The Mycota

A Comprehensive Treatise on Fungi as Experimental Systems for Basic and Applied Research

Volume 16

The fungi represent a heterogenous assemblage of eukaryotic microorganisms and have become favored organisms for research at the cellular and molecular level. Such research involvement has been stimulated by interest in the biotechnological application of fungi in processes related to industry, agriculture and ecology. Considering both yeasts and mycelial fungi, THE MYCOTA highlights developments in both basic and applied research and presents an overview of fungal systematics and cell structure. Foremost authorities in research on mycology have been assembled to edit and contribute to the volumes.

All volumes/chapters are systematically reviewed by the series editor(s) and respective volume editor(s).

Stefanie Grüttner
Krisztina Kollath-Leiß • Frank Kempken
Editors

Agricultural and Industrial Applications

Springer

Editors
Stefanie Grüttner
Botanisches Institut
Kiel University
Kiel, Germany

Krisztina Kollath-Leiß
Botanisches Institut
Kiel University
Kiel, Germany

Frank Kempken
Botanisches Institut
Kiel University
Kiel, Germany

ISSN 2945-8048 ISSN 2945-8056 (electronic)
The Mycota
ISBN 978-3-031-81903-2 ISBN 978-3-031-81904-9 (eBook)
https://doi.org/10.1007/978-3-031-81904-9

This Springer imprint is published by the registered company Springer Nature Switzerland AG
The registered company address is: Gewerbestrasse 11, 6330 Cham, Switzerland

Volume Preface

Fungi are fascinating organisms that have a significant impact on our daily lives. They are highly relevant to industry and agriculture. Their ability to produce a wide variety of secondary metabolites makes them ideal for drug production in the pharmaceutical industry and for the production of other important compounds used in various areas of biotechnology and agricultural applications. They also play an important part in the food industry, where they are utilized for food production or enrichment. As symbionts, fungi are important for healthy and robust plants in forests and agriculture. Besides all the positive impacts, they also affect our lives negatively as food contaminants or plant pathogens leading to harvest losses. As such, fungi have a tremendous impact on humanity. Therefore, it is important to understand the processes related to the possibility of fungal applications. Since the fields of industry and agriculture are interconnected, we have combined the editions of Mycota X: Industrial Applications and Mycota XI: Agricultural Applications into one volume. We are excited to have recruited some of the authors of Mycota X and XI to update or contribute new chapters for this edition. We also welcomed new authors to provide novel insights into the role of fungi in industry and agriculture. This allows us to present seven excellent chapters covering a broad area of expertise. However, due to the limited availability of authors, we are unable to cover all possible topics related to the industrial and agricultural applications of fungi in this edition. Nevertheless, we are still very delighted about the chapters we have included.

In Chap. 1, **K.-Peter Stahmann** discusses the industrial production of organic acids and vitamins using fungi. The focus is on the production of citric and glucinic acid by *Aspergillus niger* and the production of vitamins by *Mortierella alpina* and *Ashbya gossypii*. The chapter emphasizes the challenges related to increasing productivity, such as the lack of molecular tools for improving the *M. alpina* strain and the incomplete understanding of the molecular mechanisms involved in the biosynthesis of these metabolites.

Marine fungi have great potential for biotechnological applications because of their diverse nature and wide range of metabolic capabilities. In Chap. 2, **Antje Labes** discusses the industrial applications of marine fungi, which is a rapidly evolving and exciting area of research. These applications include bioremediation, biocontrol, and bioenergy.

Chapter 3, written by **Jan Dijksterhuis** and **Jos Houbraken**, discusses the issue of fungal spoilage of crops and food, which leads to significant loss in the food and feed industry: A highly relevant topic with respect to the increasing demand for food of the world population. The chapter focuses on plant-based food products and explores the relationship between fungi and living crops, as well as the contamination of processed food. Understanding these relationships can help protect crops against pathogens, thus preventing disease. Additionally, the chapter delves into the stress resistance of fungi and their involvement in conservation techniques of processed food.

Daniel Panaccione and his colleagues, in Chap. 4, address the important field of ergot alkaloids, which are metabolites of agricultural and medical significance. They focus on the biosynthesis pathways of fungi from different lineages and point out that the early steps of the ergot alkaloid biosynthesis pathways are similar across different fungi and are carried out by a common set of clustered genes. However, the diversification of the later steps leads to the production of a variety of ergot alkaloids.

Chapter 5 by **Erika Kothe** and colleagues discusses the various roles of ectomycorrhizal fungi, primarily mushroom-forming basidiomycetes, and their significance in agroforestry or afforestation in times of climate change. Ectomycorrhizal fungi can be beneficial in reclaiming degraded land, such as areas with high levels of metal contamination, for agricultural use. The chapter also focuses on the fungi's impact on soil and bacterial community composition, as well as their interaction with the host plant.

Jan van Kan and **Michele Malvestiti** discuss the important topic of necrotrophic plant pathogenic Ascomycota in Chap. 6. The successful infection of the fungi relies on necrotrophic effectors that induce programmed cell death in plant tissue. They describe the range of necrotrophic effectors used by different fungi to overcome the plant's immune system. Additionally, they highlight various studies that identified some of these effectors. Based on this, they provide an overview of the tools and methods for identifying these effectors in order to select resistant plant genotypes.

Finally, in Chap. 7, **Kerstin Voigt** and coworkers concentrate on the challenge of climate change and its impact on fungal diseases. They describe the effects of changing temperatures, humidity, and land use on the spread and occurrence of fungal pathogens, emphasizing the importance of understanding the complex relationship between climate change and fungal diseases.

We hope that these chapters will proof to be useful for scientists working in the field but also as important insight for people from other areas of research. We thank all authors for their contributions and also the Springer staff for their help.

Kiel, Germany
September 2024

Stefanie Grüttner
Krisztina Kollath-Leiß
Frank Kempken

Series Preface

The Mycota Series: 2022

The Mycota is an encyclopaedic book series published by Springer as a comprehensive treatise spanning the Fungal Kingdom through a focus on fungi as experimental systems for basic and applied research. Articles in this series present introductory information in a comprehensive manner to broaden our general knowledge, along with the state-of-the-art information on selected topics in fungal biology. Thus, these articles serve as a timeless reference source for historically relevant discoveries in fungal biology to drive further advances in the field.

The Mycota was founded in 1994 by Karl Esser and Paul Lemke. Since then 15 volumes have been published, with several now in their third edition. The steadily growing interest in *The Mycota*, reflected in particular by the high number of e-book downloads, encouraged Springer to continue this publication with a new international Series Editorial board beginning in 2022.

Historically, the study of fungi originated as a sub-discipline of botany and was a largely descriptive discipline until the early nineteenth century. Subsequent experimental research has propelled the field of mycology, and achievements in the genetics and molecular biology of fungi have benefited studies in the related fields of fungal biochemistry and biophysics, plant pathology, medical mycology, and systematics. Ground-breaking research has been carried out using fungi as recognized by the receipt of nine Nobel Prizes, including the revolutionary 1945 Nobel Prize in Physiology or Medicine for the discovery of penicillin and the foundational 2006 Nobel Prize in Chemistry for establishing the molecular basis of eukaryotic transcription. All such studies expand our knowledge of fungi and of fungal processes and improve our ability to understand, utilize, and control fungi for the benefit of humankind.

Fungi have invaded every conceivable ecological niche. Saprobic forms abound, especially in the decay of organic debris. Pathogenic forms exist with both plant and animal hosts. Fungi even grow on other fungi. They are found in aquatic as well as soil environments, and their spores have been found in the air of extreme environments. Fungi can be variously associated with plants as symbionts in the form of lichens, mycorrhizae, and endophytes and also occur as overt pathogens. Association with animal systems varies; examples include the predaceous fungi that trap nematodes, the micro-fungi that grow in the anaerobic environment of the rumen, and medically important pathogens afflicting humans. Currently, taxonomists have identified about 150,000 species, but it is thought that over 90% of fungal species remain undescribed, with conservative estimates ranging from 2 to 5 million fungal species on earth.

From this perspective, there are new topics in fungal biology to explore, which will expand the current volumes and take them in new directions. Further, there will be new volumes in areas (e.g., Cryptomycota) that were not covered before.

Our understanding of the evolution of fungi is still incomplete and mainly based on species that can be grown in culture. But recent environmental DNA analyses have revealed a highly diverse form of eukaryotic life that branches with the Fungi, and thus the resulting and highly diverse clades were named the cryptomycota. The discovery of novel intermediate forms will redefine the fungal tree of life. Other topics will consider the adaptation of fungi to climate changes, the occurrence of fungal pathogens in the environment, and the dispersal of fungi during global pandemics.

Fungal constituents of the microbiome have received much less attention thus far, yet recent findings from clinical and animal studies clearly establish fungi as a significant component of the oral, gastrointestinal, pulmonary, and skin microbiomes. Finally, the relevance of fungal biology to society is reflected by the increasing number of fungal related human diseases.

In the history of pharmacology, fungi have always been sources of useful molecules for humans, but the diversity of molecules that can be obtained from fungi is still largely under explored. However, efforts in fungal genomics provide resources for data-driven genome mining and large-scale comparisons to explore the molecular repertoires produced by fungi. The result will be new compounds with applications in the pharmaceutical and agricultural science industries.

Fungi can also help the entire planet and may, for example, be relevant in specific sectors, such as that linked to pollution from plastic. Fungi produce a wide range of enzymes that have the potential to break down the chemical bonds of plastic polymers, and in this context the potential role of marine fungi in plastic degradation may be of major relevance. Finally, new biomaterials from fungal species may open the door to alternatives to fossil-based materials, and thus reduce environmental pollution.

For consistency throughout this series of volumes, the names adopted for major groups of fungi should be followed according to the following paper, which gives an overview of all of the orders in the fungal kingdom: https://pubmed.ncbi.nlm.nih.gov/32660385/.

We are grateful to Springer for continuing *The Mycota* series and are especially thankful to all of the Volume Editors in selecting topics and assembling experts from diverse fields of fungal biology.

Sydney, Australia — Dee Carter
Delhi, India — Anuradha Chowdhary
Durham, NC, USA — Joseph Heitman
Bochum, Germany — Ulrich Kück
July 15, 2022

Contents

About the Editors and Contributors

About the Series Editors

Dee Carter is Professor of Microbiology at the University of Sydney. She has a degree in microbiology and biochemistry from Otago University, New Zealand, and a PhD from Imperial College, London, UK. After postdoctoral fellowships in Montpellier, France, and UC Berkeley, California, USA, she moved to Australia in 1995 to take up a lectureship at the University of Sydney. Dee's research has encompassed (1) the population genetics and ecology of medically important fungi, including yeast and mould pathogens; (2) responses of fungi to host and antifungal stress, including the production of variant morphological forms; and (3) the transcriptome and proteome response to antifungal therapy, particularly during synergistic interactions between antifungals and natural products. As well as research, Dee teaches microbiology including mycology, epidemiology, and molecular biology at Sydney University.

Anuradha Chowdhary is a Professor of Medical Mycology at the Vallabhbhai Patel Chest Institute, University of Delhi, India. She received her medical degree in 1992 and MD (Microbiology) degree in 1996 from the University of Delhi, India, and PhD from the Faculty of Medical Sciences, Radboud University Medical Center, Netherlands. Dr Chowdhary's research interests include molecular ecology and population genetics of pathogenic fungi, antifungal drug resistance mechanisms, especially Aspergillus and Candida spp., and the epidemiology of systemic mycoses. She is currently working on molecular epidemiology of Candida auris, terbinafine-resistant dermatophytes, and azole resistant Aspergillus fumigatus.

Joseph Heitman is James B. Duke Professor and Chair, Department of Molecular Genetics and Microbiology, Duke University, Durham, NC, USA. His research studies model and pathogenic fungi addressing fundamental questions of scientific and medical importance. Pioneering studies with Baker's yeast revealed how immunosuppressive natural products interdict signalling cascades via FKBP12-drug complexes and discovered TOR as a globally conserved nutrient sensor targeted by the immunosuppressive, antiproliferative drug rapamycin. His research discovered unisexual reproduction of pathogenic microbes, with implications for pathogen emergence, how sex generates diversity, and how sex evolved. Dr. Heitman's lab has further developed genetic and genomic approaches elucidating molecular principles of fungal virulence, identifying therapeutic targets, and illustrating convergent evolution of fungal mating-type loci with mammalian, insect, and plant sex chromosomes, defined the calcium-activated protein phosphatase calcineurin as a globally conserved fungal virulence factor, and elucidated functions of RNAi in microbial pathogen genome integrity, hypervirulent outbreak lineages, and drug resistance via epimutation.

Ulrich Kück is a Professor of General and Molecular Botany at the Ruhr-University in Bochum, Germany. He is being graduated in Biology and Chemistry, and has a long- standing experience in the molecular biology of fungi, algae, and plants. His research with fungi has focussed on two general aspects: 1st, Genetic engineering of the secondary metabolism of biotechnically relevant filamentous fungi, including functional genomics and proteomics. 2nd, Molecular genetic analysis of cellular growth in filamentous fungus, with the focus on sexual development. In particular, he is interested in the function of mating type loci, and the involvement of conserved signalling complexes, such as the striatin-interacting phosphatases and kinases (STRIPAK) complex, in the control of cellular and developmental processes.

About the Volume Editors

Stefanie Grüttner is a research associate at the University of Kiel, Germany. In her doctoral thesis, she was working on the regulation of gene expression in plant mitochondria. After completion of her doctorate in 2021, her research interest shifted to filamentous fungi with an emphasis on genome editing of the model ascomycete *Neurospora crassa*.

Krisztina Kolláth-Leiß received her doctorate in 2015 at Kiel University, Germany. She is a teaching associate at the Botanical Institute and research associate in the Department of Botanical Genetics and Molecular Biology at the CAU zu Kiel. In her research, she investigates principles of polar growth in the model ascomycete *Neurospora crassa*.

Frank Kempken is Director at the Botanical Institute and Head of the Department of Botanical Genetics and Molecular Biology at the University of Kiel (Christian-Albrechts-Universität zu Kiel), Germany. His research interests are fungal transposons, genomics of marine fungi, and auxin production of moulds. Frank Kempken was also an editor of a former Mycota edition.

Contributors

Oluwatosin Abdulrahman Abdulsalam Institute of Microbiology, Friedrich Schiller University Jena, Jena, Germany

Bain and Company, Düsseldorf, Germany

Jaime David Acosta-España Jena Microbial Resource Collection, Leibniz Institute for Natural Product Research and Infection Biology – Hans Knöll Institute, Jena, Germany

Institute of Microbiology, Friedrich Schiller University Jena, Jena, Germany

Faculty of Medicine, Pontifical Catholic University of Ecuador, Quito, Ecuador

Health Sciences Faculty, Universidad Internacional SEK del Ecuador, Quito, Ecuador

Olga Bogdanova Institute of Microbiology, Friedrich Schiller University Jena, Jena, Germany

BioNTech SE, Jena, Germany

Kyle A. Davis West Virginia University, Division of Plant and Soil Sciences, Morgantown, WV, USA

Jan Dijksterhuis Westerdijk Fungal Biodiversity Institute, Utrecht, The Netherlands

Marycolette Ndidi Ezediokpu Institute of Microbiology, Friedrich Schiller University Jena, Jena, Germany

McMaster University, Hamilton, ON, Canada

Samantha J. Fabian West Virginia University, Division of Plant and Soil Sciences, Morgantown, WV, USA

Corinne M. Hazel West Virginia University, Division of Plant and Soil Sciences, Morgantown, WV, USA

Jos Houbraken Westerdijk Fungal Biodiversity Institute, Utrecht, The Netherlands

Abigail M. Jones West Virginia University, Division of Plant and Soil Sciences, Morgantown, WV, USA

Erika Kothe Institute of Microbiology, Friedrich Schiller University Jena, Jena, Germany

Katrin Krause Institute of Microbiology, Friedrich Schiller University Jena, Jena, Germany

Antje Labes Flensburg University of Applied Science, Flensburg, Germany

Michele C. Malvestiti Laboratory of Phytopathology, Wageningen University, Wageningen, The Netherlands

Dolly Montaño Jena Microbial Resource Collection, Leibniz Institute for Natural Product Research and Infection Biology – Hans Knöll Institute, Jena, Germany

Institute of Microbiology, Friedrich Schiller University Jena, Jena, Germany

Daniel G. Panaccione West Virginia University, Division of Plant and Soil Sciences, Morgantown, WV, USA

K. -Peter Stahmann Fachgebiet Technische Mikrobiologie, Institut für Biotechnologie, Brandenburgische Technische Universität Cottbus-Senftenberg, Senftenberg, Germany

Jan A. L. van Kan Laboratory of Phytopathology, Wageningen University, Wageningen, The Netherlands

Kerstin Voigt Jena Microbial Resource Collection, Leibniz Institute for Natural Product Research and Infection Biology – Hans Knöll Institute, Jena, Germany

Institute of Microbiology, Friedrich Schiller University Jena, Jena, Germany

1 Industrial Production by Filamentous Fungi: Organic Acids and Vitamins

K. -Peter Stahmann

Abstract

Overproduction of four primary metabolites by filamentous fungi is explained from a scientific point of view. Causal connections concerning two organic acids, citrate and D-gluconate, the vitamin-like compound, arachidonic acid, and the vitamin riboflavin are explained.

While extracellular D-gluconate production by *Aspergillus niger* is already at 99% yield, recent cloning of a putative citric acid exporter gene might improve productivity in the future. Highest potential for improvement might show arachidonic acid. Its current producer, the zygomycete *Mortierella alpina*, is underdeveloped concerning molecular tools. Additionally, biochemical reactions and their regulation, e.g., the question how fatty acid chain length and degree of desaturation are controlled, are not understood. In riboflavin overproduction by the hemiascomycete *Ashbya gossypii,* metabolic engineering begins with an improved uptake of fatty acids into the hyphae. Connected induction of sporulation and riboflavin overproduction is shown and related to the second messenger cAMP. But, the molecular mechanism of induction is not investigated yet.

Keywords

Citric acid · Gluconic acid · Arachidonic acid · Riboflavin · *Aspergillus niger* · *Mortierella alpina* · *Ashbya gossypii*

1.1 Introduction

In this chapter "Industrial Production" means controlled cultivation of filamentous fungi and metabolite overproduction in 100 m^3 scale. Four processes delivering more than 10,000 tons per year are discussed. In Table 1.1 they are compared with similar fungal processes.

This chapter discusses primary metabolites. These play a role in growth or in energy metabolism. For example, citric acid is the molecule formed in the matrix of the mitochondria to get ATP by oxidation and decarboxylation *via* the citric acid cycle and the respiratory chain. Secondary metabolites are not needed directly. Penicillins and cephalosporins have no role in

---

This chapter is an updated version of "Production of Vitamin B2 and a Polyunsaturated Fatty Acid by Fungi" published in Vol. 10: Industrial Applications, 2011, ISBN 978-3-642-11457-1, https://link.springer.com/book/10.1007/978-3-642-11458-8

K. -P. Stahmann (✉)
Fachgebiet Technische Mikrobiologie, Institut für Biotechnologie, Brandenburgische Technische Universität Cottbus-Senftenberg, Senftenberg, Germany
e-mail: klaus-peter.stahmann@b-tu.de

S. Grüttner et al. (eds.), *Agricultural and Industrial Applications*, The Mycota 16,
https://doi.org/10.1007/978-3-031-81904-9_1

Table 1.1 Industrial production of selected metabolites by filamentous fungi. Natural overproduction of isolated wild types was not sufficient for economic production but the starting point for mutants generated by classical mutagenesis (CM) and metabolic engineering (ME)

Metabolite	Annual Production t/a	Species	Natural Overproducer	Mutants by CM	Mutants by ME
Organic acid					
Citric acid	> 1,000,000	*Aspergillus niger*	Yes	Yes	No
Gluconic acid[1]	> 100,000		Yes	Yes	No
L-Malic acid	> 100,000		No	Yes	Yes
Itaconic acid	> 10,000	*Aspergillus terreus*	Yes	Yes	Yes
Vitamin-like					
Arachidonic acid[2]	> 100,000	*Mortierella alpina*	Yes	Unknown	No
β-Carotene[1]	Unknown	*Blakeslea trispora*	Yes	Yes	Unknown
Vitamin					
Riboflavin[3]	> 10,000	*Ashbya gossypii*	Yes	Yes	Yes
Beta-Lactam					
Penicillins[4]	> 10,000	*Penicillium chrysogenum*	No	Yes	Unknown
Cephalosporins[4]	> 1000	*Acremonium chrysogenum*	No	Yes	Unknown

[1] produced also chemically; [2] produced as mixed triglyceride; [3] produced also by *Bacillus subtilis;* [4] modified by feeding synthetic side chain precursors

anabolism. They might play a role in the ecological niche when competitors, i.e. other microorganisms, are present.

Overproduction of metabolites needs pulling or pushing of carbon fluxes through cascades of enzymes. More than 100 g/L is required to reach a concentration range that is economically relevant for organic acids. In most cases natural overproduction of wild-type strains was the starting point for the development of mutants. Natural overproduction is evident for riboflavin. The wild-type strain of *Ashbya gossypii* can produce more than 100 mg riboflavin per gram biomass. This is 1000 times more than needed for the function as precursor of FMN and FAD. These coenzymes are found in mM concentrations intracellularly.

Industrial strains are the result of a combination of classical mutation plus metabolic engineering. Screenings for the best strains are performed in both approaches. For all processes batch fermentations are performed. Growth is limited by an essential nutrient, e.g. N, maintenance is kept on the minimal level, and production is run by a so-called overflow metabolism to reach high final concentrations. The goal for all processes is the theoretical maximum.

The ranking of the products in Table 1.1 is done with respect to orders of magnitudes produced per year. Citric acid is the largest product in that list with >1 million tons a^{-1}. With 100 million tons per year ethanol production by *Saccharomyces cerevisiae* (Hoff et al. 2021) is three orders of magnitudes larger. Because *S. cerevisiae* is a yeast it is not discussed here. As benchmarks for the production scale, penicillins and cephalosporins are listed in Table 1.1. Since they are secondary metabolites they are not further discussed. Because antibiotics are active at low concentration an overproduction is not observed in the wild-type strains *Penicillium chrysogenum* or *Acremonium chrysogenum*.

1.2 Organic Acids

In order to classify the production figures through applied microbiology, a few comparative figures from chemistry are given. The consumption of both diesel and gasoline is in the range of 1 billion tons per year. The largest products of petroleum-based organic acids are acetic acid and acrylic acid. Both are produced in an annual volume of 10 million tons (Biddy et al. 2016). Bio-based organic acids are one or two magnitudes smaller in production scale.

1.2.1 *Aspergillus niger*

A. niger is the most important filamentous fungus applied in industry. Its ability to grow at pH 2 and its fast growth with cheap substrates, e.g. sugar, keeps it at position one concerning organic acids produced worldwide. But, also its increased use for the production of robust enzymes is impressive. In the last decade the feed additive phytase, needed for the liberation of phosphoric acid from phytate in the intestine of animals, became the No.1 product, relegating proteases to second place. In the field of enzymes, the production of glucose oxidase is also worth to be mentioned. Its application as sensor is of importance for millions of blood tests performed every day to adjust insulin dosage individually to diabetes patients.

A comprehensive history of *A. niger* applications and scientific breakthroughs is published by Cairns et al. (2021).

In this subchapter citric acid and gluconic acid will be discussed. A minimum of scientific data is presented to explain overproduction.

The genotypes of industrial strains are not published. Most improvement steps performed by molecular biology methods to understand the system can be gone by random mutagenesis and subsequent screening.

1.2.1.1 Citric Acid

Citric acid was originally extracted from lemon. At the beginning of the twentieth century it was the first product of *Aspergillus niger* (Currie 1917). To get citrate titers >100 g/l extreme conditions are needed. These comprise high substrate concentrations, low pH, and traces of Mn(II), Fe(II), Cu(II), and Zn(II). The complexity of effects and the limited trace elements are discussed in a recent review (Karaffa et al. 2021). A simplified metabolic model with respect to subcellular localization is shown in Fig. 1.1. Recently, the gene encoding a putative citrate exporter was identified and its metabolic pulling effect shown (Steiger et al. 2019).

Citrate Exporter Gene *cexA*

A wild-type culture of *A. niger* can produce 15 g/l citrate within 15 hours when the concentration of Mn^{2+} is <5 ppb (Steiger et al. 2019). All trace elements, e.g., Fe^{3+}, need to be kept in solution by supplemented citrate (Steiger et al. 2019). When the gene *cexA*, encoding a putative H^+ antiporter with 12 predicted transmembrane helices, was deleted, citrate production was completely abolished. Instead, 4 g/l oxalate was found in the extracellular medium (Steiger et al. 2019). The expected side product oxalate is connected to the pathway of citrate production (Fig. 1.1).

Although an in vitro reconstitution of the transporter protein in liposomes is lacking, the probability that CexA is the transporter is high. Two arguments support that thesis, first, heterologous expression of *cexA* in *S. cerevisiae* increased citrate from detection limit <0.01 g/l to 0.4 g/l second, induced overexpression of *cexA* resulted in a citrate titer increase in *A. nidulans* (Steiger et al. 2019). For that approach, the Tet-on system, inducible with doxycycline (Meyer et al. 2011) was a convincing tool. Tiny concentrations, 0.1 mg/l or 1.0 mg/l doxycycline, caused >40 g/l or > 100 g/l citrate in the extracellular medium, respectively (Steiger et al. 2019). That citrate export can be a production rate-limiting step was suggested two decades ago (Karaffa and Kubicek 2003).

An interesting question is whether the wild type of *A. niger* secretes citrate as siderophore. Due to the fast oxidation of Fe^{2+} and the low solubility of ferric ions (Fe^{3+}) bioavailability is a problem in most environments (Odoni et al. 2017). Several complexes of iron citrate are proposed (Fig. 1.2). They play a role in the uptake of iron by *Bacillus cereus* (Fukushima et al. 2012). But, the high sugar concentrations (> 100 g/l) used for citrate overproduction are far away from starving conditions in a natural environment.

Avoiding Side Products: Gluconate and Oxalate

The above-mentioned increase in extracellular oxalate after deletion of *cexA* supports the plausible pathway connection shown in Fig. 1.1.

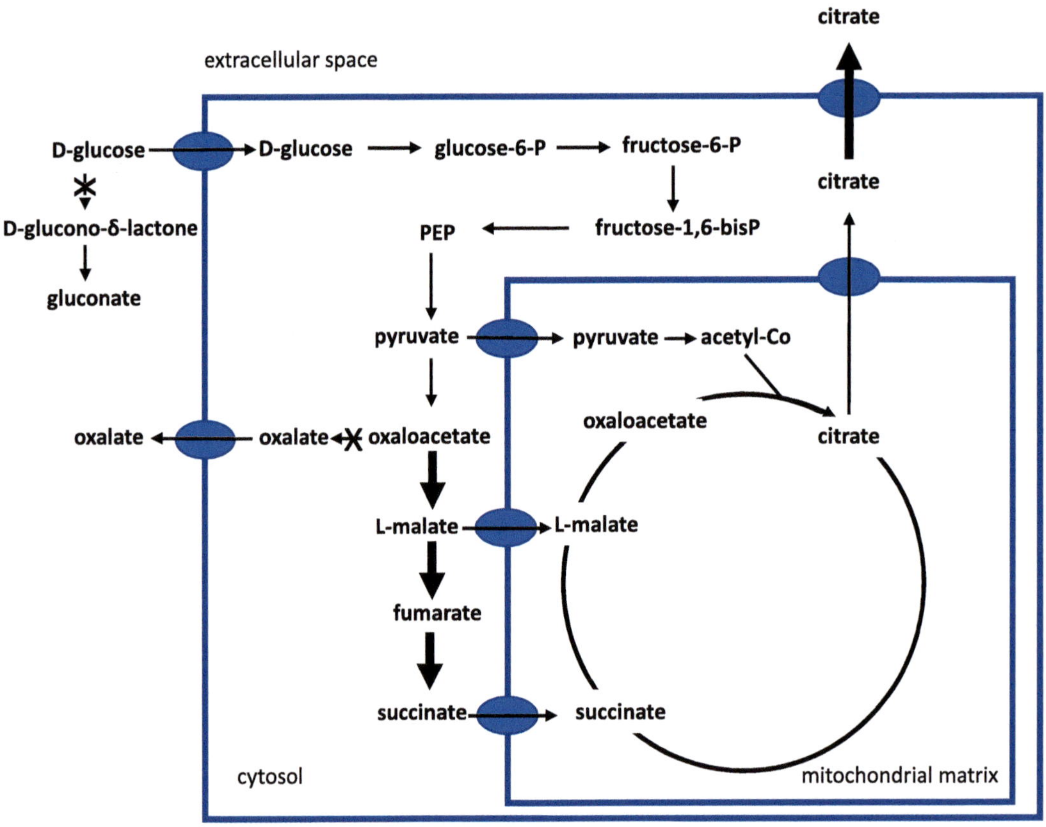

Fig. 1.1 Diagram illustrating metabolic pathways of citrate production by *Aspergillus niger*. Two membranes, the inner membrane of the mitochondrium and the cytoplasmic membrane limit the metabolic fluxe from D-glucose to the extracellular citrate. Four bold arrows indicate increased metabolic fluxes resulting in citrate overproduction. These are caused by overexpression of the four genes encoding three enzymes and one exporter. Deletions of two genes encoding enzymes responsible for unwished side products result in zero fluxes. These are indicated by crossed arrows (adapted from Chroumpi et al. 2020)

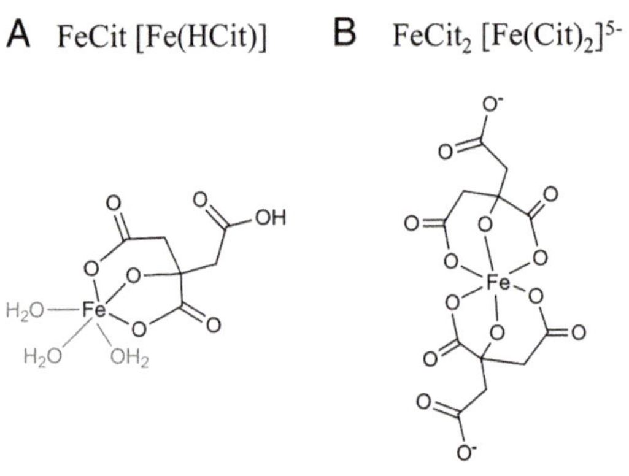

Fig. 1.2 Image showing two suggested iron citrate complexes taken from Fukushima et al. (2012). A consists of one, B of two citrate anions per ferric kation. They play a role in the iron binding and uptake by *Bacillus cereus*

Deletion of *oahA*, encoding oxaloacetate hydrolase, avoids the unwished side product (Kubicek 1988). The same is true for *goxC*, encoding an extracellular glucose oxidase. Its deletion makes a direct, partial oxidation of glucose to gluconate impossible (Kubicek 1988).

Improving Precursor Supply

In general, all enzymatic steps can be improved. Either the number of enzyme molecules per cell is increased or a heterologous gene, encoding an enzyme with better properties is used. The latter was shown by expression of the genes *mdh2* (malate dehydrogenase), *fum1s* (cytosolic fumarase), *frds1* (fumarate dehydrogenase), and *fumRs* (cytosolic fumarases) from *S. cerevisiae.*

1.2.1.2 Gluconic Acid

Gluconic acid is the result of a regioselective oxidation. The aldehyde group at C1 of D-glucopyranose is oxidized to a carboxyl group. To run an industrial process many methods, e.g. catalysis with noble metals, are described (Kornecki et al. 2020). The main reason why *A. niger* is still the preferred catalyst might be the problem of unwished side products, i.e. oxidation at C5. In contrast to citric acid the product is formed extracellularly (Fig. 1.3).

Extracellular Regioselective Oxidation

Liu et al. (2017) investigated a commercial production strain of *A. niger* to find the optimal initial starting concentration for D-glucose. When 300 g/l D-glucose was provided 99% of it was converted to D-gluconic acid within less than 24 hours fermentation time. In Fig. 1.3 a simplified scheme shows some aspects of that process. The main players are the extracellular enzymes: glucose oxidase (GoxC, EC 1.1.3.4), a flavoprotein; catalase (CatC, 1.11.1.6), a heme-dependent enzyme; and gluconolactonase (Lct 3.1.1.17), a hydrolase. Extracellular localization of these three enzymes had been shown by immuno-labeling of thin sections (Witteveen et al. 1992). At least the *goxC* gene encodes a putative N-terminal signal sequence indicating its secretion (EXPASY Enzyme database 17.6.2023).

An interesting point is that GoxC does not bind the open form of D-glucose but the ring-shaped hemiacetal, named beta-D-glucopyranose. Whether an extracellular mutarotase accelerates the conversion of alpha- to the beta-anomer is not clear. In particular the affinity of the GoxC is 20 times higher for beta-D-glucopyranose than for the alpha-anomer. This means, depending on the conditions, the final activity can be less than 1% (Leskovac et al. 2005).

The extracellular catalase pulls the flux in the direction of the oxidation and keeps the H_2O_2 low to minimize inactivating effects on extracellular proteins (Kornecki et al. 2020).

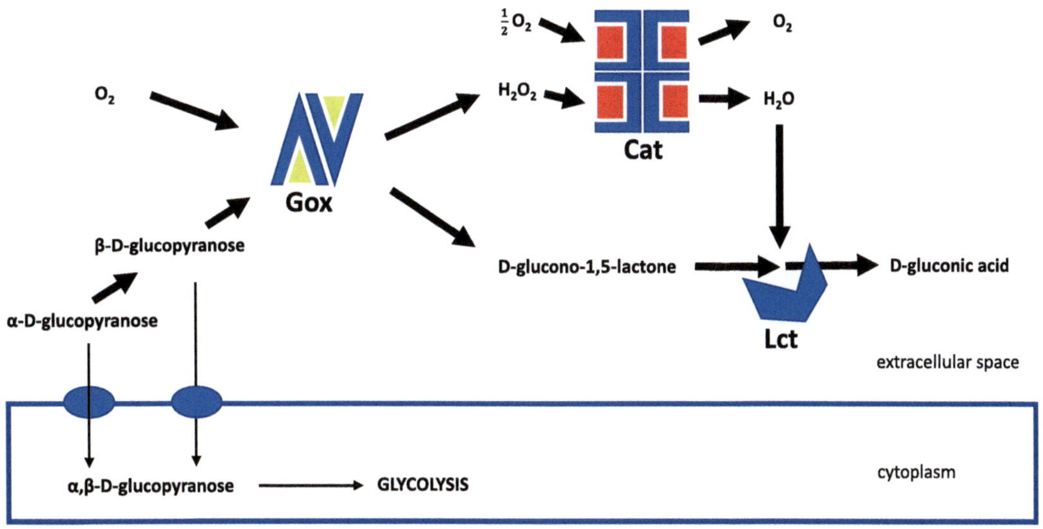

Fig. 1.3 Diagram illustrating extracellular oxidation of D-glucose to D-gluconic acid by *Aspergillus niger*. A minor flux is entering the cell and catabolized via glycolysis. Most D-glucose is oxidized to D-gluconic acid in a regioselective reaction. Extracellular enzymes are: Glucose oxidase (Gox), a dimer containing FAD (yellow) as coenzyme, catalase (Cat), a tetramer containing heme (red), and gluconolactonase (Lct), a hydrolase

1.3 Vitamins

Vitamins are organic nutrients essential for humans or domestic animals because their own anabolism does not synthesize them adequately. Since they fulfill catalytic or hormone-like functions, the intake of small amounts is sufficient. Because storage is limited they should be on the menu every day. Between 1930 and 1950 all 13 vitamin groups were determined in their structure after extraction, purification, and chemical synthesis.

A triglyceride rich in arachidonic acid, vitamin B_2, and vitamin B_{12} is produced by fermentation exclusively. Today technical production of thousands of tons per year is performed by chemistry, extraction, fermentation or bioconversion, or combined processes (Table 1.2). A replacement of chemical synthesis by Industrial Microbiology, also called White Biotechnology, is encouraged by increasing costs for disposal of waste and a shift to renewable educts to reach sustainability. Many countries follow a policy to reach the so-called BioEconomy.

This chapter discusses two fungal metabolites, arachidonic acid and riboflavin. The vitamin-like long-chain polyunsaturated fatty acid (PUFA) arachidonic acid is quantitatively leading (Table 1.1), followed by beta-carotene, also called provitamin A, and in the third place, riboflavin, also called vitamin B_2. Only the most important physiological properties and strategies for their improved production are discussed.

1.3.1 *Mortierella alpina*

Mortierella alpina, a filamentous zygomycete, produces lipids rich in arachidonic acid (ARA) when grown at 18 °C (Lindberg and Molin 1993). Its presence in the spores and preferred degradation during germination suggest a role as a store for carbon and energy (Lounds et al. 2007). The melting point of ARA is −49 °C. Because of four *cis* double bonds the linked alkyl tails prevent tight packing. This leads to a higher membrane fluidity in the phospholipid bilayer (Ernst et al. 2016). This fits to *M. alpina* psychrophilic phenotype.

1.3.1.1 Arachidonic Acid

Polyunsaturated fatty acids (PUFAs) are vitamin-like compounds with structural functions in cell membranes and lipid tissue. They serve as precursors for the biosynthesis of hormones like leukotrienes, prostaglandins, and thromboxanes. PUFAs are aliphatic monocarboxylic acids with two to six double bonds in biologically active *cis* configuration within their C18 to C24 backbone. They are categorized as ω-3 or ω-6, depending on the position of the first double bond, counted from the terminal carbon atom of the fatty acid. Their biosynthesis starts from oleic acid (18:1) and is continued by a sequence of desaturase and elongase reactions (Fig. 1.4; Marszalek and Lodish 2005). Since humans cannot introduce double bonds at the ω-3 or ω-6 position, linoleic acid (18:2, ω-6) and α-linolenic acid (18:3, ω-3) are essential. They belong to the vitamin F group.

Table 1.2 Scale and methods for industrial production of vitamins and vitamin-like compounds with an impact of microorganisms. Production processes use either/and chemical synthesis (Chem. Synth.), extraction from plants (Extrac. Chem.), microbial fermentation, or combined methods

	Production Method				
Compound	Chem. Synth.	Extract. Chem.	Microbial	Combined	World Production (tons $year^{-1}$)
Vit C				+	>100,000
Vit B_2			+		>10,000
Carotenoids	+	+	+	+	>1000
PUFAs		+	+		>100,000
Vit B_{12}			+		>10

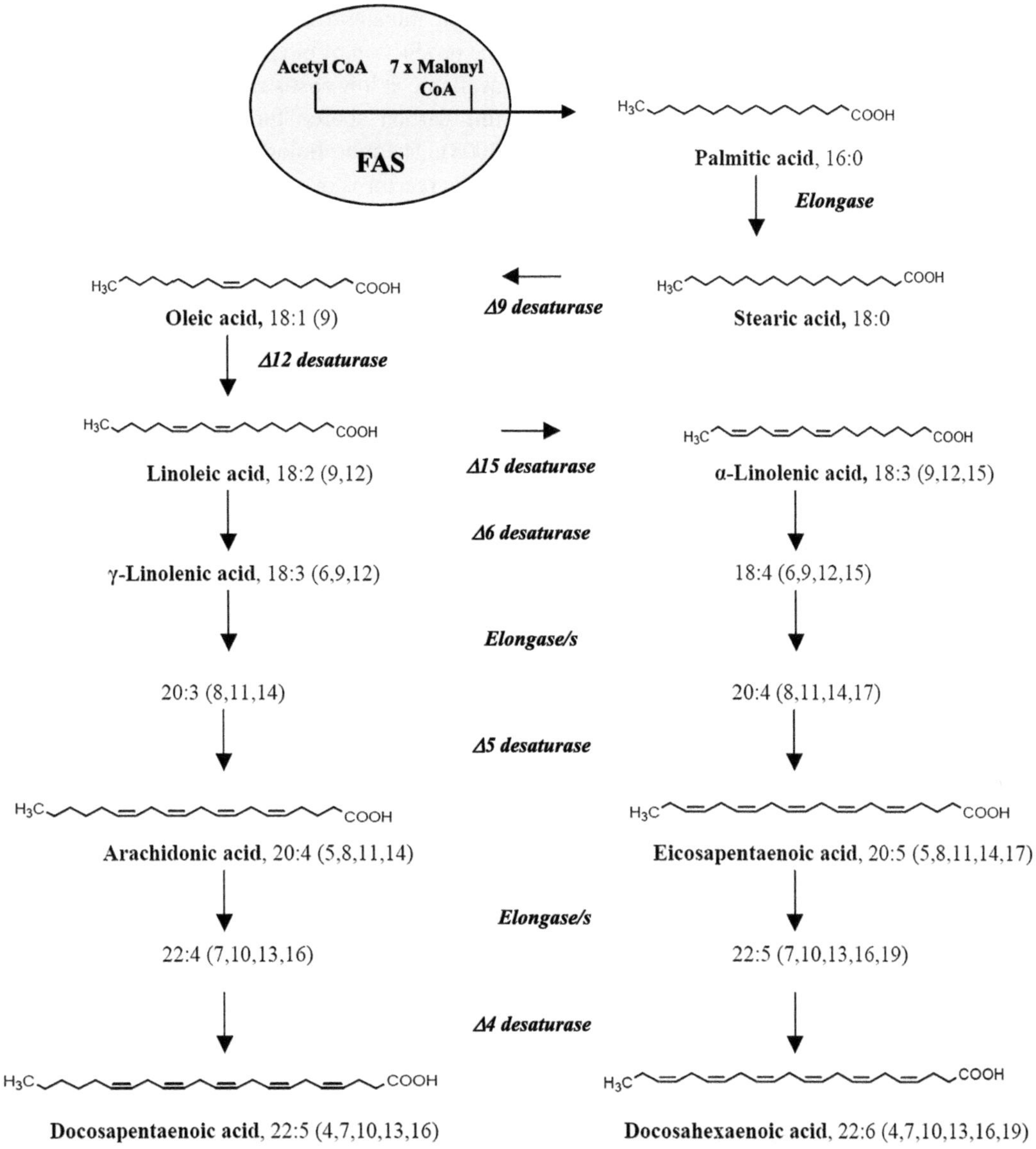

Fig. 1.4 Diagram illustrating the eukaryotic biosynthesis of fatty acids. It starts with fatty acid synthase (FAS) in the cytosol (Ratledge 2004). Parallel elongation and desaturation steps take place in the membrane of the endoplasmic reticulum. Endproducts shown here are docosapentaenoic acid (20:5) and docosahexaenoic acid (22:6). Arachidonic acid (20:4) is an intermediate metabolite

Arachidonic acid (20:4, ω-6; ARA) and docosahexaenoic acid (22:6, ω-3; DHA) are not essential but of commercial interest. They are needed for the developing brain, the neuronal tissue, and retina of the human fetus and are therefore contained in human breast milk. Studies on supplemented infant formula suggested a correlation between PUFAs content in the cerebral cortex of breastfed infants and better cognitive and visual functions of infants fed with infant formula supplemented with long-chain PUFAs (Heird and Lapillonne 2005).

ARA supports the neonatal eye and brain development (Alessandri et al. 2004). Breastfed infants have higher ARA blood levels than formula-fed infants. Infant formulas fortified with ARA and DHA are sold in Europe for more than 20 years. After permission by the FDA they became introduced in the USA in 2002. In contrast to DHA, which can be obtained from fish oil or microbial processes, industrial ARA production solely relies on fermentation. The first publications about the mycelial fungus *Mortierella alpina* as ARA producer appeared in the late 1980s.

The pattern of the fatty acids produced by *M. alpina* could be influenced by the addition of a specific delta5-fatty acid desaturase inhibitor, i.e. dioxabicyclo [3.3.0] octane derivatives like sesamin, or a selection of mutants with absent or reduced delta5-fatty acid desaturase activity (Kawashima et al. 1992). A number of *M. alpina* genes encoding enzymes which are involved in the synthesis of polyunsaturated fatty acids were cloned, e.g. delta5-fatty acid (Knutzon et al. 1998), delta6-fatty acid (Sakuradani et al. 1999a), delta9-fatty acid (Sakuradani et al. 1999b), and delta12-fatty acid (Huang et al. 1999) desaturase. Verification of the genes functionality by expression, e.g. in *Aspergillus oryzae*, resulted in a shift to more than 70% of linoleic acid (18:2, ω-6) in the cell's fatty acid composition (Sakuradani et al. 1999c). Two transformation systems were established, one working with uracil auxotrophs (Takeno et al. 2004) and a second using zeocin resistance as dominant selective marker (Takeno et al. 2005). By use of the latter ARA production was increased after transformation with a gene encoding a rate-limiting elongase (Takeno et al. 2005). More details concerning breeding of *M. alpina* are presented by Sakuradani et al. (2009).

The accumulation of lipid starts in *M. alpina* cells when growth becomes limited, e.g. after consumption of the nitrogen source, and depends on high activity of malic enzyme needed for NADPH supply in fatty acid biosynthesis (Zhang and Ratledge 2008). Highest titers with glucose as carbon source reported so far are 20 g l^{-1} of PUFAs with up to 70% ARA produced by *M. alpina* mutants (Sakuradani et al. 2009). Since the production of biodiesel today makes glycerol available at low cost strains were screened to use this carbon source for ARA production (Hou 2008). The morphology of the fungal mycelium in the reactor is of specific concern as only small pellets with a diameter of 1–2 mm allow sufficient mass transfer and keep the broth viscosity at an acceptable level. After fermentation the mycelium is harvested by filtration, dried, and solvent extracted. Alternatively, the mycelium is homogenized as an aqueous suspension and extracted with a water-immiscible organic solvent.

Chang et al. (2022) conclude in a recent review that to shorten fermentation time (8–10 days) and to improve the conversion ratio from carbon source to the lipid are current research goals. In granted patent applications, initial glucose concentrations (Chen et al. 2022) or efficient extraction of lipids from dried biomass (Adugna 2022) are claimed.

1.3.2 *Ashbya gossypii*

Ashbya gossypii is a hemiascomycete related to *Saccharomyces cerevisiae* (Wendland 2020). Comparative genomics suggest that the filamentous *A. gossypii* is evolutionarily closer to a common ancestor with *S. cerevisiae* being the younger relative after whole genome duplication. Convincing arguments for that idea are the number of chromosomes (*A. gossypii*: 7 chromosomes, *S. cerevisiae*: 16 chromosomes) and the synteny of many genes (Dietrich et al. 2004). The life cycle of *A. gossypii* starts with an ascospore (Fig. 1.5). It germinates with hyphae and forms asci again. This means *A. gossypii* is homothallic (self-fertile). The ascus opens and releases the spores (Fig. 1.5). In a sporulating culture only a minor part of the hyphal cells are asci. The majority produces and stores or excretes riboflavin (Nieland and Stahmann 2013, Fig. 1.5).

Fig. 1.5 (continued) C) The photograph shows a typical mixture of three types of single cells: white cells, yellow cells, and asci. D) The ascus shows release of one of eight needle-shaped spores. E) White and yellow cells can be converted into protoplasts by lysing enzyme. Fluorescent light shows that the yellow protoplasts contain riboflavin (taken from Nieland and Stahmann 2013)

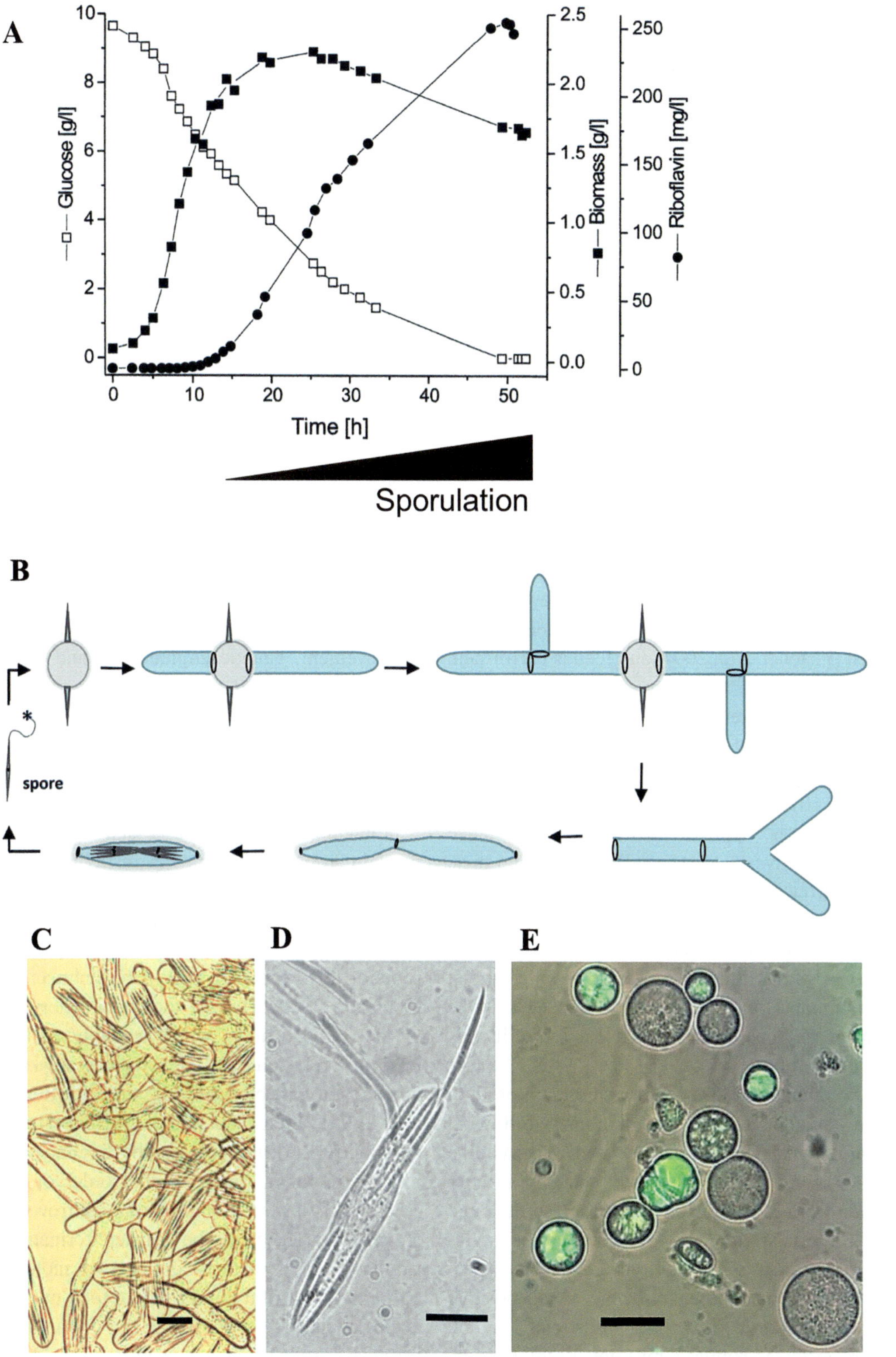

Fig. 1.5 Time course of growth and sporulation in a culture of the *Ashbya gossypii*. A) Riboflavin production starts when growth rate declines. A medium was used where yeast extract but not glucose limited growth. Some cells differentiate into asci and liberate ascospores. Because they aggregate it is not possible to count them. B) The life cycle (taken from Wendland 2020) shows a fragmentation into single cells, when nutrients are declining.

1.3.2.1 Riboflavin

Riboflavin is the name of 7,8-dimethyl-10-(D-1′-ribityl) isoalloxazine, indicating its yellow (*flavus* lat.) color. Riboflavin or its derivatives have key functions in energy metabolism, maintenance of healthy skin and muscles, support of immune and nervous system, and promotion of cell growth and division. It is the precursor for the coenzymes FMN (flavin mononucleotide) and FAD (flavin adenine dinucleotide), important electron carriers in redox reactions. Additionally, both coenzymes participate in bioluminescence, light sensing, phototropism, DNA protection against UV-light damage, and reset of the circadian clock (Cashmore et al. 1999).

Since higher eukaryotes cannot produce riboflavin it has the status of a vitamin. Because of light sensitivity and poor resorption riboflavin deficiency is often a result of surveys on the nutritional status. Therefore, supplementation of food (Fletcher et al. 2004) and feed is performed. Overdosing due to dietary supplementation is impeded by excretion of riboflavin into the urine. In industrialized countries processed food is often fortified by the use of riboflavin as a colorant (E 101). About 70% of industrially produced riboflavin is applied in animal feed, since productive livestock, especially poultry and pigs, show growth retardation and diarrhea in case of riboflavin deficiency. Deficiency symptoms like dermatitis can be avoided by a nutritional intake of 0.3–1.8 mg/d for humans and 1–4 mg/kg diet for animals (Eggersdorfer and Adam 1996). One cup of milk per day is sufficient to fulfill this requirement.

For about half of a century riboflavin production was a domain of chemical industry following a multi-step process starting with 3,4-xylidine and D-ribose. The latter was provided by fermentation from glucose using transketolase-deficient *Bacillus* sp. strains. The first commercial single-step riboflavin fermentations came up in the 1940s employing the anaerobic Gram-positive bacterium *Clostridium acetobutylicum* or the more efficient hemiascomycetes *Eremothecium ashbyi* or *Ashbya gossypii* (Wickerham et al. 1946; Perlman 1979). All companies manufacturing riboflavin by fermentation shut down their production plants in the 1960s being competitively disadvantaged to the chemical processes (Lago and Kaplan 1981). Today, microbial riboflavin production is the paradigm of environmentally superior White Biotechnology replacing chemistry. After Merck (USA) resumed fermentation by *A. gossypii* in 1974, the success story started with an *A. gossypii* production plant of BASF in Ludwigshafen (Germany) in 1990. The market share of riboflavin produced by the microbial processes increased from 5% in 1990 to 75% in 2002. Initially the two naturally vitamin B_2 overproducing fungi *A. gossypii* and *Candida famata*, and later the metabolically engineered Gram-positive bacterium *Bacillus subtilis* were used (Stahmann et al. 2000). Today *C. famata* is not used anymore and chemical riboflavin production came to an end leaving only *A. gossypii*- and *B. subtilis*-based production methods. Both processes are the outcome of classical strain improvement, genetic engineering, and process optimization. The *B. subtilis* process was compared with the *A. gossypii* process by Hohmann and Stahmann (2010).

The Riboflavin Biosynthesis Pathway

Riboflavin biosynthesis was found in plants and most microorganisms but not in higher eukaryotes. Two precursors, guanosine triphosphate, also needed for RNA, and ribulose-5-phosphate, provided by pentose phosphate pathway, are the roots of this anabolism (Fig. 1.6; Stahmann and Hohmann 2020). Despite the dephosphorylation of the ribityl residue, which is catalyzed by an unspecific phosphatase, all enzymes are known and all genes are cloned for *A. gossypii* (Revuelta et al. 1995). While reduction of the ribosyl residue comes prior to deamination of the diamino pyrimidinone in fungi, the sequential order of these reactions is inverted in bacteria (Burrows and Brown 1978). Because the last step, formation of riboflavin, is a dismutation of two molecules 6,7-dimethyl-8-ribityllumazine, in total two molecules of ribulose-5-phosphate and one molecule guanosine triphosphate are needed. In total half of the carbon origins from ribulose-5-phosphate (Fig. 1.6). Because of that stoichiometry activity of 3,4-dihydroxy-2-butanone-4-phosphate syn-

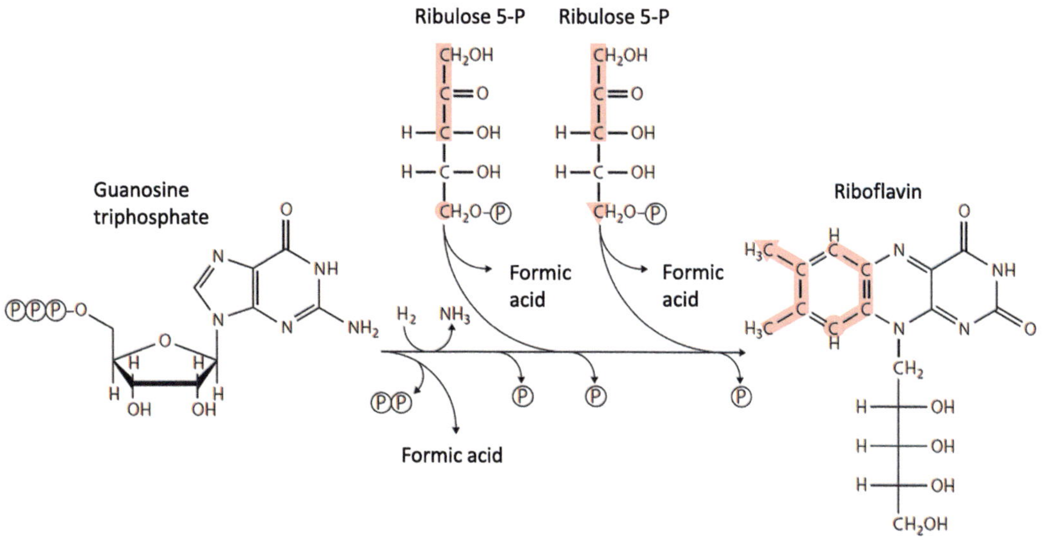

Fig. 1.6 Riboflavin biosynthesis, an overview. One guanosine triphosphate and two ribulose 5-P are needed for the synthesis of riboflavin in fungi as well as in bacteria (Stahmann and Hohmann 2020). The carbon atoms originating from ribulose 5-P are highlighted in the precursors and in the product

thase must be double in comparison with the other enzymes. Detailed insights into riboflavin biosynthesis and the catalytic mechanisms of the involved enzymes are described by Fischer and Bacher (2005).

Isolation of the *RIB* Genes

The yeast *Saccharomyces cerevisiae* played an important role in the genetic analysis of riboflavin biosynthesis. Mating between riboflavin auxotrophic mutants obtained by random mutagenesis and screening of haploid strains revealed six complementation groups (Oltmanns and Bacher 1972). The *S. cerevisiae* genes *RIB1* to *RIB5* and *RIB7* were then cloned by functional complementation of the respective auxotrophic mutants (Revuelta et al. 1994). The *RIB* genes of *A. gossypii* could not be identified based on the *S. cerevisiae RIB* sequence information because neither the genome of *S. cerevisiae* nor that of *A. gossypii* was available at that time. Functional expression of *A. gossypii* cDNA or genomic DNA libraries in the *S. cerevisiae* riboflavin auxotrophs allowed cloning of the *A. gossypii RIB* genes (Revuelta et al. 1995).

Regulation of Riboflavin Overproduction

Overproduction of riboflavin by *A. gossypii* cultivated on glucose as carbon and energy source was found to be linked to the formation of ascospores (Fig. 1.5). Supplementation of the medium with cAMP or non-metabolizable derivatives of cAMP arrested the fungus in the state of a vegetative mycelium without riboflavin overproduction. This phenomenon is an argument for a cAMP-dependent signaling cascade controlling riboflavin overproduction (Schlösser et al. 2001). Wendland (2020) studied the role of cAMP in sporulation. Wendlands research group suggested that sporulation and riboflavin overproduction are under mating type locus and nutritional control (Wasserstrom et al. 2017).

An independent approach to study regulation was performed by chemostatic cultivation. *A. gossypii* did not overproduce riboflavin at constant dilution rates. Dilution rate down-shifts causing nutritional stress resulted in sporulation of the culture and riboflavin overproduction. Increased *RIB3*, *RIB4*, and *RIB5*, but not *RIB2* and *RIB7* mRNA levels were detected after the nutritional downshift (Schlösser et al. 2007). In batch culture the DHBP synthase activity increased 50-fold when growth rate declined due to an increased transcription initiation as revealed by promoter-reporter fusion experiments (Schlösser et al. 2001). Whereas the molecular mechanism of *rib* gene regulation in *A. gossypii* has still to be elucidated glucose repression could be excluded (Schlösser et al. 2001). Since the *RIB* genes are not clustered but distributed over the chromosomes the mechanism of induction or derepression might be new. Using *lacZ* as reporter for the strength of the fused *RIB3* promoter up to 30-fold increase in beta-galactosidase specific activity was reported (Nieland and Stahmann 2013).

Since the initiation of *RIB* gene expression in *A. gossypii* correlates with sporulation a linked regulatory mechanism for asci development and the induction of riboflavin overproduction can be suggested. *A. gossypii*, originally described as severe but today a negligible plant pathogen, was isolated from plant tissues like cotton balls (*Gossypium hirsutum*). They were reported to become yellowish (Batra 1973) probably due to pigmentation with riboflavin when infected by the fungus. Riboflavin can protect the hyaline *A. gossypii* spores against UV-light induced damage (Stahmann et al. 2001). *A. gossypii* is unable to penetrate the epidermis of plant tissue and therefore depends on insects with piercing-sucking mouthparts like *Dysdercus nigrofasciatus*. They take up and distribute the spores during feeding at the plants. A riboflavin pigmentation of *A. gossypii* infected tissue, probably also taking place in citrus fruits (Dammer and Ravelo 1990) might attract the insects to the infected plants.

Classical and Molecular Tools for Strain Improvement

The wild type of *A. gossypii*, isolated by Ashby and Nowell (1926), is a natural overproducer of riboflavin, accumulating 100 mg/g biomass under growth-limiting cultivation conditions (Schlösser et al. 2001). The selection of improved production strains is possible by visual inspection of colonies obtained by cultivation of a randomly mutagenized *A. gossypii* population on agar plates. Submerged production of haploid monokaryotic spores, which can be easily harvested by extraction with lipophilic solvents, allows the generation of classical mutants at high throughput.

A broad spectrum of molecular tools is available including plasmid transformation, replacement mutagenesis with dominant marker genes, restriction enzyme-mediated chromosomal integration, and transposon mutagenesis (Wright and Philippsen 1991). Replacement mutagenesis with a high site specificity is possible with less than 50 bp homology regions flanking the DNA fragment, which has to be integrated into the genome (Wendland et al. 2000). Pop-out of dominant markers, e.g. the *kan* gene which provides geneticin resistance, is possible via short repetitive DNA sequences. This option allows a repeated use of the same marker for consecutive chromosomal DNA manipulations.

Today the *A. gossypii* genome sequence is publicly available (Gattiker et al. 2007). Previously, functional complementation of *Saccharomyces cerevisiae* mutants with *A. gossypii* genomic libraries was the method of choice to isolate genes. This was possible because introns are rare in the *A. gossypii* genome.

Metabolic Engineering

Riboflavin Pathway

Additional copies of *RIB3*, *RIB4*, and *RIB5* were integrated into the genome of *A. gossypii* (Althöfer et al. 1999) using a geneticin resistance cassette controlled by the *A. gossypii TEF* promoter (Steiner and Philippsen 1994). The original *RIB* promoters caused more than a doubling in productivity when the itaconate-resistant mutant Ita-GS-01 (Schmidt et al. 1996b) was used as parental strain (Althöfer et al. 1999). Integration of additional copies of the *RIB1*, *RIB2*, *RIB 4*, and *RIB7* into the genome of *A. gossypii* LU21 also resulted in a significant increase in riboflavin production (Althöfer et al. 2001). Flanking of the geneticin resistance gene with two TEF promoter elements facilitated the pop-out of the resistance marker. The marker-free strains fulfill the legal requirements for self-cloning strains according to EC Council directive

98/81. The above-mentioned two examples are applications of restriction enzyme-mediated integration. Although successful the method has the disadvantage to risk a disruption of loci involved in product formation (Casas-Flores et al. 2004). Site-specific insertion of linear DNA by double cross-over into the *A. gossypii* genome is possible in *A. gossypii* (Steiner et al. 1995). The low frequency of homologous recombination needs larger screening efforts but provides strains with a defined molecular structure. Today, CRISPR/Cas9 and CRISPR/Cpf1 systems are developed enabling gene deletions and the introduction of mutations (Jiménez et al. 2020).

Riboflavin Precursor Supply

In the purine biosynthesis pathway, phosphoribosylamine synthesis, catalyzed by phosphoribosylpyrophosphate amidotransferase, is negatively regulated at the transcriptional level by adenine in *A. gossypii*. For a deregulated expression of this amidotransferase the encoding *AgADE4* gene was functionally linked to the constitutive promoter of the glyceraldehyde-3-phosphate dehydrogenase gene. Furthermore, a feed-back-resistant amidotransferase mutein designed by three amino acid exchanges was constructed by site-directed mutagenesis. Overexpression of the mutein gene in the wild type of *A. gossypii* led to an increased riboflavin production (Jiménez et al. 2005).

Glycine supplementation is known to stimulate riboflavin overproduction of *A. gossypii* (Demain 1972) presumably by overcoming a metabolic limitation in de novo biosynthesis of the riboflavin precursor GTP. This idea is supported by the effect of the glycine analog aminomethylphosphonic acid, which did not increase but competitively reduced riboflavin production (Monschau et al. 1999), suggesting that glycine did not induce riboflavin production via a regulatory function but alleviated a possible glycine shortage as a substrate.

Overexpression of *GLY1* encoding a threonine aldolase increased riboflavin production only if the medium was supplemented with L-threonine. Since *A. gossypii* prefers uptake of L-threonine and excretion of glycine the boosting effect exceeds simple supplementation of the medium with the precursor of purine biosynthesis (Monschau et al. 1998).

Elevated glycine supply also explains the enhanced riboflavin overproduction of *SHM2*-disrupted *A. gossypii* mutants. *SHM2* encodes the cytosolic serine hydroxymethyltransferase catalyzing conversion of glycine into L-serine by consumption of methylene tetrahydrofolate (Schlüpen et al. 2003). In vivo ^{13}C-labeling experiments confirmed that this cytosolic serine hydroxymethyltransferase was a carbon-one compound producer during growth but a glycine-consuming enzyme during the riboflavin production phase.

Isocitrate Lyase and Isocitrate Dehydrogenase

Since technical production uses plant triglycerides as carbon source the glyoxylate pathway came into consideration to be flux limiting. Indeed, *A. gossypii* mutants resistant to inhibitors of isocitrate lyase, e.g. itaconate, showed increased riboflavin production (Schmidt et al. 1996a, b). Isocitrate lyase and malate synthase constitute the anaplerotic glyoxylate shunt providing malate from acetyl-CoA, which is derived, e.g., from beta-oxidation of fatty acids. The genes of both enzymes are induced upon glucose depletion (Maeting et al. 1999) and assimilation of the intracellular reserve lipids (Stahmann et al. 1994) or upon growth on plant lipids. The mutation leading to itaconate resistance increased isocitrate lyase specific activity (Schmidt et al. 1996b) probably favoring anaplerosis but also 3,4-dihydroxy-2-butanone 4-phosphate synthase specific activity was significantly increased (Schlösser et al. 2001). The itaconate resistance might be a result of a pleiotropic mutation, affecting upregulation of the glyoxylate shunt as well as upregulation of the riboflavin pathway. A first transcription factor identified to have an impact on riboflavin production was Ag*BAS1*. A truncation of amino acids 632–743 led to an enhanced productivity but avoided an auxotrophy for adenine, which was seen when the gene was replaced with a dominant marker (Mateos et al. 2006).

Isocitrate lyase and malate synthase are transported into the peroxisomes (Fig. 1.7) of *A. gos-*

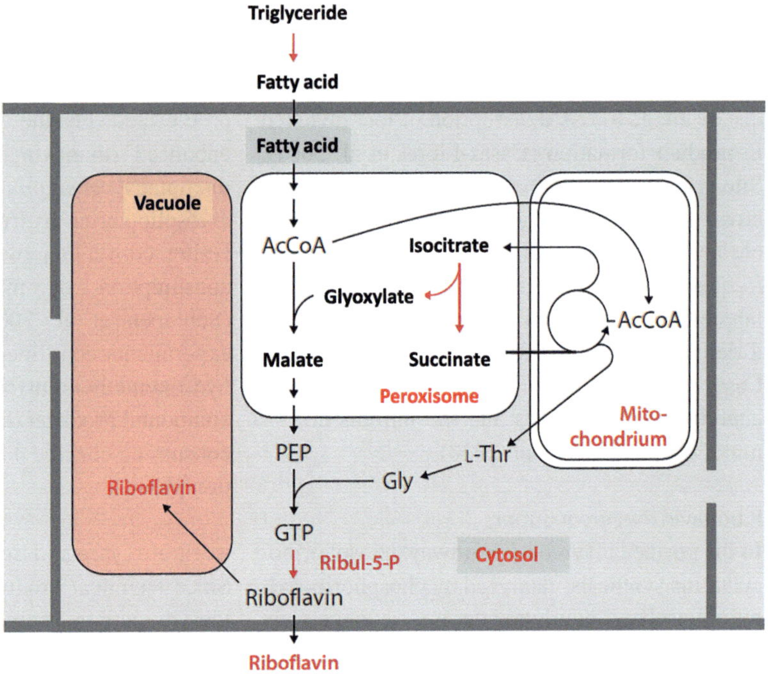

Fig. 1.7 Diagram illustrating simplified metabolic pathways and subcellular compartments relevant for riboflavin formation in a cell of *Ashbya gossypii*; subcellular compartments printed red contain anabolic steps, arrows printed red indicate regulated enzymatic steps. L-Thr, L-threonine; Gly, glycine; GTP, guanosine triphosphate; AcCoA, acetyl coenzyme A; PEP, phosphoenol pyruvate

sypii (Maeting et al. 1999), an organelle in which beta-oxidation takes place (Maeting et al. 2000). Interestingly, a knock-out of the gene encoding a peroxisomal isocitrate dehydrogenase led to a reduced production of riboflavin. Overexpression of the gene using the constitutive *TEF* promoter led to an increase in riboflavin production. A higher activity of the peroxisomal NADP-dependent isocitrate dehydrogenase might lift the intracellular levels of NADPH, which might be limiting in *A. gossypii* cells assimilating unsaturated fatty acids liberated from plant oils (Althöfer et al. 2001).

Genome, Transcriptome, Proteome, Metabolome

The genome of *A. gossypii* contains less than 6000 genes. Concerning growth and production the relevant transcriptome, proteome, and metabolome are smaller. Therefore, metabolic models of *A. gossypii* are now used as a scaffold to differentially analyze microarray data from both, growth and production phase. In total, less than 50 reactions showed correlated increases. Transcript levels and metabolic fluxes in studied, but also in not-studied catalytic pathways were found. All the genes identified as up- or down-regulated are interesting (Ledesma-Amaro et al. 2014; Schwechheimer et al. 2018).

Riboflavin Excretion and Vacuolar Accumulation

Kinetic studies of riboflavin influx and efflux with riboflavin auxotrophic or overproducing *A. gossypii* mutants indicated the presence of active riboflavin transport systems (Förster et al. 2001). Inhibition of riboflavin uptake by extracellular FMN or FAD led to an increase in the apparent riboflavin efflux in the early production phase indicating the presence of a separate excretion carrier.

A. gossypii transports the yellow pigment not only into the cultivation medium. A significant amount is transported into the vacuole and is accumulated there so that crystals can be formed. Disruption of *VMA1* encoding a subunit of the vacuolar ATPase did not interfere with the viability of *A. gossypii*, but no vacuolar riboflavin was detectable and instead all riboflavin was excreted into the culture medium (Förster et al. 1999). The genes encoding the riboflavin transporter of the vacuolar membrane and probably two others for import and export via the cytoplasmic membrane are not identified yet.

Production and Downstream Processing

The industrial riboflavin production process utilizes soybean oil and soybean meal as carbon and energy source. Since these are present in the fermentation broth as osmotically inactive emulsified lipid droplets or solid particles a high nutrient load is possible. *A. gossypii* secretes a 35 kDa lipase (Stahmann et al. 1997) hydrolyzing the triglycerides in the culture medium into free fatty acids, which are then ingested by the fungus. In stirred or shaken systems the lipase is inactivated within minutes due to aggregation at the gas/water and the lipid/water interface. Furthermore, elevated concentrations of free fatty acids interfere with lipase production (Stahmann et al. 1997). Recently, the fatty acid uptake system was identified. The overexpression of its gene increases the overproduction of riboflavin (Hoff et al. 2015).

Riboflavin overproducing cells accumulate significant amounts of the product as intracellular crystals. A heating step after completion of the fermentation run induces self-lysis and liberates riboflavin from the biomass (Kurth 1992). Heating of the fermentation broth and slow cooling over several hours promotes riboflavin crystals formation in the broth facilitating the separation of the crystals from a supernatant by decantation (Faust et al. 1991). A partial purification is carried out by resuspension of the riboflavin containing precipitate in diluted acid, heating, and decantation (Grimmer et al. 1992). Since *A. gossypii* production strains with increased copy numbers of homologous genes are classified as self-clones remains of the production microorganism in the feed application product are tolerated.

1.4 Conclusions and Outlook

Mineral-oil-based chemical riboflavin production, successfully used for half a century, was replaced by two microbial processes, one with a bacterium, used by many companies and the other by a filamentous fungus, applied by one company, now for three decades. These microbial processes save half of the costs, reduce waste and energy requirements, and use renewable resources like sugar or plant oil.

A main point is that more than a hundred biochemical conversions take place in a single cell and these grow and produce in a single vessel. A complex multi-step plant was replaced by a single-step process. The key to this remarkable success were decades of strain development. A combination of selection after random mutagenesis and genetic engineering based on rational design resulted in strains improved not only by the "removal of a bottleneck" but also by the generation of a "network of anabolic highways" and the deletions of unwished reactions. A sensible use of "high tech" like NMR or genome sequencing has allowed to explain phenomena but simple techniques like the comparison of the yellow color of different strains on agar plates allowing to predict which one will be better in the hundred cubic meter vessel are sometimes more powerful.

The coexistence of fast bacterial processes using *Bacillus subtilis* and a much slower fungal process applying *A. gossypii* means that both are more efficient than chemistry and not under pressure because the market is growing and the selling price is high. This is probably true for both, riboflavin and ARA, because they are needed in food or feed in low dosages. The design of both *A. gossypii* and *M. alpina* strains is not at an end. Many catalytic steps have potential for improvement and some steps are not yet known.

An open question is the transport of riboflavin and its precursors across membranes. Neither the transporter accumulating the pigment in the vacuole nor the carrier excreting it against a concentration gradient is known. Little is known about the facilitation of shuttle processes in the cell, e.g. between peroxisomes and mitochondria.

A problem in studying fatty acid biosynthesis is the lack of in vitro assays for desaturases and elongases, which are probably working on membranes. Overexpression and in vivo data do help indirectly but determination of specific activities would shift the research on the most relevant level for their direct comparison. Mamani et al. (2019) calculated a demand for arachidonic acid of 1 million t per year by 2025. Since the price for arachidonic acid is high even with the current productivity of *Mortierella alpina* industrial production will rise.

Vitamin C (L-ascorbic acid) might be a future product for *Aspergillus niger*. It is produced since the 1930ies by the Reichstein seven-step process. The process starts with D-glucose as precursor. Six steps are chemically performed and only one step, the regioselective oxidation of D-sorbitol to 2-keto-L-gulonic acid is catalyzed extracellularly by *Gluconobacter oxydans*. Concentrations are >100 g per liter. *A. niger*, engineered in three enzymatic steps, can directly produce 100 mg per liter vitamin C starting from D-galacturonate. The uptake system for D-galacturonate was identified (Sloothaak et al. 2014) and its conversion to L-galactonate is catalyzed by the reductase GaaA (Martens-Uzunova and Schaap 2008). Heterologous expression of the genes encoding L-gulono-1,4-lactone lactonase and L-galactono-lactone dehydrogenase from *Malpighia glabra*, a tropical fruit tree, resulted in ascorbate production. The third step is deletion of the gene encoding L-galactonate dehydrogenase (Gaab; Martens-Uzunova and Schaap 2008). A source of the precursor D-galacturonate can be pectin, a polysaccharide of the plant cell wall (Mohnen 2008).

A comparison of all processes discussed in this chapter leads to the question of robustness. Most probably, the citric acid production by *A. niger* is leading because of its acidic pH. Recent studies with *Aspergillus fumigatus* showed that this species followed Arrhenius' rule when the temperature is shifted from 30 °C to 40 °C (Nieland et al. 2021). This means that a microbial process, where temperature and nothing else is limiting, is possible. Now, it might become a tool to check the potential for further strain improvement. Second, the optimum process temperature for a eukaryotic system might be between 40 °C and 50 °C. To get closer to that optimum will reduce fermentation time, reduce cooling costs, and contribute to less contaminations by other microorganisms.

References

Adugna N (2022) Methods of obtaining lipids from a microbial cell composition. U.S. Patent Application Nr. 17/439, 134

Alessandri JM, Guesnet P, Vancassel S, Astorg P et al (2004) Polyunsaturated fatty acids in the central nervous system: evolution of concepts and nutritional implications throughout life. Reprod Nutr Dev 44(6):509–538

Althöfer H, Seulberger H, et al. (1999) Genetic method for producing riboflavin. WO 9961623

Althöfer H, Zelder O, et al. (2001) Monocellular or multicellular organisms for production of riboflavin. WO0111052

Ashby SF, Nowell N (1926) The fungi of stigmatomycosis. Ann Bot 40:69–83

Batra LR (1973) Nematosporaceae (Hemiascomycetidae): taxonomy, pathogenicity, distribution and vector relations. USDA Technol Bull 1469:1–71

Biddy MJ, Scarlata C, Kinchin C (2016) Chemicals from biomass: a market assessment of bioproducts with near-term potential (No. NREL/TP-5100-65509). National Renewable Energy Lab (NREL), Golden, CO

Burrows RB, Brown GM (1978) Presence in *Escherichia coli* of a deaminase and a reductase involved in biosynthesis of riboflavin. J Bacteriol 136:657

Cairns TC, Barthel L, Meyer V (2021) Something old, something new: challenges and developments in *Aspergillus niger* biotechnology. Essays Biochem 65(2):213–224

Casas-Flores S, Rosales-Saavedra T et al (2004) Three decades of fungal transformation: novel technologies. Methods Mol Biol 267:315–325

Cashmore AR, Jarillo JA, Wu Y, Liu D (1999) Cryptochromes: blue light receptors for plants and animals. Science 284:760–765

Chang L, Lu H, Chen H, Tang X et al (2022) Lipid metabolism research in oleaginous fungus *Mortierella alpina*: current progress and future prospects. Biotechnol Adv 54:107794

Chen H, Yongquan C, Chen W, Chengfeng GE et al. (2022). U.S. Patent No. 11, 266, 167

Chroumpi T, Mäkelä MR, de Vries RP (2020) Engineering of primary carbon metabolism in filamentous fungi. Biotechnol Adv 43:107551

Currie JN (1917) The citric acid fermentation of *Aspergillus niger*. J Biol Chem 31(1):15–37

Dammer KH, Ravelo HG (1990) Verseuchung von *Leptoglossus gonagra* (Fabr.) mit *Nematospora coryli* Peglion und *Ashbya gossypii* (Ashby et Novell) Guilliermond in einer Zitrusanlage der Republik Kuba. Arch Phytopathol Pflanzenschutz, Berlin 26:71–78

Demain AL (1972) Riboflavin oversynthesis. Ann Rev Microbiol 26:369–388

Dietrich FS, Voegeli S et al (2004) The *Ashbya gossypii* genome as a tool for mapping the ancient

Saccharomyces cerevisiae genome. Science 304(5668):304–307

Eggersdorfer M, Adam G (1996) Vitamins., Chapter 1. In: Ullmann's encyclopedia of industrial chemistry, vol A27. VCH Verlagsgesellschaft, Weinheim, pp 521–530

Ernst R, Ejsing CS, Antonny B (2016) Homeoviscous adaptation and the regulation of membrane lipids. J Mol Biol 428(24):4776–4791

Faust T, Meyer J, et al. (1991) Process for the separation of riboflavin from a fermentation suspension. EP0438767

Fischer M, Bacher A (2005) Biosynthesis of flavocoenzymes. Nat Prod Rep 22(3):324–350

Fletcher RJ, Bell IP, Lambert JP (2004) Public health aspects of food fortification: a question of balance. Proc Nutr Soc 63:605–614

Förster C, Santos MA, Ruffert S, Krämer R, Revuelta JL (1999) Physiological consequence of the disruption of the *VMA1* gene in the riboflavin overproducer *Ashbya gossypii*. J Biol Chem 274:9442–9448

Förster C, Revuelta JL et al (2001) Carrier-mediated transport of riboflavin in *Ashbya gossypii*. Appl Microbiol Biotechnol 55(1):85–89

Fukushima T, Sia AK, Allred BE, Nichiporuk R et al (2012) *Bacillus cereus* iron uptake protein fishes out an unstable ferric citrate trimer. Proc Natl Acad Sci 109(42):16829–16834

Gattiker A, Rischatsch R et al (2007) *Ashbya* genome database 3.0: a cross-species genome and transcriptome browser for yeast biologists. BMC Genomics 8:9

Grimmer J, Kiefer H, et al. (1992) Process for the purification of riboflavin obtained by fermentation. EP0464582

Heird and Lapillonne (2005) The role of essential fatty acids in development. Annu Rev Nutr 25:23.1–23.23

Hoff B, Molt A, Haefner S, Zelder O (2015) Patent WO2015/086427

Hoff B, Plassmeier J et al (2021) Unlocking Nature's biosynthetic power—metabolic engineering for the fermentative production of chemicals. Angew Chem Int Ed 60(5):2258–2278

Hohmann HP, Stahmann KP (2010) Biotechnology of riboflavin production. In: Lew M, Hung WL (eds) Comprehensive natural products chemistry II. Elsevier, Oxford

Hou CT (2008) Production of arachidonic acid and dihomo-gamma-linolenic acid from glycerol by oil-producing filamentous fungi, *Mortierella* in the ARS culture collection. J Ind Microbiol Biotechnol 35(6):501–506

Huang YS, Chaudhary S, Thurmond JM, Bobik EG Jr, Yuan L, Chan GM, Kirchner SJ, Mukerji P, Knutzon DS (1999) Cloning of delta12- and delta6-desaturases from Mortierella alpina and recombinant production of gamma-linolenic acid in Saccharomyces cerevisiae. Lipids 34:649–659

Jiménez A, Santos MA et al (2005) Metabolic engineering of the purine pathway for riboflavin production in *Ashbya gossypii*. Appl Environ Microbiol 71(10):5743–5751

Jiménez A, Hoff B, Revuelta JL (2020) Multiplex genome editing in *Ashbya gossypii* using CRISPR-Cpf1. New Biotechnol 57:29–33

Karaffa L, Kubicek CP (2003) *Aspergillus niger* citric acid accumulation: do we understand this well working black box? Appl Microbiol Biotechnol 61:189–196

Karaffa L, Fekete E, Kubicek CP (2021) The role of metal ions in fungal organic acid accumulation. Microorganisms 9(6):1267

Kawashima H, Akimoto K, Yamada H et al. (1992) Process for production of 8,11-eicosadienoic acid. Patent EPA 0 535 941 A1

Knutzon DS, Thurmond JM, Huang YS, Chaudhary S, Bobik EG Jr, Chan GM, Kirchner SJ, Mukerji P (1998) Identification of Delta5-desaturase from *Mortierella alpina* by heterologous expression in Bakers' yeast and canola. J Biol Chem 273:29360–29366

Kornecki JF, Carballares D, Tardioli PW, Rodrigues RC et al (2020) Enzyme production of D-gluconic acid and glucose oxidase: successful tales of cascade reactions. Cat Sci Technol 10(17):5740–5771

Kubicek CP (1988) Regulatory aspects of the tricarboxylic acid cycle in filamentous fungi—a review. Trans Br Mycol Soc 90(3):339–349

Kurth R (1992) Process for the enhancement of riboflavin levels in spray-dried riboflavin fermentation extracts. EP0487985

Lago BD, Kaplan L (1981) Vitamin fermentations: B_2 and B_{12}. Adv Biotechnol 3:241–246

Ledesma-Amaro R, Kerkhoven EJ, Revuelta JL, Nielsen J (2014) Genome scale metabolic modeling of the riboflavin overproducer *Ashbya gossypii*. Biotechnol Bioeng 111(6):1191–1199

Leskovac V, Trivić S et al (2005) Glucose oxidase from *Aspergillus niger*: the mechanism of action with molecular oxygen, quinones, and one-electron acceptors. Int J Biochem Cell Biol 37(4):731–750

Lindberg AM, Molin G (1993) Effect of temperature and glucose supply on the production of polyunsaturated fatty acids by the fungus *Mortierella alpina* in fermenter cultures. Appl Microbiol Biotechnol 39:450–455

Liu X, Tian X, Hang H, Zhao W et al (2017) Influence of initial glucose concentration on seed culture of sodium gluconate production by *Aspergillus niger*. Bioresour Bioprocess 4(1):1–14

Lounds C, Eagles J, Carter AT, MacKenzie DA, Archer DB (2007) Spore germination in *Mortierella alpina* is associated with a transient depletion of arachidonic acid and induction of fatty acid desaturase gene expression. Arch Microbiol 188(4):299–305

Maeting I, Schmidt G, Sahm H, Revuelta JL, Stierhof YD, Stahmann KP (1999) Isocitrate Lyase of *Ashbya gossypii* - transcriptional regulation and peroxisomal localization. FEBS Lett 444:15–21

Maeting I, Schmidt G, Sahm H, Stahmann KP (2000) Role of a peroxisomal NADP-specific isocitrate dehydrogenase in the metabolism of the riboflavin overproducer *Ashbya gossypii*. J Mol Catal 10(1–3):335–343

Mamani LDG, Magalhães AI Jr, Ruan Z, de Carvalho JC, Soccol CR (2019) Industrial production, patent landscape, and market trends of arachidonic acid-rich oil of *Mortierella alpina*. Biotechnol Res Innov 3(1):103–119

Marszalek JR, Lodish HF (2005) Docosahexaenoic acid, fatty acid–interacting proteins, and neuronal function: breastmilk and fish are good for you. Annu Rev Cell Dev Biol 21:633–657

Martens-Uzunova ES, Schaap PJ (2008) An evolutionary conserved D-galacturonic acid metabolic pathway operates across filamentous fungi capable of pectin degradation. Fungal Genet Biol 45(11):1449–1457

Mateos L, Jiménez A, Revuelta JL, Santos MA (2006) Purine biosynthesis, riboflavin production, and trophic-phase span are controlled by a Myb-related transcription factor in the fungus *Ashbya gossypii*. Appl Environ Microbiol 72(7):5052–5060

Meyer V, Wanka F, van Gent J, Arentshorst M et al (2011) Fungal gene expression on demand: an inducible, tunable, and metabolism-independent expression system for *Aspergillus niger*. Appl Environ Microbiol 77(9):2975–2983

Mohnen D (2008) Pectin structure and biosynthesis. Curr Opin Plant Biol 11(3):266–277

Monschau N, Sahm H, Stahmann KP (1998) Threonine aldolase overexpression plus threonine supplementation enhanced riboflavin production in *Ashbya gossypii*. Appl Environ Microbiol 64:4283–4290

Monschau N, Stahmann KP, Sahm H, Zelder O (1999) Uni- or multicellular microorganisms for the production of riboflavin. Patent EP 0930 367 A2

Nieland S, Stahmann KP (2013) A developmental stage of hyphal cells shows riboflavin overproduction instead of sporulation in *Ashbya gossypii*. Appl Microbiol Biotechnol 97:10143–10153

Nieland S, Barig S, Salzmann J, Gehrau F et al (2021) *Aspergillus fumigatus* AR04 obeys Arrhenius' rule in cultivation temperature shifts from 30 to 40° C. Microb Biotechnol 14(4):1422–1432

Odoni DI, van Gaal MP, Schonewille T, Tamayo-Ramos JA et al (2017) *Aspergillus niger* secretes citrate to increase iron bioavailability. Front Microbiol 8:1424

Oltmanns O, Bacher A (1972) Biosynthesis of riboflavin in *Saccharomyces cerevisiae*: the role of genes rib1 and rib7. J Bacteriol 110(3):818–822

Perlman D (1979) Microbial process for riboflavin production. In: Peppler H, Perlman D (eds) Microbial technology, microbial processes, vol 1, 2nd edn. Academic Press, New York, pp 521–527

Ratledge C (2004) Fatty acid biosynthesis in microorganisms being used for single cell oil production. Biochimie 86(11):807–815

Revuelta DJL, Sanios Garcia MA, Garcia Ramirez JJ, Gonzales-Hernandez GA, Buitrago Sema MJ (1994) *Saccharomyces cerevisiae* recombinant riboflavin production. Patent WPI 94-177113

Revuelta DJL, Buitrago SMJ, Santos GMA (1995) Riboflavin synthesis in fungi. Patent WO 9526406-A

Sakuradani E, Kobayashi M, Shimizu S (1999a) Delta6-fatty acid desaturase from an arachidonic acid-producing *Mortierella* fungus. Gene cloning and its heterologous expression in a fungus, *Aspergillus*. Gene 238:445–453

Sakuradani E, Kobayashi M, Shimizu S (1999b) Delta9-fatty acid desaturase from arachidonic acid-producing fungus. Unique gene sequence and its heterologous expression in a fungus, *Aspergillus*. Eur J Biochem 260:208–216

Sakuradani E, Kobayashi M, Ashikari T, Shimizu S (1999c) Identification of Delta12-fatty acid desaturase from arachidonic acid-producing mortierella fungus by heterologous expression in the yeast Saccharomyces cerevisiae and the fungus *Aspergillus oryzae*. Eur J Biochem 261:812–820

Sakuradani E, Ando A, Ogawa J, Shimizu S (2009) Improved production of various polyunsaturated fatty acids through filamentous fungus *Mortierella alpina* breeding. Appl Microbiol Biotechnol 84(1):1–10

Schlösser T, Schmidt G et al (2001) Transcriptional regulation of 3,4-dihydroxy-2-butanone 4-phosphate synthase. Microbiology 147(Pt 12):3377–3386

Schlösser T, Wiesenburg A et al (2007) Growth stress triggers riboflavin overproduction in *Ashbya gossypii*. Appl Microbiol Biotechnol 76(3):569–578

Schlüpen C, Santos MA, Weber U et al (2003) Disruption of the SHM2 gene, encoding one of two serine hydroxymethyltransferase isoenzymes, reduces the flux from glycine to serine in *Ashbya gossypii*. Biochem J 369(2):263–273

Schmidt G, Stahmann KP, Sahm H (1996a) Isolation and characterization of isocitrate lyase from the riboflavin-producing fungus *Ashbya gossypii*. Microbiology 142:411–417

Schmidt G, Stahmann KP, Kaesler B, Sahm H (1996b) Correlation of isocitrate lyase activity and riboflavin formation in the riboflavin overproducer *Ashbya gossypii*. Microbiology 142:419–426

Schwechheimer SK, Becker J et al (2018) Improved riboflavin production with *Ashbya gossypii* from vegetable oil based on 13C metabolic network analysis with combined labeling analysis by GC/MS, LC/MS, 1D, and 2D NMR. Metab Eng 47:357–373

Sloothaak J, Schilders M, Schaap PJ, de Graaff LH (2014) Overexpression of the *Aspergillus niger* GatA transporter leads to preferential use of D-galacturonic acid over D-xylose. AMB Express 4:1–9

Stahmann KP, Hohmann HP (2020) Vitamins, Nukleotides und Carotenoids. In: Wilson DB, Sahm H, Koffas M (eds) Industrial microbiology. Wiley-Verlag, Weinheim, pp 161–183

Stahmann KP, Kupp C, Feldmann SD et al (1994) Formation and degradation of lipid bodies found in the riboflavin-producing fungus *Ashbya gossypii*. Appl Microbiol Biotechnol 42:121–127

Stahmann KP, Böddecker T, Sahm H (1997) Regulation and properties of a fungal lipase showing interfacial inactivation by gas bubbles, or droplets of lipid or fatty acid. Eur J Biochem 244:220–225

Stahmann KP, Revuelta JL, Seulberger H (2000) Three biotechnical processes using *Ashbya gossypii*,

Candida famata, or *Bacillus subtilis* compete with chemical riboflavin production. Appl Microbiol Biotechnol 53:509–516

Stahmann KP, Arst HN et al (2001) Riboflavin, overproduced during sporulation of *Ashbya gossypii*, protects its hyaline spores against ultraviolet light. Environ Microbiol 3(9):545–550

Steiger MG, Rassinger A, Mattanovic D, Sauer M (2019) Engineering of the citrate exporter protein enables high citric acid production in *Aspergillus niger*. Metab Eng 52:224–231

Steiner S, Philippsen P (1994) Sequence and promoter analysis of the highly expressed TEF gene of the filamentous fungus *Ashbya gossypii*. Mol Gen Genet 242:263–271

Steiner S, Wendland J, Wright MC et al (1995) Homologous recombination as the main mechanism for DNA integration and cause of rearrangements in the filamentous ascomycete *Ashbya gossypii*. Genetics 140:973–987

Takeno S, Sakuradani E, Murata S, Inohara-Ochiai M, Kawashima H, Ashikari T, Shimizu S (2004) Establishment of an overalltransformation system for an oil-producing filamentous fungus, *Mortierella alpina* 1S-4. Appl Microbiol Biotechnol 65(4):419–425

Takeno S, Sakuradani E, Tomi A, Inohara-Ochiai M, Kawashima H, Shimizu S (2005) Transformation of oil-producing fungus, *Mortierella alpina* 1S-4, using Zeocin, and application to arachidonic acid production. J Biosci Bioeng 100(6):617–622

Wasserstrom L, Dünkler A, Walther A, Wendland J (2017) The APSES protein Sok2 is a positive regulator of sporulation in *Ashbya gossypii*. Mol Microbiol 106(6):949–960

Wendland J (2020) Sporulation in *Ashbya gossypii*. J Fungi 6(3):157–171

Wendland J, Ayad-Durieux Y et al (2000) PCR-based gene targeting in the filamentous fungus *Ashbya gossypii*. Gene 242(1–2):381–391

Wickerham LJ, Flickinger MH, Johnsten RM (1946) The production of riboflavin by *Ashbya gossypii*. Arch Biochem 9:95–98

Witteveen CF, Veenhuis M, Visser J (1992) Localization of glucose oxidase and catalase activities in *Aspergillus niger*. Appl Environ Microbiol 58(4):1190–1194

Wright P, Philippsen P (1991) Replicative transformation of the filamentous fungus *Ashbya gossypii* with plasmids containing *Saccharomyces cerevisiae* ARS elements. Gene 109:99–105

Zhang Y, Ratledge C (2008) Multiple isoforms of malic enzyme in the oleaginous fungus *Mortierella alpina*. Mycol Res 112(Pt 6):725–730

Industrial Applications of Marine Fungi

2

Antje Labes

Abstract

Marine fungi, a diverse group of microorganisms found in marine environments, have gained increasing attention in recent years due to their potential industrial applications. This chapter provides a comprehensive overview of the current knowledge regarding the industrial applications of marine fungi. It explores their biotechnological potential and discusses the challenges and prospects of harnessing the full potential of marine fungi for sustainable industrial development.

Keywords

Marine diversity · Molecular studies · Untapped resource · Strain development

2.1 Introduction: Marine Fungi

Over the last five decades, fungal biotechnology, known as "mycotechnology," has witnessed significant progress, particularly in terrestrial environments. Terrestrial fungi have been made instruments for the production of extracellular enzymes, organic acids, antibiotics, and anti-cholesterolemic statins (Magan 2002). They have served as expression hosts and sources of new genes, playing a pivotal role in modern molecular genetic tools and acting as "cell factories" for the production of heterologous proteins, including those of human origin (Liu et al. 2023; Punt et al. 2002). The diversity in the fungal tree of life and their manifold ecological functionalities are not yet reflected in the applied number of taxonomic groups and the diversity of products on the market. Hence, new fungal taxa and diverse strains can serve as a source for new products and processes. So far, non-touched ecological niches and habitats have been focused on increasing the available (i.e., cultivated) fungal diversity, including the world's oceans (Manohar and Raghukumar 2013). Marine fungi comprise a group of organisms that have been neglected for a long time; therefore, knowledge is still relatively rare. Early screening campaigns revealed marine fungi are an excellent source for new natural products, indicating their potential for biotechnological application. Especially fungi from unique and extreme marine environments, such as the deep sea, are assumed to be a rich source of unique and unusual compounds (Fenical and Jensen 2006; Vignesh et al. 2011; Vignesh et al. 2011).

After the discovery of fungi from marine resources (Kohlmeyer 1979), a scientific dispute on the ecological role and true origin of marine isolates was ongoing, especially as a majority of culture-dependent and culture-independent fungi

A. Labes (✉)
Flensburg University of Applied Science, Flensburg, Germany
e-mail: antje.labes@hs-flensburg.de

S. Grüttner et al. (eds.), *Agricultural and Industrial Applications*, The Mycota 16,
https://doi.org/10.1007/978-3-031-81904-9_2

recovered from marine resources have homology to terrestrial species. The presence of novel fungal sequences with less than 97% similarity to previously identified fungal sequences in public databases and culture studies demonstrating obligate marine culture conditions have proven the multiphyletic ecological group of marine fungi (for review, see, e.g., Raghukumar (2008)). Today, marine fungi are classified as obligate or facultative: obligate marine fungi grow and sporulate exclusively in a marine or estuarine habitat, depending on the saline condition and salt composition. In contrast, facultative marine fungi derive from freshwater or terrestrial origin but can grow in marine environments (Chi et al. 2009; Kohlmeyer and Volkmann-Kohlmeyer 2003; Li et al. 2007). More generally, the term "marine-derived fungi" is often used, especially when the ecological functionality was not the focus of the original research approach. During the last years, several sampling campaigns dedicated to obtaining culturable marine fungi were performed, but not necessarily with proof of marine activity (Reich and Labes 2017).

Recent molecular studies on marine mycology revealed environmental clusters within the major fungal phyla and assigned marine fungi to occupy a central position in many marine habitats (Manohar and Raghukumar 2013). The occurrence of marine fungi has been reported on different living and decaying substrates (e.g., sponges, algae, wood, tunicates, sediments, mollusks, corals, plants, and fish) but also as endophytes and in sediments, even in deep-sea habitats. Many sampling sites have been along coastlines, such as mangrove, sand, beach, river, and estuarine habitats, mainly due to sampling efficacy and accessibility of such habitats. This restriction remains a substantial limitation in describing the complete marine fungal abundance and ecofunctionality. Fortunately, with technological advancements and decreasing sequencing costs, genomic studies have become feasible on a global scale, and genetic data has contributed to a broader understanding of fungal biodiversities in the oceans (Compson et al. 2020). Despite the increasing efforts in isolating and discovering new species, marine fungi are still an understudied group compared to other marine microorganisms (Rotter et al. 2021; Tisthammer et al. 2016). Nevertheless, the uncultivable fungi described by genomic techniques still represent the major component of the marine fungal community (Comeau et al. 2016; Rämä et al. 2014; Xu et al. 2018).

2.2 Focus of Research on the Application Potential

Acknowledging the existing application of terrestrial fungi, the potential of the marine counterparts has become increasingly attractive in recent years. The unique ecological niches of the marine habitats implied additional and novel applications of marine fungi in "mycotechnology." The unique features may be applied in biotechnology for new products but also for redesigning existing biotechnological processes with more robust culture conditions, as some marine fungi did adapt to harsh conditions, including high pressure, high osmolarity, cold, and toxic heavy-metal concentrations. Most fungi are found in the environment in coastal environments and seasonal variability, especially concerning decomposing species, which follow the abundance of organic matter and its seasonal variability (Wang et al. 2012). Screening campaigns for marine fungi have been selectively focused on specific habitats with respect to their intended later application.

2.2.1 Plant-Associated Fungi

Fungi associated with marine algae, seagrasses, and mangroves, either as associates or endophytes, were explored for their properties and potential applications with respect to lignin and polymer degradation but also concerning plant–fungus interaction (Patyshakuliyeva et al. 2019). Fungi isolated from living and decaying marine plant material are capable of degrading both the ubiquitous plant macromolecules, like cellulose, cellobiose, lipids, pectin, starch, xylan, and tannic acid (Gessner 1980), and the marine-specific carbohydrates such as alginates, sulfated polysaccharides, laminarin, fucoidan, and others (Balabanova et al. 2018). These enzymes can be applied in transforming these sugars for the energetic use of algal remaining (e.g., for biogas production), especially from aquacultures. Also, fungal polygalacturonases are valuable enzymes for clarifying fruit juices in the food industry (Sandri et al. 2011).

2.2.2 Animal-Associated Fungi

Particular attention has been given to fungi associated with marine invertebrates, particularly corals and sponges, as potential sources of bioactive compounds with pharmaceutical or industrial relevance. Marine natural product discovery started with examining sponges and, subsequently, some other marine invertebrates. During the course of the discovery, it became apparent that associated microorganisms, including fungi, are the actual producers of these bioactive compounds (Amend et al. 2019). A wide variety of biological activities were described, and many marine fungal compounds are filling the preclinical pipeline (Gonçalves et al. 2022).

2.2.3 Deep-Sea Fungi

Fungi from extreme marine environments, such as the deep sea, hypersaline waters, and oxygen-deficient sediments, were highlighted for their adaptability and potential applications. Deep-sea sediments especially represent the sink for organic matter, creating a habitat where fungi are the dominant eukaryotic microbes (Amend et al. 2019; Nicoletti and Andolfi 2018). Deep-sea hot vent systems are often characterized by high heavy-metal concentration. Fungi, thriving in these habitats, must cope by producing detoxification compounds, often extracellular sugars. These can be applied in the cosmetic industry. Also, enzymes from these fungi show high resistance against heavy-metal intoxication (Alvarado-Campo et al. 2023). Sampling access remains the major bottleneck in gaining these organisms, along with appropriate cultivation conditions that do not favor fast-growing abundant species but specifically adapted ones (Rédou et al. 2015).

2.3 Existing Industrial Processes

Despite the growing number of studies highlighting the biotechnological and industrial potential, few marine fungi are employed as production organisms. The most prominent ones have made their way into pharmaceutical applications. Not many marine-derived compounds made their way to the pharmaceutical market. Marine fungal compounds face the same challenges in drug discovery and development as any other fungal compounds. However, already in the 1950s, cephalosporin C, a β-lactam type natural antibiotic, was discovered from a *Cephalosporium* (later reclassified as *Acremonium*) strain obtained at the Sardinian coast (Newton and Abraham 1955). Gliotoxin was identified as a new type of antibiotic, diketopiperazine, produced by a marine deep-sea sediment *Aspergillus* sp. strain (Okutani 1977). Marine fungi also proved to be effective producers of antitumoral compounds (Pejin and Karaman 2017).

The compound halimid (and its semisynthetic analogon with the trade name plinabulin) is the only anti-cancer compound of marine fungal origin already being clinically tested in phases 2 and 3 for different indications (Natoli et al. 2021). The preclinical pipeline contains a high number of candidates. However, a prediction of market success is not appropriate. A steadily increasing number of new, active fungal natural products have been identified from the marine environment, proving them to be a prolific source of bioactive compounds. Blunt et al. listed over 200 natural products from marine-sourced fungi (excluding those from mangroves (Blunt et al. 2015)). Cosmetic and cosmeceutical applications might also be considered for marine fungal secondary metabolites (Agrawal et al. 2018). Biotechnological approaches can help close the developmental gap between discovery and production to solve compound availability issues in drug development (as reviewed for antibiotics by Silber et al. (2016)).

2.4 Novel Aspects of Industrial Application of Marine Fungi

2.4.1 Bioremediation and Environmental Applications

Exceptional degradative properties dominate the fungal ecosystemic function. They are the lignin degraders; they metabolize hydrocarbons, and new evidence suggests that fungi degrade plastics (Ekanayaka et al. 2022). Plastic degradation is already a topic as plastic surfaces are becoming an increasing habitat in the ocean (Wayman and Niemann 2021). Accordingly, fungi can play an additional role in the industry as remediators rather than producers. Considering that the circular economy and zero waste approaches are increasingly important and will be the future of any industrial process if humankind wants to survive, fungi may play a crucial role in cycling carbon. Mycoremediation, the use of fungi for environmental remediation, has emerged as a promising approach to cleaning up various pollutants. While most research in mycoremediation has focused on terrestrial fungi, recent studies have highlighted the potential of marine fungi in this field. Marine fungi possess unique adaptations and enzymatic capabilities that enable them to degrade and detoxify a wide range of contaminants present in marine environments. State-of-the-art research on mycoremediation using marine fungi has demonstrated their effectiveness in degrading hydrocarbons, heavy metals, pesticides, and other pollutants. Several marine fungal species have been identified as efficient degraders of crude oil, polycyclic aromatic hydrocarbons (PAHs), and petroleum-derived compounds (for review, see Stiebeling and Labes (2022)). These fungi can produce diverse enzymes, such as ligninolytic enzymes (laccases, peroxidases), cellulases, and lipases, which are crucial in the breakdown of complex pollutants.

In addition to pollutant degradation, marine fungi can contribute to the immobilization and biosorption of heavy metals through their mycelial networks and extracellular polysaccharides. This feature makes them potential candidates for the remediation of metal-contaminated sediments and wastewater. Studies have identified various marine fungal species, including *Aspergillus*, *Penicillium*, *Fusarium*, and *Trichoderma*, that can biosorb and sequester heavy metals. These fungi utilize their mycelial networks and extracellular polysaccharides to bind and immobilize metal ions. Furthermore, they can produce metal-binding compounds and enzymes, such as chitin, melanin, and extracellular polymeric substances (EPS), which enhance metal uptake and detoxification (Silva et al. 2019; Stiebeling and Labes 2022).

Challenges still need to be addressed despite the potential of mycoremediation using marine fungi. Cultivation techniques and scale-up processes for marine fungi are relatively underdeveloped compared to their terrestrial counterparts. Furthermore, selecting and optimizing suitable fungal strains for specific pollutants and environmental conditions require further investigation.

2.4.2 Biocontrol Agents Against Pathogens

Marine fungi possess bioactive compounds and mechanisms that enable them to antagonize and inhibit the growth of pathogens. They can produce secondary metabolites such as antibiotics, enzymes, and peptides that exhibit antimicrobial properties against various pathogens, including bacteria, viruses, and fungi. Research on marine secondary metabolites soon revealed additional application fields: biocontrol agents against plant pathogens can be developed as eco-friendly alternatives for disease management in various industries, including agriculture, aquaculture, and human health.

Studies have identified several marine fungal species with biocontrol potential. Cocultures with plant pathogens (Oppong-Danquah et al. 2018) were especially proven to be effective in stimulating marine fungi to counteract as biocontrol organisms. The fungi can act as direct antagonists by competing for nutrients and space with pathogens or by producing antifungal compounds that inhibit pathogen growth. Additionally, some marine fungi can induce plant defense responses, enhancing the resistance of plants to pathogenic attacks. Efforts are also being made to optimize the production and formulation of biocontrol agents derived from marine fungi. Strategies such as solid-state fermentation, submerged fermentation, and co-cultivation have been investigated to enhance the production of bioactive metabolites. Formulation techniques, including encapsulation and biofilm formation, have been explored to improve the stability and efficacy of biocontrol agents (Rojas-Sánchez et al. 2022).

Despite progress, challenges remain in the practical application of marine fungi as biocontrol agents. Optimization of cultivation techniques, formulation strategies, and scale-up processes are necessary to ensure the commercial viability of these agents. Additionally, the ecological impact of introducing marine fungi into different ecosystems and the potential for resistance development in pathogens require careful evaluation.

2.4.3 Bioenergy and Biofuel Production

Marine fungi have recently emerged as promising candidates for the degradation of lignocellulosic materials, thereby contributing to the production of biofuels through biomass conversion. Lignocellulose, a complex matrix composed of cellulose, hemicellulose, and lignin, represents the most abundant renewable resource on Earth and is available in many marine habitats (Jex et al. 2014; Rojas-Sánchez et al. 2022). The utilization of marine fungi for biomass conversion aligns with the growing emphasis on sustainable and eco-friendly biofuel production. The abundance of lignocellulosic biomass in marine environments, often considered waste, can be harnessed as a valuable resource for biofuel production, thereby addressing both environmental and energy security concerns. Moreover, the marine fungi-mediated biomass conversion process offers the potential for integrated biorefineries, where multiple value-added products can be obtained from the same feedstock (Cheng et al. 2022).

Unlocking its potential for biofuel production is imperative for sustainable energy solutions, and marine fungi exhibit unique enzymatic capabilities that make them valuable contributors to this endeavor. Their ability to secrete an array of lignocellulolytic enzymes, such as cellulases, hemicellulases, and ligninases, enables the efficient breakdown of complex plant polymers into fermentable sugars. Recent studies have identified numerous marine fungi, belonging to genera like *Aspergillus*, *Penicillium*, and *Trichoderma*, exhibiting remarkable lignocellulolytic activities. The characterization of their enzymatic profiles has revealed novel and potent enzymes, expanding the repertoire available for biomass degradation compared to their terrestrial counterparts (Bonugli-Santos et al. 2012).

In addition to their enzymatic prowess, marine fungi contribute to biomass degradation through the modification of lignin, a recalcitrant polymer that hinders access to cellulose and hemicellulose. Fungal species like *Marasmius* sp. and *Corollospora maritima* have been found to produce lignin-modifying enzymes, facilitating the breakdown of lignin and enhancing overall lignocellulosic biomass conversion efficiency (Bonugli-Santos et al. 2010; Bonugli-Santos et al. 2012).

2.5 Challenges and Prospects

The bioprospecting of marine fungi has revealed their immense biodiversity and the potential for novel bioactive compounds, enzymes, and application in ecosystem services. However, several challenges, including cultivation techniques, genetic manipulation, and regulatory issues, must be addressed to translate marine fungal research into practical applications successfully. With further exploration and advancements, marine fungi hold promise for contributing to sustainable industrial development and addressing various societal and environmental challenges.

Today, the potential of marine fungi in mycotechnology is untapped despite opportunities for innovative biotechnological processes with applications in various industries. Few examples of industrial processes using marine fungi as production organisms are available. However, the number of studies highlighting the potential is imposing for enzymes, secondary metabolites, and other possible products. Industrial-scale biotechnological processes should be feasible from scientific, technical, and economic perspectives. This viability must be demonstrated by developing process concepts based on fermentation, including metabolic engineering, purification, molecular design, and synthesis data. Some obstacles remain, which need to be solved. Especially, the reduction of the salinity in culture media is a technological challenge.

The availability of novel marine genera is still limited concerning the cultivated diversity. The high proportion of members of well-known genera such as *Penicillium* and *Aspergillus* may reflect the high abundance of representatives of these two genera in the samples; however, the commonly used isolation media favor the growth of these fast-growing groups, reducing the probability of isolating fungi present only in low abundance and other nutrient requirements. Accordingly, culture techniques need to be adapted for the isolation of the so far unknown species. Optimization of nutrient composition for culture media is well established for numerous fungi from terrestrial environments and was confirmed for marine fungi (Imhoff 2016; Kjer et al. 2010). Marine fungal strains of genera also described from terrestrial habitats have approximately the same optimum conditions for growth and use the same primary pathways for energy and biomass production. However, during the screening and the later production of products from marine samples, the presence of salt can be crucial as a consequence of adaptation to ocean salinity (e.g., Chen et al. (2011); D'Souza-Ticlo et al. (2009)) for enzymes from marine fungi. In this sense, salt means not only the presence of sodium chloride in the appropriate concentration. Sulfate, one of the major components of sea water, and minor elements, including heavy metals, are essential for proper growth media composition of many, especially the obligate marine fungi. Accordingly, many researchers use "artificial sea water" to cultivate these strains (Kjer et al. 2010). Cultivating marine fungi for industrial applications poses challenges, from slow growth rates to contamination risks. Bioreactor systems offer controlled environments, optimizing growth conditions for targeted fungi. Co-culturing, involving symbiotic interactions between different microorganisms, enhances productivity and promotes the expression of unique compounds. Synthetic biology approaches enable the modification of fungal genomes, tailoring strains for specific industrial needs.

The diversity of marine fungi is not adequately represented in the cultured diversity. Many studies have focused on just a few genera, mainly *Penicillium*, *Aspergillus*, *Fusarium*, and *Cladosporium*. This focus might be due to these groups' relative easiness of isolation, which also dominated isolation campaigns from terrestrial habitats. Interestingly, when comparing marine strains of these groups to terrestrial relatives, different secondary metabolite biosynthesis pathways can be found (Imhoff 2016). However, many of the novel and unknown environmental fungal taxa identified in molecular studies are likely to be difficult to propagate in culture, either because spores of terrestrial fungi, also present in the environmental samples, outcompete them, or because their life cycle depends on a symbiotic interaction (Del Campo et al. 2014). Targeted bioprospecting of marine fungi for industrial applications seems to be a major necessity for a successful exploitation. Isolating marine fungi is the first crucial step in targeted bioprospecting. Traditional methods, such as dilution plating and direct plating, remain essential, but recent advancements enhance efficiency. Single-cell isolation techniques, leveraging microfluidics and fluorescence-activated cell sorting, enable the isolation of rare and unculturable fungi (Oberpaul et al. 2022). Additionally, high-throughput culturing systems expedite the screening process, accelerating the discovery of fungi with industrially valuable traits.

Metagenomics revolutionizes the exploration of the genomic potential of entire microbial communities, including marine fungi. By extracting

DNA directly from environmental samples, researchers gain insights into the genetic diversity of fungi within complex ecosystems, future studies need to apply metagenomics in identifying novel genes and pathways with industrial significance (Culligan et al. 2014). Metagenomic data will not only accelerate the discovery of bioactive compounds but also provide a broader understanding of the functional capabilities encoded in the collective genomes of marine fungi and help to design processes along the physiological demands of the fungi. Omics are also relevant on the level of the metabolites: metabolomics approaches offer a powerful tool for characterizing the chemical diversity of marine fungi. By analyzing the metabolites produced, compounds with industrial potential can be identified and prioritized. From secondary metabolites to enzymatic products, metabolomics provides valuable insights into the metabolic pathways and bioactive molecules that hold promise for industrial applications (Palermo 2023).

For the optimal and sustainable use of marine fungal sources, it is essential to store the microbial cultures. Culture collections play a vital role in the conservation and sustainable use of microbial resources, including fungi (Daniel and Prasad 2010; Reich and Labes 2017). Culture collections keep strains available for further investigations and production processes. Conservation of life cells while reducing metabolism, e.g., by cryopreservation, is crucial in order to prevent genetic and phenotypic changes induced by repeated passaging of the cultures.

The approach to bring marine fungi into biotechnological application is similar to other microbial processes. Bonugli-Santos et al. (2015) reviewed the biotechnological production of marine enzymes, including bioprocess strategies adopted for the cultivation of marine organisms for enzyme production. Large-scale production (e.g., in bioreactors) of glucoamylase, superoxide dismutase, lignin peroxidase, chitinase, protease, and glutaminase by marine strains is reported in the literature (Sarkar et al. 2010). These enzymes are produced in bioreactors largely through submerged-state fermentation, and the conditions related to bioreactor production are comparable to processes using terrestrial strains. Specific studies for establishing strategies for marine products are scarce (for review, see Kramer and Labes (n.d.)). A consideration of the unique properties of the marine environment is vital for marine biotechnology: Biotechnological production processes are influenced by the particular adaptations of organisms to their environment. Interestingly, only a few reports have investigated the impact of varying salt concentrations on the production of marine fungal metabolites. Bioprocess engineering in marine biotechnology follows the path from discovery to commercialization with various possible starting points. However, an entire value chain remains mainly theoretical, as quantitative biotechnological engineering studies on marine fungi are virtually nonexisting in the literature. Silber et al. (2016) reviewed the biotechnological realization of marine fungal antibiotics and demonstrated that biotechnology has a vast potential for the sustainable production of antibiotics. Biotechnology, as studied and developed with terrestrial representatives, may help to expand and understand the chemical space in a targeted manner, provide classical full fermentative and semi-synthetic processes, and may include metabolic engineering manipulating the genetic background as a basis for the generation of "biological" derivatives. For terrestrial fungi, the technical and economic feasibility of large-scale production has been proven for many processes (Silber et al. 2016). Sarkar et al. (2010) provided bioprocess data for fermentations of marine fungi for enzyme production. Based on their analysis, there is a demand for further adaptations of scale and inclusion of knowledge on the biology of the fungi into the production processes (e.g., salinity adaptations or pressure for deep-sea organisms). Process development of complete fermentative processes is required (Sarkar et al. 2010).

2.6 Conclusion

The bioprospecting of marine fungi has unveiled a rich reservoir of biodiversity with significant potential for novel bioactive compounds, enzymes, and applications in ecosystem services. Despite these promising discoveries, the translation of marine fungal research into practical applications faces several challenges, including cultivation techniques, genetic manipulation, and regulatory hurdles. The untapped potential of marine fungi in mycotechnology remains despite opportunities for innovative biotechnological processes across various industries. Industrial-scale biotechnological processes with marine fungi are feasible from scientific, technical, and economic perspectives, but obstacles such as the reduction of salinity in culture media and slow growth rates need to be addressed. The current focus on well-known genera like *Penicillium* and *Aspergillus* limits the representation of the true diversity of marine fungi. Furthermore, the challenges of isolating and culturing novel and unknown species require adaptation of traditional methods and the development of advanced techniques.

Bioreactor systems, co-culturing, synthetic biology, metagenomics, and metabolomics are powerful tools that enhance the exploration and utilization of marine fungal resources. These approaches expedite the discovery of bioactive compounds, provide insights into genetic diversity, metabolic pathways, and facilitate the identification of molecules with industrial potential. Cultivating marine fungi for industrial applications requires targeted bioprospecting, and metagenomic data can aid in identifying novel genes and pathways.

While there is potential for the sustainable production of antibiotics and other valuable products through biotechnology, a comprehensive understanding of marine fungal biology and the development of complete fermentative processes are essential. The current literature on quantitative biotechnological engineering studies for marine fungi is limited, emphasizing the need for further research and development in this field. Overall, marine fungi hold significant promise for contributing to sustainable industrial development and addressing various societal and environmental challenges, but overcoming the existing hurdles requires concerted efforts from researchers, industry, and regulatory bodies.

References

Agrawal S, Adholeya A, Barrow CJ, Deshmukh SK (2018) Marine fungi: an untapped bioresource for future cosmeceuticals. Phytochem Lett 23:15–20. https://doi.org/10.1016/j.phytol.2017.11.003

Alvarado-Campo KL, Quintero M, Cuadrado-Cano B, Montoya-Giraldo M, Otero-Tejada EL, Blandón L, Sánchez O, Zuleta-Correa A, Gómez-León J (2023) Heavy metal tolerance of microorganisms isolated from coastal marine sediments and their lead removal potential. Microorganisms 11:2708. https://doi.org/10.3390/microorganisms11112708

Amend A, Burgaud G, Cunliffe M, Edgcomb VP, Ettinger CL, Gutiérrez MH, Heitman J, Hom EFY, Ianiri G, Jones AC, Kagami M, Picard KT, Quandt CA, Raghukumar S, Riquelme M, Stajich J, Vargas-Muñiz J, Walker AK, Yarden O, Gladfelter AS (2019) Fungi in the marine environment: open questions and unsolved problems. MBio:10. https://doi.org/10.1128/mBio.01189-18

Balabanova L, Slepchenko L, Son O, Tekutyeva L (2018) Biotechnology potential of marine fungi degrading plant and algae polymeric substrates. Front Microbiol 9:1527. https://doi.org/10.3389/fmicb.2018.01527

Blunt JW, Copp BR, Keyzers RA, Munro MHG, Prinsep MR (2015) Marine natural products. Nat Prod Rep 32:116–211. https://doi.org/10.1039/C4NP00144C

Bonugli-Santos RC, Durrant LR, Sette LD (2010) Laccase activity and putative laccase genes in marine-derived basidiomycetes. Fungal Biol 114:863–872. https://doi.org/10.1016/j.funbio.2010.08.003

Bonugli-Santos RC, Durrant LR, Sette LD (2012) The production of ligninolytic enzymes by marine-derived basidiomycetes and their biotechnological potential in the biodegradation of recalcitrant pollutants and the treatment of textile effluents. Water Air Soil Pollut 223:2333–2345. https://doi.org/10.1007/s11270-011-1027-y

Bonugli-Santos RC, Dos Santos Vasconcelos MR, Passarini MR, Vieira GA, Lopes VC, Mainardi PH, Dos Santos JA, de Azevedo Duarte L, Otero IV, da Silva Yoshida AM, Feitosa VA, Pessoa A Jr, Sette LD. (2015) Marine-derived fungi: diversity of enzymes and biotechnological applications. https://doi.org/10.3389/fmicb.2015.00269 Front Microbiol. Apr 10;6:269. PMID: 25914680; PMCID: PMC4392690.

Chen H-Y, Xue D-S, Feng X-Y, Yao S-J (2011) Screening and production of ligninolytic enzyme by a marine-derived fungal *Pestalotiopsis* sp. J63. Appl Biochem Biotechnol 165:1754–1769. https://doi.org/10.1007/s12010-011-9392-y

Cheng P, Li Y, Wang C, Guo J, Zhou C, Zhang R, Ma Y, Ma X, Wang L, Cheng Y, Yan X, Ruan R (2022) Integrated marine microalgae biorefineries for improved bioactive compounds: a review. Sci Total Environ 817:152895. https://doi.org/10.1016/j.scitotenv.2021.152895

Chi Z, Chi Z, Zhang T, Liu G, Li J, Wang X (2009) Production, characterization and gene cloning of the extracellular enzymes from the marine-derived yeasts and their potential applications. Biotechnol Adv 27:236–255. https://doi.org/10.1016/j.biotechadv.2009.01.002

Comeau AM, Vincent WF, Bernier L, Lovejoy C (2016) Novel chytrid lineages dominate fungal sequences in diverse marine and freshwater habitats. Sci Rep 6:30120. https://doi.org/10.1038/srep30120

Compson ZG, McClenaghan B, Singer GAC, Fahner NA, Hajibabaei M (2020) Metabarcoding from microbes to mammals: comprehensive bioassessment on a global scale. Front Ecol Evol 8:581835. https://doi.org/10.3389/fevo.2020.581835

Culligan EP, Sleator RD, Marchesi JR, Hill C (2014) Metagenomics and novel gene discovery: promise and potential for novel therapeutics. Virulence 5:399–412. https://doi.org/10.4161/viru.27208

D'Souza-Ticlo D, Sharma D, Raghukumar C (2009) A thermostable metal-tolerant laccase with bioremediation potential from a marine-derived fungus. Mar Biotechnol 11:725–737. https://doi.org/10.1007/s10126-009-9187-0

Daniel H-M, Prasad GS (2010) The role of culture collections as an interface between providers and users: the example of yeasts. Res Microbiol 161:488–496. https://doi.org/10.1016/j.resmic.2010.01.011

Del Campo J, Sieracki ME, Molestina R, Keeling P, Massana R, Ruiz-Trillo I (2014) The others: our biased perspective of eukaryotic genomes. Trends Ecol Evol 29:252–259. https://doi.org/10.1016/j.tree.2014.03.006

Ekanayaka AH, Tibpromma S, Dai D, Xu R, Suwannarach N, Stephenson SL, Dao C, Karunarathna SC (2022) A review of the fungi that degrade plastic. J Fungi (Basel) 8:772. https://doi.org/10.3390/jof8080772

Fenical W, Jensen PR (2006) Developing a new resource for drug discovery: marine actinomycete bacteria. Nat Chem Biol 2:666–673. https://doi.org/10.1038/nchembio841

Gessner RV (1980) Degradative enzyme production by salt-marsh fungi. Bot Mar 23:133–139. https://doi.org/10.1515/botm.1980.23.2.133

Gonçalves MFM, Esteves AC, Alves A (2022) Marine fungi: opportunities and challenges. Encyclopedia 2:559–577. https://doi.org/10.3390/encyclopedia2010037

Imhoff JF (2016) Natural products from marine fungi—still an underrepresented resource. Mar Drugs 14:19. https://doi.org/10.3390/md14010019

Jex CN, Pate GH, Blyth AJ, Spencer RG, Hernes PJ, Khan SJ, Baker A (2014) Lignin biogeochemistry: from modern processes to quaternary archives. Quatern Sci Rev 87:46–59. https://doi.org/10.1016/j.quascirev.2013.12.028

Kjer J, Debbab A, Aly AH, Proksch P (2010) Methods for isolation of marine-derived endophytic fungi and their bioactive secondary products. Nat Prot 5:479–490. https://doi.org/10.1038/nprot.2009.233

Kohlmeyer J, Kohlmeyer E (1979) Marine mycology—The higher fungi. Academic Press, New York.

Kohlmeyer J, Volkmann-Kohlmeyer B (2003) Fungi from coral reefs: a commentary. Mycol Res 107:386–387. https://doi.org/10.1017/s0953756203227775

Kramer A, Labes A (n.d.) 16 Biotechnology of marine fungi: new workhorses and applications. In: Kramer A (ed) Genetics and biotechnology. Springer, Cham, pp 399–412

Li H, Chi Z, Wang X, Duan X, Ma L, Gao L (2007) Purification and characterization of extracellular amylase from the marine yeast *Aureobasidium pullulans* N13d and its raw potato starch digestion. Enzym Microb Technol 40:1006–1012. https://doi.org/10.1016/j.enzmictec.2006.07.036

Liu D, Garrigues S, de Vries RP (2023) Heterologous protein production in filamentous fungi. Appl Microbiol Biotechnol 107:5019–5033. https://doi.org/10.1007/s00253-023-12660-8

Magan N (2002) Bio-Exploitation. Mycol Res 106:863–864. https://doi.org/10.1017/S0953756202216652

Manohar CS, Raghukumar C (2013) Fungal diversity from various marine habitats deduced through culture-independent studies. FEMS Microb Lett 341:69–78. https://doi.org/10.1111/1574-6968.12087

Natoli M, Herzig P, Pishali Bejestani E, Buchi M, Ritschard R, Lloyd GK, Mohanlal R, Tonra JR, Huang L, Heinzelmann V, Trüb M, Zippelius A, Kashyap AS (2021) Plinabulin, a distinct microtubule-targeting chemotherapy, promotes M1-like macrophage polarization and anti-tumor immunity. Front Oncol 11:644608. https://doi.org/10.3389/fonc.2021.644608

Newton GG, Abraham EP (1955) Cephalosporin C, a new antibiotic containing sulphur and D-alpha-aminoadipic acid. Nature 175:548. https://doi.org/10.1038/175548a0

Nicoletti R, Andolfi A (2018) The marine-derived filamentous fungi in biotechnology. In: Rampelotto PH, Trincone A (eds) Grand challenges in marine biotechnology. Springer, Cham, pp 157–189

Oberpaul M, Brinkmann S, Marner M, Mihajlovic S, Leis B, Patras MA, Hartwig C, Vilcinskas A, Hammann PE, Schäberle TF, Spohn M, Glaeser J (2022) Combination of high-throughput microfluidics and FACS technologies to leverage the numbers game in natural product discovery. Microb Biotechnol 15:415–430. https://doi.org/10.1111/1751-7915.13872

Okutani K (1977) Gliotoxin produced by a strain of *Aspergillus* isolated from marine mud. Nippon Suisan Gakkaishi 43:995–1000

Oppong-Danquah E, Parrot D, Blümel M, Labes A, Tasdemir D (2018) Molecular networking-based metabolome and bioactivity analyses of marine-adapted fungi co-cultivated with phytopathogens. Front Microbiol 9:2072. https://doi.org/10.3389/fmicb.2018.02072

Palermo A (2023) Metabolomics- and systems-biology-guided discovery of metabolite lead compounds and druggable targets. Drug Discov Today 28:103460. https://doi.org/10.1016/j.drudis.2022.103460

Patyshakuliyeva A, Falkoski DL, Wiebenga A, Timmermans K, de Vries RP (2019) Macroalgae derived fungi have high abilities to degrade algal polymers. Microorganisms 8:52. https://doi.org/10.3390/microorganisms8010052

Pejin B, Karaman M (2017) Antitumor natural products of marine-derived fungi. In: Mérillon JM, Ramawat K (eds) Fungal metabolites. Reference series in phytochemistry. Springer, Cham, pp 1–28. https://doi.org/10.1007/978-3-319-19456-1_25-1

Punt PJ, van Biezen N, Conesa A, Albers A, Mangnus J, van den Hondel C (2002) Filamentous fungi as cell factories for heterologous protein production. Trends Biotechnol 20:200–206. https://doi.org/10.1016/S0167-7799(02)01933-9

Raghukumar C (2008) Marine fungal biotechnology: an ecological perspective. Fungal Divers 31:19–35

Rämä T, Nordén J, Davey ML, Mathiassen GH, Spatafora JW, Kauserud H (2014) Fungi ahoy! Diversity on marine wooden substrata in the high north. Fungal Ecol 8:46–58. https://doi.org/10.1016/j.funeco.2013.12.002

Rédou V, Navarri M, Meslet-Cladière L, Barbier G, Burgaud G (2015) Species richness and adaptation of marine fungi from deep-subseafloor sediments. Appl Environ Microbiol 81:3571–3583. https://doi.org/10.1128/AEM.04064-14

Reich M, Labes A (2017) How to boost marine fungal research: a first step towards a multidisciplinary approach by combining molecular fungal ecology and natural products chemistry. Mar Genomics 36:57–75. https://doi.org/10.1016/j.margen.2017.09.007

Rojas-Sánchez B, Guzmán-Guzmán P, Morales-Cedeño LR, Del Orozco-Mosqueda MC, Saucedo-Martínez BC, Sánchez-Yáñez JM, Fadiji AE, Babalola OO, Glick BR, Santoyo G (2022) Bioencapsulation of microbial inoculants: mechanisms, formulation types and application techniques. Appl Biosci 1:198–220. https://doi.org/10.3390/applbiosci1020013

Rotter A, Barbier M, Bertoni F, Bones AM, Cancela ML, Carlsson J, Carvalho MF, Cegłowska M, Chirivella-Martorell J, Conk Dalay M, Cueto M, Dailianis T, Deniz I, Díaz-Marrero AR, Drakulovic D, Dubnika A, Edwards C, Einarsson H, Erdoğan A, Eroldoğan OT, Ezra D, Fazi S, FitzGerald RJ, Gargan LM, Gaudêncio SP, Gligora Udovič M, Ivošević DeNardis N, Jónsdóttir R, Kataržytė M, Klun K, Kotta J, Ktari L, Ljubešić Z, Lukić Bilela L, Mandalakis M, Massa-Gallucci A, Matijošytė I, Mazur-Marzec H, Mehiri M, Nielsen SL, Novoveská L, Overlingė D, Perale G, Ramasamy P, Rebours C, Reinsch T, Reyes F, Rinkevich B, Robbens J, Röttinger E, Rudovica V, Sabotič J, Safarik I, Talve S, Tasdemir D, Theodotou Schneider X, Thomas OP, Toruńska-Sitarz A, Varese GC, Vasquez MI (2021) The essentials of marine biotechnology. Front Mar Sci 8:629629. https://doi.org/10.3389/fmars.2021.629629

Sandri IG, Fontana RC, Barfknecht DM, Da Silveira MM (2011) Clarification of fruit juices by fungal pectinases. LWT 44:2217–2222. https://doi.org/10.1016/j.lwt.2011.02.008

Sarkar S, Pramanik A, Mitra A, Mukherjee J (2010) Bioprocessing data for the production of marine enzymes. Mar Drugs 8:1323–1372. https://doi.org/10.3390/md8041323

Silber J, Kramer A, Labes A, Tasdemir D (2016) From discovery to production: biotechnology of marine fungi for the production of new antibiotics. Mar Drugs 14:137. https://doi.org/10.3390/md14070137

Silva A, Delerue-Matos C, Figueiredo S, Freitas O (2019) The use of algae and fungi for removal of pharmaceuticals by bioremediation and biosorption processes: a review. Water 11:1555. https://doi.org/10.3390/w11081555

Stiebeling DA, Labes A (2022) Mycoremediation of sewage sludge and manure with marine fungi for the removal of organic pollutants. Front Mar Sci 9:946220. https://doi.org/10.3389/fmars.2022.946220

Tisthammer KH, Cobian GM, Amend AS (2016) Global biogeography of marine fungi is shaped by the environment. Fungal Ecol 19:39–46. https://doi.org/10.1016/j.funeco.2015.09.003

Vignesh S, Raja A, James RA (2011) Marine drugs: implication and future studies. Int J Pharmacol 7:22–30

Wang G, Wang X, Liu X, Li Q (2012) Diversity and biogeochemical function of planktonic fungi in the ocean. In: Raghukumar C (ed) Biology of marine fungi [recurso electrónico]. Springer Healthcare Ltd, Alemania, pp 71–88

Wayman C, Niemann H (2021) The fate of plastic in the ocean environment - a mini review. Environ Sci Process Impacts 23:198–212. https://doi.org/10.1039/D0EM00446D

Xu W, Gong L, Pang K-L, Luo Z-H (2018) Fungal diversity in deep-sea sediments of a hydrothermal vent system in the southwest Indian ridge. Deep-Sea Res I Oceanogr Res Pap 131:16–26. https://doi.org/10.1016/j.dsr.2017.11.001

Fungal Spoilage of Crops and Food

3

Jan Dijksterhuis and Jos Houbraken

Abstract

Food products are a rich nutrient source that will attract both bacterial and fungal colonizers. Food spoilage is a major threat for our food stock and is responsible for enormous losses worldwide, which makes it a research area that is very relevant with respect to the increasing demand on food during the next decennia.

This chapter will highlight fungal spoilage dealing mostly with plant-based food products. At first, the relationship between the living crop and fungi will be illustrated. Between host and pathogen there exists a surprisingly complex communication and successful intervention may result in novel strategies to prevent post-harvest diseases.

Then the association of fungi with different types of processed food will be described. In time, different preservation techniques are developed with the aim to discourage fungal development in the food product. These include fermentation, addition of salts or high concentrations of sugars, pickling, drying, cooling, the addition of preservatives, or a heating treatment before packaging. Only fungi that can survive these adverse conditions can successfully spoil processed food. Different aspects of stress resistance are addressed in this chapter including osmo- and xerotolerance, protective compounds inside cells, and heat-resistant structures. Food parameters are often surprisingly restrictive to the spectrum of fungal species which are able to grow and thus spoil the individual food types. Normally, less than ten and often one to three species are responsible for spoilage. Careful elucidation of the specific aspects of food spoilage may lead to novel tailor-made methods to decrease the enormous portion of our food that is lost.

Keywords

Food Spoilage · Fungi · Food losses · Post-harvest diseases · *Colletotrichum* · *Penicillium* · *Aspergillus* · Water activity · Heat-resistant fungi · *Zygosaccharomyces*

3.1 Food Losses, Its Magnitude, and Carbon Footprint

The astronomical amount of food products that are produced worldwide account for approximately one-third of all anthropogenic greenhouse

This chapter is an updated version of 'Fungal Spoilage of Crops and Food' published in Vol. 11: Agricultural Applications, 2013, ISBN 978-3-642-36820-2, https://link.springer.com/book/10.1007/978-3-642-36821-9

J. Dijksterhuis (✉) · J. Houbraken
Westerdijk Fungal Biodiversity Institute,
Utrecht, The Netherlands
e-mail: j.dijksterhuis@wi.knaw.nl

S. Grüttner et al. (eds.), *Agricultural and Industrial Applications*, The Mycota 16,
https://doi.org/10.1007/978-3-031-81904-9_3

gas emissions (Crippa et al. 2021). In 2015, these emissions were estimated to be 14–22 (18) Gt CO_2 equivalents of which 27% was emitted by industrialized countries. This accounts for approximately 2.4 ton of CO_2 emission per capita per year. Emissions from solid waste disposal and waste water treatment comprise approximately 10% of this amount.

Is it possible to estimate the amount of food produced and determine the losses? These calculations are far from simple and include the combination of many different data. Caldeira et al. (2019) calculated a primary production of 638 Mt food (plant and animal wet mass) in the European Union during 2011. During the food production chain, a total of 129 Mt food waste is generated of which 60 Mt is realized during the consumption stage. The authors stated that: "The food groups presenting the highest amount of food waste (considering the total food waste generated in all the supply chain stages) are fruit and vegetables (total produce 160 Mt, losses nearly 44%)." Finally, 306 Mt of all the produce is eaten.

In 2015, the United Nations defined the Sustainable Development Goals (SDG) aiming to reduce per capita food waste at the retail and consumer levels with 50% by 2030. Additionally, food losses along production and supply chains, including post-harvest losses must be reduced. In 2011, the food loss per capita is estimated to be 257 kg/year.

Food waste is not all edible food waste, for instance we remove the peel of an orange. However, the observation that perishable crops as fruits and vegetables show the highest losses may indicate that microbial attacks (by bacteria and fungi) during all stages of production and storage of food products are a notable part of food losses. For fruits and vegetables, in Latin America and sub-Saharan Africa these losses were as high as 16.5% (9.1% for cereals) and 11% (7.9% for cereals) of commodity availability. For high-income countries (HIC) and middle-and low-income countries (LMIC) these figures are 6% and 9.7%, respectively (1.3 and 6.4% for cereals). Therefore, food spoilage is one of the threats for our food stock (see Fig. 3.1) and might be responsible for enormous losses worldwide.

The consumption of processed, packaged, and prepared foods has increased in all but the most isolated rural communities (FAO, 2018, 2019). This makes the study on food spoilage a research area that is very relevant with respect to the increasing demand on food production during the next decennia. Knowledge on the specific mechanisms that occur during food spoilage might generate novel insights that result in increased net amounts of food without an increase of land use. Such a reduction would play a role in making food production more sustainable (FAO, 2018).

3.2 Food Losses and Microbial Attack

Food products are rich in nutrients and will attract both bacterial and fungal colonizers. After successful colonization the nutritional properties of a product are altered. When the nutritional value, structure, and taste of the product are deteriorated, this colonization is called food spoilage. Food spoilage can result in the production of toxic secondary metabolites resulting in grave medical problems and is an issue that needs our continuous awareness with respect to food safety.

If the nutritional quality and storage longevity of a food product is improved due to microbial action, we have a fermented food product. These two aspects of food colonization are two sides of the same coin.

This chapter will highlight fungal spoilage, which deals in many cases with plant- based food products. Fungi are the main degraders of the sturdy plant cell-wall components that otherwise would accumulate within the ecosystems of the world. Prior to spoilage, the fungi are present on or inside the crop in low numbers or as survival structures.

Food products include two main groups, namely living crops and processed food. Colonization of food products, henceforth, is very diverse. This chapter will evaluate different fungal-food relationships. At first, the relationship between the living crop and fungus will be illustrated (Fig. 3.1). Subsequently, the association of fungi with different types of processed

Fig. 3.1 A collage of various fruits and vegetables in different stages of decay. The top section shows a collection of produce including pumpkins, melons, onions, bananas, and berries, some with visible mold. The middle section features close-ups of moldy and rotten spots on different fruits, highlighting textures and colors of decay. The bottom section includes detailed images of mold growth on raspberries and a blackened banana. The overall theme emphasizes food spoilage and decomposition (Pictures made by Jos Houbraken)

food will be described. Different preservation techniques make the food product a difficult environment to colonize, although it is also a rich medium. Only fungi that can survive certain adverse conditions including high osmolarity and heat treatment can successfully spoil processed food. Different aspects of stress resistance are addressed in this chapter including osmotolerance, protective compounds inside cells, and heat-resistant structures.

Several books on food spoilage fungi summarize many different aspects of fungi and food. Pitt and Hocking (2009) and Samson et al. (2019) provide overviews on the taxonomic description and specificity of food spoilage fungi and Dijksterhuis and Samson (2007) highlight numerous aspects of the relation between food and fungi including spoilage and fermentation.

3.3 Developments in Fungal Taxonomy Are Important for the Study of Fungal-Food Spoilage

During the last decade, the taxonomy of fungi has developed profoundly. Initially, the ITS region (the internal transcribed spacer region) was seen as the most promising marker for successful identification for the broadest range of fungi (Schoch et al. 2012). It became clear that DNA sequencing of specific multiple loci in the genome is necessary to identify fungal species within the genera. However, it has to be stressed that species identification within in fungi is based on a combination of approaches including phenotypic (the appearance of the fungus), biochemical (specific biomolecules), sequencing (of several loci in the genome which are specific for the fungal taxon). This is dubbed as integrative taxonomy or the polyphasic species concept among other terminologies (Lücking et al. 2021). It remains to be stated that a complete and coherent picture of fungal taxonomy does not exist and approaches may have a pragmatic element.

The level of accuracy and maximal precision in fungal taxonomy is particularly critical in plant pathology, food safety, and clinical mycology (Lücking et al. 2021). This is related to issues as quarantine in the case of plant diseases, production of mycotoxins in food by specific food-spoiling fungi and antifungal resistance in the case of human pathogenic fungi. In this case even specific molecular markers or transcriptomics could be used to discern virulent strains or actively synthesized mycotoxin clusters.

With the onset and influx of DNA sequencing as a tool to identify fungal species it was observed that both sexual (teleomorph) and asexual (anamorph) stages of the fungal life cycle often clustered in one clade. These stages represent different phenotypes which were traditionally named and classified separately under the concept of "dual nomenclature." Many mycologists favored to adopt the principle of "one fungus = one name," which took several decades to be accepted (Hawksworth et al. 2011) and since then this principle is implemented throughout the fungal taxonomy.

For example, the nomenclature of the Eurotiales that harbor many fungal strains important as food spoilers is treated in Houbraken et al. (2020) and is also very relevant in the case of plant pathogenic fungi (including post-harvest species, Wingfield et al. 2012). This includes that the food-relevant and heat-resistant fungal species *Byssochlamys spectabilis*, *B. nivea,* and *B. fulva* are now designated as *Paecilomyces variotii*, *P. niveus,* and *P. fulvus*. Further, the xerophilic species *Eurotium chevalieri* is known as *Aspergillus chevalieri* (Chen et al. 2017). The characteristic ascomata and ascospores are designated as *Aspergillus chevalieri* (eurotium morph). Similar are the heat-resistant species *Neosartorya fischeri* and *N. spinosa* now mentioned as *Aspergillus fischeri* and *A. spinosus* (neosartorya morph).

3.4 Infection of Living Crops: Post-Harvest Diseases

The relation between fungi and living agricultural crops is plant pathogenic in nature and involves a complex communication between parasite and host.

This is characterized by the following themes and variations: 1.) The living crop can be entered in a passive mode as wounds or natural openings in the epidermis or via active penetration of the plant cell wall by means of specialized cells. 2.) Establishment of a temporary fungus–host interface that can exhibit (prolonged) survival in a quiescent state, followed by a necrotrophic stage in which plant tissue is killed and lesions develop. 3.) Creation of a microenvironment in which pH changes are imposed in the crop.

Even within fungal genera, and depending on the plant tissue, pathogenic species vary strongly in the length of this quiescent stage (De Silva et al. 2017). However, for crops this is important as losses occur during transport or even at later stages, such as during ripening at the consumer's location. This quiescent stage occurs in different post-harvest fungi among them the genera *Colletotrichum*, *Alternaria,* and *Botrytis* (Prusky et al. 2013). Understanding the onset, duration, and breaking of quiescence is potentially vital for developing novel ways to ensure disease control. The total number of fungal species involved in post-harvest diseases is by far larger than what ever can be covered in this short overview; we would like to illustrate some principles of infection and the fungi involved.

3.4.1 Anthracnose

Members of the fungal genus *Colletotrichum* (teleomorph; *Glomerella*) cause anthracnose disease in a wide range of fruits and vegetables (and flowers). During the last decade, the taxonomy of the genus was intensively studied and currently over 257 species in 15 species complexes are described (Talhinhas and Baroncelli 2021). With respect to many crops these fungi are known for their latent infections that become active after fruit ripening and in a necrotrophic fashion destroy crop, making it mentioned as one of the ten most important fungal pathogens (Dean et al. 2012). They affect crops as banana, avocado, papaya, strawberries, tomato, coffee, apple, and melon among many more (Ciofini et al. 2022). These fungi produce highly specialized structures called appressoria that provide fungal entry into healthy plant tissue (Fig. 3.2).

Appressoria generate enormous turgor pressure, which enables the penetration peg formed from it, to breach the sturdy plant cell walls with pure mechanical force (Bechinger et al. 1999; Ryder et al. 2022). These cells have melanized cell walls and accumulate glycerol to generate such pressures, although the precise function of melanization is not clear (Lin et al. 2012; Ludwig et al. 2014). Actin cables and the actin-crosslinking protein fimbrin are markedly present before and during formation of the penetration peg (Wang and Shaw 2016; Zhang et al. 2022). In the case of infection of the climacteric fruit of the tomato with *C. gloeosporioides* the fungus actually waits for the right moment to infect. The plant hormone ethylene produced by the tomato during senescence is an important trigger for proper infection (Flaishman and Kolattukudy 1994).

On avocado, appressoria form on unripe fruit and penetrate the plant cell wall. The subcuticular hyphae then become quiescent until ripening of the fruit has occurred (Prusky and Lichter 2007; Prusky et al. 2013). Quiescence is correlated with the presence of an antifungal compound (AFD) that is degraded by the enzyme lipoxygenase. Indirectly, AFD levels are controlled by the antioxidant epicatechin that is present in the avocado peel and acts as a lipoxygenase inhibitor (Ardi et al. 1998).

Synthesis of the antifungal compound correlates with the expression of a Δ12-fatty acid desaturase (Wang et al. 2004). Interestingly, a cold treatment that resulted in increase of unsaturated fatty acid also gave rise to increased antifungal diene (AFD). *C. gloeosporioides* also reacts to other chemical triggers, namely long chain fatty acid alcohols in the host surface wax in the case of anthracnose of avocado fruit (Podila et al. 1993). Another basic requirement for appressorium formation is the presence of a surface (Kim et al. 1998). Thus, different thigmotropic (sense-reactive) and chemical signals (at least partly sensored by G-proteins, Ryder et al.

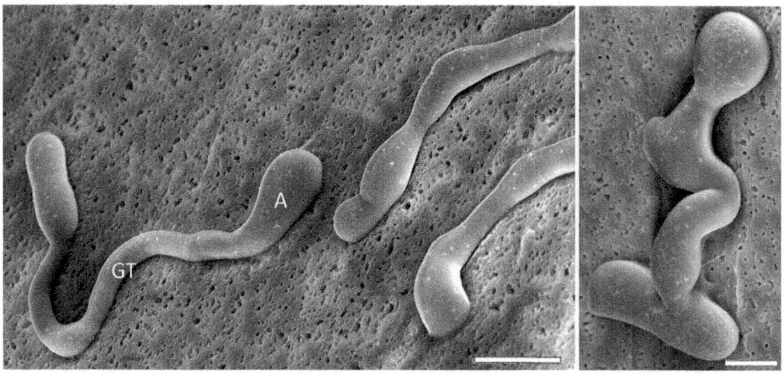

Fig. 3.2 Conidia (C) of a *Colletotrichum* species germinate on an avocado surface. The germ tube (GT) differentiates into an appressorium (A) that is firmly attached to the plant cell wall (micrograph made by Jan Dijksterhuis, Westerdijk Fungal Biology Institute, The Netherlands). Bars are 10 and 5 μm, respectively

2022) cooperate during differentiation and infection. The central signal protein GcPmk1 needs to be phosphorylated and is vital for appressorium formation and melanization (He et al. 2017). Wax- and ethylene-dependent signaling pathways are not identical, but share two proteins that must be phosphorylated (Kolattukudy et al. 1995). Unripe avocado tissue is also able to react on the presence of fungal elicitor with the formation of ROS, during ripening this ability is almost absent (Beno-Moualem and Prusky 2000).

Necrotrophic hyphae are thinner than biotrophic hyphae and produce a variety of plant cell-wall degrading enzymes and other factors as reactive oxygen species or secondary metabolites that lead to cell death. This is characteristic for a hemibiotrophic lifestyle (as reviewed in Prusky and Lichter 2007; Prusky and Kolattukudy 2007; Münch et al. 2008; Prusky et al. 2013).

Colletotrichum genomes are relatively large compared to many other fungi, typically above 50 Mb or even higher (Crouch et al. 2014) with 12–16.000 predicted proteins. Expanded gene families include carbohydrate-active enzymes (CAZymes), secondary metabolism (SM) enzymes, secreted proteases, and putative secreted effectors.

3.4.2 True Necrotrophs, *Botrytis cinerea* as an Example

Necrotrophs kill plant tissue of the host and some of them have developed into broad spectrum pathogens destroying vegetables and fruits upon harvesting. *Botrytis cinerea* (the sexual stage was known as *Botryotinia fuckeliana)* is a fungus that causes widespread infection of grapes, (straw) berries, and other fruits as well as vegetables and flowers, but with a preference for soft and mature tissues (Dean et al. 2012). The fungus enters the host by means of appressorium-like cells or infection cushions or via wounds.

The appressoria require the presence of tetraspanins (the gene *BcPls1*) for successful penetration (Gourgues et al. 2004). Homologues of these specialized membrane proteins are also found in the plant parasite *Magnaporthe grisea* that forms similar appressoria as *Colletotrichum.*

After entrance into the plant tissue the fungus can exhibit a (short) quiescent stage before it starts to kill host cells. Toxic secondary metabolites as botrydial and botcinic acid were thought to play a role, however a strain without the transcription factor of the biosynthetic gene cluster for botrydial did not show reduced infection on bean leaves (Porquier et al. 2016). There is an interesting interplay between induced cell death (by cell death inducing proteins) or use of the host hypersensitivity response (where cell death is induced as a plant response) or killing of cells by overwhelming cell stress caused by reactive oxygen, oxalic acid, or cell-wall degrading enzymes (Choquer et al. 2007; Bi et al. 2023). It is of interest that the fungus itself has to remain insensitive for apoptotic signals.

The fungal genome of *B. cinerea* contains 275 plant-cell-wall-degrading enzymes (CWDE) and up to twelve different lipases are identified (van Kan 2006; Bi et al. 2023). Especially, endopolygalacturonases (pectin degradation) are important while *B. cinerea* induces ripening of the host, which produces ethylene, CWDE, and coloration-related molecules. It has to be noted that *B. cinerea* produces the plant maturation hormone ethylene by itself (Guo et al. 2020). Strikingly, the presence of two polygalacturonases produced by the fungus is vital for virulence, causing some pectin degradation during very early infection establishment (Silva et al. 2021, 2023). Further cellobiohydrolase and endoglucanase are related to reduced fungal virulence when removed from the fungus. In tomato fruit, the activity of expansins and polygalacturonases produced by the host loosens the plant cell wall during ripening. Transgenic tomato strains that did not have these activities showed clearly reduced and delayed infection development after inoculation with *B. cinerea* (Cantu et al. 2008). Thus cell-wall modulation and degradation by both fungus and host are vital for infection.

These findings show that the interaction between necrotroph and the host is considerably more elaborated than thought before (Amselem et al. 2011) and that there exists a balance between host and pathogen.

3.4.3 Opportunistic Fungi

Opportunistic fungi can grow well without plant hosts, acting as saprotrophs on decaying plant material or in soil. They infect crops via natural structures (such as lenticels) or cracks and wounds in the outer layer (exocarp) of the crop. These fungi develop into true pests of harvested crops. In citrus fruit, the fungi *Penicillium italicum* and *P. digitatum* (blue and green rot of citrus, respectively) cause the most serious and widespread rots of these crops, with 90% of the losses caused by the latter species (Kanashiro et al. 2020).

Other opportunistic fungi on these fruits are *Alternaria alternata*, *Aspergillus niger*, *A. flavus*, *Fusarium* spp., *Geotrichum candidum,* and *Trichoderma viride* (Snowdon 1990). In apple, *P. expansum* is a post-harvest problem of similar magnitude. Opportunists can also enter via the dying leaves of the flower before the fruit is fully grown (Snowdon 1990). Remnants of leaves present on the fruit (the so-called calix) contain more fungal cells and become a springboard for colonization of the tissues of a fruit (Smid et al. 1996).

Growing hyphae of the opportunistic fungi release enzymes that degrade the plant cell wall and define if dry or wet rot of the food crop develops. Cellulolytic enzymes do not disrupt the pectin middle lamella and therefore do not dissociate plant cells, which results in a more preserved structure of the tissue during dry rot (as reviewed in Prusky and Kolattukudy 2007). Pectin degrading enzymes destroy this connection between the cells resulting in maceration and wet rot.

Prevention of secretion of many enzymes by disruption of the *Blistering1* gene of *P. expansum* showed reduced lesion formation in apple (Jurick et al. 2020).

There is a suite of microorganisms living on the exocarp of fruits and vegetables, also known as the carposphere (Bartholomew et al. 2021). Upon lesion formation, these organisms have the opportunity to thrive due to the release of many nutrients. Patulin, a mycotoxin produced by *P. expansum* in lesions, is extremely active as an antibiotic (in fact against many multi-resistant bacteria!). This suggests that the component is active in keeping the lesion area clean of competing organisms (Bartholomew et al. 2022a, 2022b).

Interestingly patulin itself also causes lesions on apple and other fruit, which suggests that it is a generic toxin, with *P. expansum* cells protecting themselves for the effect. Interestingly, Li et al. (2015, see also Luciano-Rosario et al. 2020) describe that strains that were not able to form patulin did not show decreased pathogenicity to fruit. Similarly, *Aspergillus carbonarius* mutants that do not produce ochratoxin A show no effect in infection of nectarines (Maor et al. 2021).

The variety of the secreted polysaccharide-degrading enzymes is large. The variability of these enzymes has been reviewed (De Vries 2000; De Vries and Visser 2001; Pel et al. 2007) in case of the fungus *Aspergillus niger*. The genome of this fungus contains ORFs of 131 secreted carbohydrate-active enzymes (CAZymes), which illustrates the versatility of the tool box to degrade plant cell walls. Recently over 900 secreted proteins are predicted for *A. niger*, including over 300 small secreted proteins (often involved in host–pathogen interactions of which 150 were recognized as effector like SSPs (Vivek-Ananth et al. 2018). *A. niger* is a cosmopolitan fungus and causes serious opportunistic infections in onions and hyacinth bulbs. The fungi *P. expansum*, *P. digitatum,* and *P. italicum* (Kanashiro et al. 2020) are computed to contain 640, 460, and 394 putative CAZymes, respectively (Li et al. 2015).

That opportunistic fungi are markedly adapted to infection of harvested crops is illustrated by conidia (asexual spores) of *P. digitatum* that germinate quicker and in higher numbers in the presence of volatiles that surround wounded oranges (Eckert and Ratnayake 1994; Droby et al. 2008; Tao et al. 2018). Interestingly, the strongest stimulation was observed when the "authentic" volatile mixture was applied and was invariably lower in preparations of single compounds or mixtures with concentrations above and below that of the wounded oranges. Later studies showed that there are fruit specific effects and that in addition *P. italicum* shows germination activation, but to a lesser extent. Further, conidia of *B. cinerea* and *P. expansum* are inhibited by the same components (Droby et al. 2008).

Germination of conidia in the presence of one of the components, myrcene (10 μl droplet), had increased from <10% to 60% after 24 hours incubation for *P. digitatum* and from approx. 20 to 60% for *P. italicum*. Interestingly, *P. expansum* and *B. cinerea* show approx. 60% germination in the control. This suggests a germination block instead of an activation signal.

Microconidia of *Fusarium oxysporum* f. sp. *tulipae* cause a devastating dry rot in tulip bulbs and only germinate in wounds on the surface of the bulbs and are not able to grow on undamaged epidermis even when it is very close to a wound (Fig. 3.3).

Hyphae grow underneath the epidermis and a lesion is formed. At later stages, numerous microconidia are formed on the surface of the bulbs. Dependent on environmental factors as humidity and temperature fungal cells stop developing but remain alive and can enter a quiescent stage. When tulip bulbs are planted out upon storage, fungal development resumes after many months and lead to infection of plant tissues and subsequent death (Dijksterhuis, van der Lee and de Boer, unpublished results). The quiescent stage, also called latency, is an important aspect of post-harvest diseases and the mechanisms of prolonged survival of fungal cells during this stage are a potentially important research topic in order to develop novel strategies to prevent damage to food products.

3.4.4 pH Modulation

A typical phenomenon which is correlated with the early onset of the necrotrophic stage is the accumulation of ammonium at the leading edge of the developing lesion (Alkan et al. 2008, 2009) as is observed with *C. coccodes* (on tomato), *C. gloeosporioides* (on avocado), and *Alternaria alternata* (on persimmon fruit). This alkalinization is induced by the low pH of the host fruit and is a prerequisite for the activation of a host NADPH oxidase, an ROS-producing enzyme. The presence of ROS increases local cell death, a hallmark of necrotrophic growth. Further, ammonia via the NADPH oxidase activates salicylic acid- dependent genes and suppresses the expression of jasmonic acid dependent genes (Alkan et al. 2012). These fungi realize quiescence via a balance with the plant host antifungal factors and waves of expression of genes occur toward the necrotrophic infection stage (O'Connell et al. 2012; Alkan et al. 2015).

Tissue alkalinization in *C. gloeosporioides* also results in the secretion of a pectate lyase (encoded by *PelB*) via increased expression of the transcription factor PacC. Loss of function mutants of the latter factor showed strongly reduced pathogenicity and pectate lyase secretion (Miyara et al. 2008). In addition, *C. gloeosporioides* forms a laccase in the avocado peel that is able to degrade epicatechin and as such shortens the period of quiescence in case of some active isolates of the fungus (Guetsky et al. 2005).

The fungus *Alternaria alternata*, a post–harvest pathogen of many vegetables and fruits (Thomma 2003), can attack apples via the calycine tube and causes core rot in susceptible cultivars (Niem et al. 2007). *A. alternata* also causes rot of persimmon fruit, pear, peach, melon, and tomato, forms appressoria that partly penetrate the cell wall toward the pectin lamella (Suzuki et al. 2003; Morita et al. 2013), and is also a fungus that alkalinizes the host tissue.

An important factor in the disease is an endo-1,4-β-glucanase that is more highly expressed in the presence of cell-wall polymers and a higher pH (above 6, Eshel et al. 2002). The fungus can survive in a quiescent state in plant material and enters plant tissues that are weakened due to senescence or wounding, but the formation of small appressoria is not ruled out.

B. cinerea is an avid producer of oxalic acid inside the lesion and belongs to the acidic fungi. This organic acid stimulates cell-wall degrading enzymes and also has a strong calcium chelating activity that helps to destabilize the pectin net-

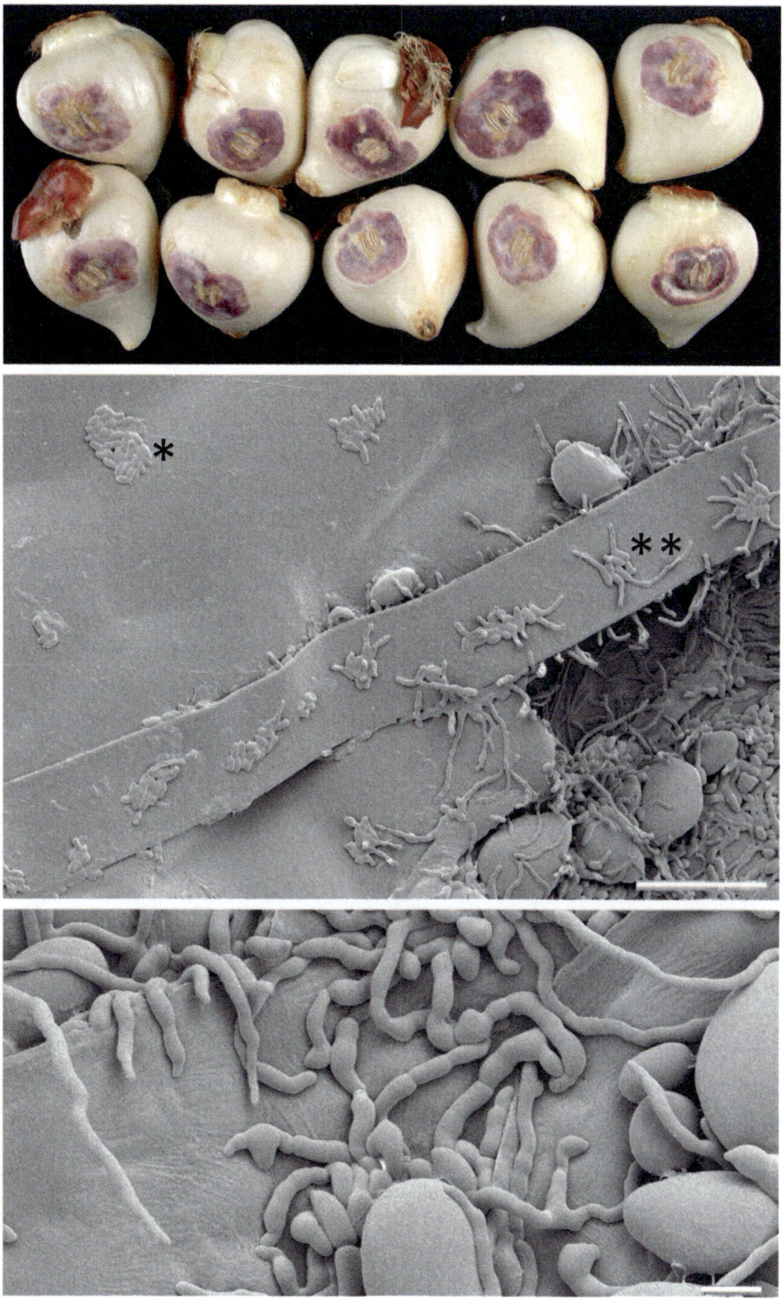

Fig. 3.3 Post-harvest infection of tulip bulbs. (A) Tulip bulbs are infected with conidia of *Fusarium oxysporum* f. sp. *tulipae* and develop purple stained lesions after 120 h following inoculation. (B) Microconidia do not germinate outside a wound in the epidermis of the bulb (*), but do so near the wound (**). (C) Germination of conidia on plant cell wall and starch granules (photographs made by Jan Dijksterhuis, Westerdijk Fungal Biodiversity Centre, The Netherlands). Bars are 20 and 5 μm, respectively

work in which calcium ions are embedded (van Kan 2006; Prusky and Lichter 2007). Oxalic acid production is even more a hallmark of infection by another widespread necrotrophic pathogen *Sclerotinia sclerotiorum* (Kim et al. 2007; Hegedus and Rimmer 2005) that is related to *B. cinerea* (Amselem et al. 2011). This fungus is notorious as a post-harvest pathogen of carrot, sunflower seeds, and bean pods among 400 plant species, most of them dicots.

Here, oxalic acid also modulates the hypersensitivity response including programmed cell death around the pathogen in delaying the oxidative burst and prevention of callose deposition at the leading edges of the lesions (Williams et al. 2011).

P. italicum, digitatum, and *expansum* all are able to acidify the host tissue and produce citric acid in liquid culture (Prusky et al. 2004). In citrus and apple fruit, citric acid and gluconic acid accumulate, and the expression of several polygalacturonases as pepg1 at pH 3.5–4.0 contributes to pectin depolymerization and tissue maceration. Ammonium produced at the leading edge of the lesions formed by *P. expansum* in apple was found to activate expression of the PacC transcription factor as well as the genes of the patulin biosynthesis cluster (Barad et al. 2016).

The production of gluconic acid by this fungus was accompanied with the expression of a glucose oxidase gene (*gox2*) and virulent isolates showed more of both (Hadas et al. 2007). Interestingly, GOX activity, gluconic acid accumulation and decay dropped significantly when oxygen levels dropped to 10% or lower. This indicates that gluconic acid and not citric acid is an important factor for disease.

The fungus *Aspergillus carbonarius* infects peaches, nectarines, grapes and produces gluconic acid and ochratoxin A, but gluconic acid was important to decrease the pH of fruit tissue from 3.5–3.7 to 3.1–3.2 (Maor et al. 2021).

The Δpks mutant, not able to produce ochratoxin, was slightly more virulent as the wildtype on nectarine, and the Δgox mutant that could not produce gluconic acid showed only some reduction in lesion formation. In the latter case citric acid, also produced in the wildtype, acidified the tissue to pH 3.2.

Finally, some other species are mentioned as *Fusarium oxysporum* and *F. proliferatum* that produce fumaric acid in watermelon, tomatoes among others. The gluconic acid producing *Phomopsis mangiferae* rots mango (Jiao et al. 2022).

3.5 Processed Foods: Spoilage as Colonization of a Medium

Many processed foods contain vegetables, fruits, and other plant material mixed and treated to make the nutrients more available for the human digestive system. In a certain way, these foods are comparable to the plant-based media used in an microbiological laboratory and are very prone to spoilage when incubated. In addition, animal-based products, such as milk, yoghurts, and cheeses can be spoiled by yeasts and filamentous fungi (Garnier et al. 2017).

The routing of spoilage fungi into the food product can be via the different constituents of the product. For example, some spoilage fungi enter the product via added spices (small pieces of plant material). Alternatively, fungi are introduced during the food production chain or subsequent storage. Especially, airborne spores enter food products if they are not effectively shielded. The density of fungal spores in the (indoor) air varies greatly and is dependent on the ability of certain fungal species to form large numbers of them. Fungal spoilage organisms can build up considerable biomass in certain areas of the food production chain and when not sufficiently cleaned act as a recurrent source of contamination.

In the case of *Penicillium expansum* infecting apple, high spore densities in the air are probably caused by growth of the mold in high concentration in rotten organic material in orchards (Borner 1963). Fungi can build up in the production environment on surfaces of machinery (even "slime building") as *Geotrichum candidum*, the "machinery mold" or "dairy mold" (Wildman and Clark 1947; Snyder and Worobo 2018). Previous indoor food spoilage may grossly enhance the inoculum pressure on newly introduced food products. For example, in Dutch cheese warehouses, *Penicillium discolor* can cause serious spoilage when poor hygienic conditions increase the sporulation of this fungus. In such a way a "house flora" can develop inside certain factories; e.g., *Penicillium roqueforti* that causes spoilage in rye bread factories and *Fusarium oxysporum* in dairy products. Fungi also develop inside storage facilities after leakage, flooding, condensation and at high humidity. Occupants inside homes contribute to enhance humidity because of many activities such as cooking and breathing, especially in combination with reduced ventilation of the building caused by, for instance, effective insulation.

Massive production of asexual mitotic spores (conidia) can be regarded as a vital strategy for dispersion of a number of important food-borne fungi. The order *Eurotiales* includes many relevant food spoilage fungi (Samson et al. 2019; Pitt and Hocking 2009; Houbraken et al. 2020) with an emphasis on the genera *Paecilomyces, Penicillium,* and *Aspergillus*. As a rule of thumb, *Aspergillus* seems to be more related to spoilage in tropical areas compared to *Penicillium*, which is observed more in temperate areas.

In time, different preservation techniques are developed with the aim to discourage fungal development in the food product. The classical methods include fermentation, addition of salts or high concentrations of sugars, pickling, drying, cooling and the addition of preservatives or a heating treatment before packaging. Further, modified atmosphere packaging and the application of high pressure treatment (Barbosa-Cànovas et al. 1998; Smelt 1998) of the food product reduce fungal outgrowth. Although heat-resistant ascospores survive pasteurisation treatments (Butz et al. 1996; Palou et al. 1998), elevated temperature in combination with high pressure is more effective (Timmermans et al. 2020).

Conidia of *P. expansum* and *Penicillium buchwaldii* show a nearly 4-log reduction after a heat treatment at 58 ° C for 20 sec at and after a high pressure treatment at 400 MPa for 1 min in orange juice (Nierop-Groot et al. 2019). A concentration of 500 ppm free chlorine from ClO_2 reduced viable numbers after 5 min treatment below the detection limit, while sucrose protected the spores resulting in 4.5-log and 3.5-log reduction for both species, respectively. UV radiation (254 nm) at 100 mJ/cm^2 on conidia dried on filter resulted in a 2-log reduction, while cold atmospheric plasma for 30 min resulted in a similar reduction. Pulsed electric fields resulted in reduction of the viability with 5-log reduction of *P. expansum* conidia using monopolar pulses of 2 µs at 17 kV/cm leading to a temperature of 60 °C at 15 ml/min of orange juice.

In certain aspects, the ecological niche of processed food products therefore can be regarded as an extreme environment containing rich nutrients. This is of evolutionary interest; the fungi that are able to overcome these stresses are heavily rewarded.

3.5.1 Association of Fungal Species with Food Products

It was already recognized by Johanna Westerdijk (1949) that there might be an association between specific fungal species with certain materials, food products, or crops.

As she stated: "Referring to the use of the same concept in plant sociology, the term association is proposed for the specific combination of microorganisms, found in decaying biological substrates" and also: "A very striking example of an association involved by selection is that on moldy oranges. We always found the two species *Penicillium italicum* and *Penicillium olivaceum*, independent of the type of the fruit and the country, where they were grown. On rotten apples in cellars we often found one single fungus only, i.e. *Penicillium expansum*. This same fungus is nearly always to be found on decaying bulb scales of tulips."

The adaptability of the fungal species to overcome the restrictions of the crop or the limitations introduced by preservation techniques determines the dominance of the species on the relevant food product.

Food parameters, dedicated to stability and shelf-life extension, are restrictive to the spectrum of species which are able to grow and thus spoil the individual food types (see also Rico-Munoz et al. 2019; Snyder et al. 2019).

The spoilage etiology was identified in 127 different products (originating from more than 50 companies) with one dominant spoiling organism. Nine bacterial contaminants, 26 yeast species and 92 filamentous fungi belonging to 22 genera were observed. A number of 107 species were Ascomycetes containing *Penicillium* (23 isolates), while *Aspergillus* and *Paecilomyces* were also important (Snyder et al. 2019).

Normally, less than ten and often one to three species are responsible for a spoilage case (Frisvad and Filtenborg 1988, 1993; Filtenborg et al. 1996, 2004; Frisvad et al. 2007a). Table 3.1

Table 3.1 Most common associated fungal species (from Frisvad et al. 2007a, b, with courtesy of Taylor and Francis, CRC Press)

Crop	Product	Fungal species
Beans & peas	Black beans, cowpeas	*Alternaria alternata, Aspergillus flavus,* Asp. *ochraceus,* Asp. *parasiticus, Fusarium proliferatum, Penicillium citrinum*
Cereal	Maize	*Asp. flavus, Asp. niger, Asp. ochraceus, F. graminearum, F. proliferatum, F. verticillioides, P. citrinum*
	Rice	*Asp. flavus, Asp. niger, P. citrinum*
	Rye bread	*Eurotium repens, Eur. rubrum, P. carneum, P. paneum, P. roqueforti*
	Sorghum	*Alt. alternata, Asp. flavus, F. verticillioides, F. semitectum, P. citrinum*
	Wheat bread	*Asp. flavus, Eur. repens, Eur. Rubrum*
	Wheat, rye, barley, oat	*Alt. tenuissima* and *infectoria* sp.-grps., *Asp. flavus, Asp. parasiticus, F. avenaceum, F. culmorum, F. graminearum, P. aurantiogriseum, P. cyclopium, P. freii, P. melanoconidium, P. polonicum, P. verrucosum*
Cheeses	Hard cheese	*Asp. versicolor, P. commune, P. discolor, P. nalgiovense, P. solitum*
Coffee	Coffee—monsoon	*Asp. candidus, Asp. niger, Asp. tamarii*
	Coffee—traditional	*Asp. carbonarius, Asp. steynii, Asp. westerdijkiae, P. citrinum*
Fruit	Citrus	*Alt. tangelonis, Alt. tenuissima* sp.-grp., *Alt. turkisafria, P. digitatum, P. italicum*
	Dried fruits	*Asp. carbonarius, Asp. flavus, Asp. niger, Asp. ochraceus, Xeromyces bisporus, Wallemia sebi*
	Fruit juice	*Byssochlamys nivea, B. spectabilis (= Paecilomyces variotii), Eupenicillium* spp., *Neosartorya* spp,. *Talaromyces* spp.
	Grapes	Asp. *carbonarius,* Asp. *niger, Asp. tubingensis, P. expansum*
	Pomaceous & stone	*Alt. arborescens* sp.-grp., *Alt. tenuissima* sp.-grp., *F. lateritium, P. crustosum, P. expansum, P. solitum*
Meat	Sausages	*P. nalgiovense, P. nordicum, P.olsonii, P. chrysogenum, Eurotium* spp.
Nuts	Almonds, hazelnuts, pistachio, walnuts	*Alt. arborescens* sp.-grp., Asp. *flavus,* Asp. *niger, Asp. tamari, F. acuminatum, F. avenaceum, F. semitectum, P. crustosum, P. discolor*
Oil crop	Olives	*Alt. alternata, Asp. versicolor, P. citrinum, P. expansum*
	Peanuts	*Asp. flavus, Asp. niger*
	Sunflower	*Alt. alternata, Asp. flavus, Asp. niger, Asp. parasiticus, F.verticilloides, F. semitectum*
Vegetables	Ginger	*P. brevicompactum*
	Onion & garlic	*P. allii, P. glabrum, Petromyces alliaceus*
	Pepper—bell	*Alt. alternata*
	Pepper—black	*Asp. flavus, Asp. parasiticus, Asp. tamarii*
	Potatoes	*Alt. alternata, Alt. solani, F. coeruleum., F. sambucinum*
	Tomatoes	*Alt. alternata, Alt. subtropica, Alt. tenuissima* sp.-grp., *P. expansum, P. olsonii, P. tularense, Stemphylium eturmiunum, Stemphylium solani*
	Yams	*Botryosphaeria rhodina, F. verticillioides, P. sclerotigenum*
	Yams chips	*Asp. flavus, Asp. niger*

shows a survey of different classes of food products and their associated fungi (see also Filtenborg et al. 2004; Scholte et al. 2004; Frisvad et al. 2007b). Frisvad et al. (2007a) have stressed the importance of accurate identification of spoilage fungi. The wrong identification will blur the development of a conclusive scheme of food spoilage and disturb the research on the (cellular) mechanisms responsible for this specificity as well as the specificity of mycotoxin production (see also Frisvad, 2015). Knowledge of these parameters and comparison of strains may lead to novel tailor-made preservation strategies.

Fungal culture collections may play an important role in this development. As an example, Samson and Frisvad (2004) started a taxonomy of *Penicillium* subgenus *Penicillium* based on a polyphasic approach that includes macro- and microscopic morphology, growth characteristics, DNA sequences, and the excretion of secondary metabolites. A flurry of large taxonomic studies have redefined *Eurotiales* taxonomy (including Samson and Varga 2007, Samson and Houbraken 2011, Visagie et al. 2014; Yilmaz et al. 2014; Chen et al. 2017; Sklenář et al. 2017; Houbraken et al. 2020). *Penicillium* and *Aspergillus* are important genera within the *Eurotiales* and belong to the most abundant fungi in air and are predominant with respect to food spoilage. The correct classification and deposition of the identified strains in a culture collection may become an important element of our struggle to prevent food spoilage (see also Fig. 3.4).

3.5.2 Food-spoiling Fungi

During history, there has been a continuous race between humankind and food-spoilage organisms about food crops and stored food. As water availability is vital for the growth of the fungal cell, manipulation of this parameter has been one of the major classical approaches to enhance food shelf-life. It has to be stressed that water-biomolecule relations are surprisingly subtle (Ball and Hallsworth 2015) and gaining more knowledge about them may lead to novel preservation techniques. Increasing osmolarity by adding sugar or salt is a classical approach (Dakal et al. 2014). Several yeasts, namely *Debaryomyces hansenii*, *Zygosaccharomyces rouxii,* and *Z. bailii* are very osmotolerant and can grow in environments with (extremely) high concentrations of sugar and salt. *D. hansenii* spoils brine foods and contaminates fruit powders, but is also the dominant yeast species isolated from dry cured meat products and plays a role in taste development by the production of specific volatile compounds (Andrade et al. 2010). It is also prominent on cheese and sausages.

Ecologically, *D. hansenii* is a marine fungus (Clipson and Jennings 1992) but it is also isolated from penguin droppings, anthills, and soils. It is observed in hypersaline habitats and subglacial ice (Gunde-Cimerman et al. 2009).

Zygosaccharomyces species, including the extremely osmotolerant species *rouxii* and *bailii*, cause major food spoilage on many different products including fruit juices, sauces (salad dressings), carbonated soft drinks, ketchup (Pitt and Hocking 2009), and also honey, jams, syrups, fruit juices, fruit-juice concentrates (e.g., apple, Wang et al. 2016), chocolate candies as well as salty food including soy sauce and miso paste (Dakal et al. 2014). Further, they spoil wine and ciders and mustard sauce (Kuanyshev et al. 2017).

The yeast is virtually unknown outside the food and beverage context. Candidates of natural sources are fermenting fruits in wine yards and orchards, mummified fruits, gummy exudates from (fruit) trees. Also dried fruits including dates and dried prunes contain fungus. Very low numbers of cells result in spoilage during long storage times.

Other filamentous fungi can also grow at low water activities (such as *Aspergillus montevidensis* and *Aspergillus penicillioides, Wallemia sebi,* and *Xeromyces bisporus*). *A. montevidensis* is known for spoilage in corn silos where it can develop at 15–16% moisture levels and *A. penicillioides* is probably the pioneer species for fungal spoilage of stored grains (Pitt and Hocking 2009). *W. sebi* grows on dried figs, dates, chocolate, and fruit bars; *X. bisporus* was stated to be the most xerophilic fungus known to date and develops on pure marzipan (Williams and Hallsworth 2009; Vinnere-Pettersson et al. 2011; Leong et al. 2011). This is recently debated as *A. penicillioides* is a fungus very slowly growing at similar low water activity media (0.640, Stevenson et al. 2015, 2016, 2017). *A. chevalieri* was the most dominant fungus isolated from several samples of poultry feed (Dijksterhuis et al. 2019).

Fungi that are present on food products or are introduced via the food production chain are inactivated by heat treatments. Spoilage can occur after pasteurization treatments as heat-resistant fungi survive high temperature- or

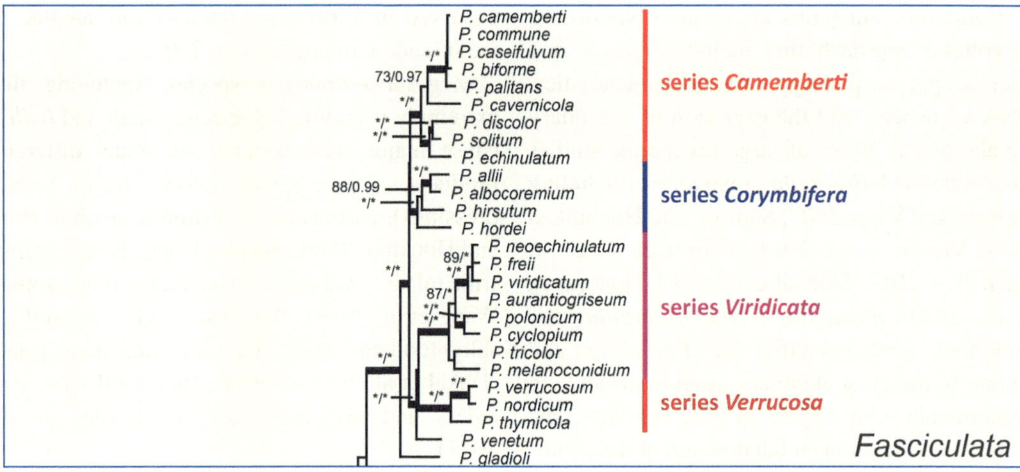

Fig. 3.4 Different clades of *Penicillium* section *Fasciculata* illustrate specificities that can be linked to different food products or growth environments. This illustrates the importance of correct identification of the fungal species. For example, the series *Camemberti* can be linked to proteinaceous and lipid-containing foods. The series *Corymbifera* occurs as post-harvest disease on onions, tap root plants, and flower bulbs. The series *Viridicata* is associated with cereals. The series *Verrucosa* grows on nitrite sole as an N-source

high-pressure treatments. Fungi that cause damage worth millions of dollars in the fruit-juice industry include *Paecilomyces niveus, Pae. fulvus,* and *Pae. variotii, as* well as Aspergillus species with a neosartorya-morph such as *Aspergillus fischeri* (as reviewed by Tournas 1994). Soil-borne fungi and fruits that develop in contact with soil (like strawberries) are more prone to contamination (Santos et al. 2018). These fungi can survive temperatures of 85 °C for time intervals that are markedly longer than those used for pasteurization treatments. The heat resistance is conveyed by sexual ascospores that are candidates to be the most stress-resistant eukaryotic described to date (Dijksterhuis 2007, recently reviewed by Wyatt et al. 2013 and Dijksterhuis 2019).

Storage of food products in refrigerators is often accompanied with the presence and growth of the so-called psychrotolerant fungi (cold-tolerant fungi). These fungi often belong to the genera *Alternaria*, *Fusarium*, *Penicillium,* and *Cladosporium.* In addition, *Botrytis cinerea* is also a fungus that develops well at a surprisingly low temperature (Hoogerwerf et al. 2002).

Several fungi (including filamentous species and yeasts) are able to degrade the important food preservative sorbate by the action of enzymatic activity. Especially, the species *Penicillium roqueforti* and *Pae. variotii* are notorious for the spoilage of rye bread, drinks, and margarine that contain sorbic acid, benzoic acid, and propionic acid.

Aspergillus niger is also capable in degrading sorbic acid, but not a major spoiler of these products.

Modified atmosphere packaging is used to keep products free of spoilers. However, some fungi have traits that make them suitable for development under very low oxygen and/or high carbon dioxide. These include *Saccharomycopsis fibuligera* and *Hyphopichia burtonii,* the "chalk molds," well-known from products as pre-backed bread.

In addition, the fungi *P. roqueforti* and *Fusarium oxysporum* are able to grow at low levels of oxygen. The main question is: Are these fungi able to grow anaerobic or are they microaerophilic?

The fungus *Geotrichum candidum (Galactomyces candidus)* is an important spoilage organism (dairy, vegetables, and fruit) without any particular stress resistance (Cai and Snyder 2019; Eliskases-Lechner et al. 2022). It can grow at relatively low temperatures as in fridges, it is salt and heat sensitive (not growing above 7%, absolute minimal $a_w = 0.91$) and may be able to deal with relatively high (0.35 mM/L) concentrations of carbonic acid at pH 7.0 (Konuchová and Valík 2021; Valle et al. 2022, 2023). It is also sensitive for high pressure treatment, with a $\log_5$ inactivation 300 Mpa for 90 seconds. Yet, the fungus can heavily contaminate food production chains. It forms numerous one-celled arthrospores and cultures show some resemblance with yeasts in development and morphology.

Arabitol, a sugar alcohol, accumulates in *Geotrichum* species in 1 M NaCl up to 30–40% dry weight decreasing slightly at the stationary phase of culturing. One species of *Geotrichum* (out of five), however, accumulated mannitol as a compatible solute in similar amounts. Mannitol is also present during the stationary growth phase without salt (Luxo et al. 1993).

3.6 Coping with Adverse Conditions

3.6.1 Osmotolerant and Halotolerant Yeasts

There is only a relatively small number of yeast species that are involved in food spoilage, these include extreme osmotolerant organisms. Halotolerance and osmotolerance are two overlapping but not identical traits (Dakal et al. 2014). Lages et al. (1999) summarize 33 yeast species that have a maximum tolerance above 2 M NaCl. The yeast *D. hansenii* grows at sodium chloride concentrations up to 2.5 M and growth is stimulated in 0.5 M (as reviewed in Prista et al. 2005). During mid-exponential growth in medium containing 1,5 M NaCl, *D. hansenii* accumulates high concentrations of sodium (approx. 400 mM) and potassium (130 mM, Herrera et al. 2017).

The amount of potassium accumulated was markedly higher at lower NaCl concentrations in the medium, and 60% of the sodium ions were located in the vacuole. *D. hansenii* can be regarded as a "sodium includer" in contrast to "sodium excluders" that have a strategy of keeping the internal concentration of the sodium ion low (Prista et al. 2005). For example, *S. cerevisiae* is very sensitive to sodium ions (Herrera et al. 2017) and these are virtually absent in the cytoplasm.

During salt stress, different transporter proteins located in the plasma and vacuolar membrane (Prista et al. 2005, 2016) are active in both efflux and influx of protons, sodium and potassium ions.

D. hansenii expresses two P-type ATPase, DhENA1 and DhENA2, one specific for sodium efflux at higher pH, but probably not involved in salt tolerance (Almagro et al. 2001). DhKha1, a putative intracellular Na+/H+ antiporter was identified to provide sodium tolerance after heterologous expression in *S. cerevisiae* and another important intracellular Na+/H+ transporter (DhNhx1) sequestered Na^+ to vacuoles. Alternatively, the DhHAK1 and DhTRK1 transporters enable the cell to take up monovalent cations, especially potassium (Martínez et al. 2011).

In addition, another major process is important for the salt tolerance of these cells, namely the accumulation of compatible solutes (Jennings and Burke 1990). The term compatible means that high intracellular concentrations of certain solutes are compatible with enzyme functioning. For fungi, the most common solutes are the polyols glycerol, erythritol, arabinitol, and mannitol as well as the disaccharide trehalose.

During mid-exponential growth in 16% (w/v) NaCl, this yeast accumulates glycerol up to approximately 35% of the dry weight of the cells (Adler et al. 1985). At the beginning of the stationary phase, a major portion of the glycerol leaks out of the cells. Simultaneously, glycerol is actively taken up by the cell and metabolized. This could be the result of the activity of a proton/glycerol symporter or even a, still putative, sodium/glycerol symporter (Lucas et al. 1990; Prista et al. 2005, 2016). The latter makes sense as glycerol has the lowest molecule weight and thus highest concentration to counteract high osmolarity. *D. hansenii* preferably accumulates trehalose at low salt levels and glycerol at high salt (2.0 M or higher, González-Hernández et al.

2005). Further, the type of solute may change with the growth phase; arabinitol (arabitol) is the major solute present in the cells of *D. hansenii* at the stationary growth phase (Adler and Gustafsson 1980). *D. hansenii* has 8 copies of *DhSTL1*, encoding for a glycerol proton symporter suggesting that this process has a role in osmotolerance (Palma et al. 2007).

The now 12 species of the genus *Zygosaccharomyces* have a reputation to be able to grow on a very high concentration of sugars (e.g., 5 M glucose [72% w/v] for *Z. rouxii*, Martorell et al. 2007) and form high amounts of carbon dioxide, leading to explosion of cans and bursting of packs. *Z. rouxii* also accumulates high levels of glycerol, but is better able to retain it inside the cell (Hosono 2000). *Z. rouxii* can still grow in 875 g sugar/l and at pH 2.5 (Membre et al. 1999) or in 3.1 M NaCI (Hosono 2000). Is their ability to ferment sugars at high rates even in the presence of oxygen, a factor in their survival (Leyva et al. 1999)? Under anaerobic conditions, the yeast also grows exponentially with vigorous fermentation, given that the medium is complex (as food and beverages are, Rodriguez et al. 2001).

Z rouxii employs two unique uptake systems for fructose molecules (Leandro et al. 2011) and is called fructophilic as it can transport these molecules with a higher capacity.

Z. bailii is extremely tolerant to organic acids such as acetic (Palma et al. 2018), sorbic, and benzoic acid (Steels et al. 2000; Martorell et al. 2007). The minimal inhibitory concentration varies between 375 and 550 mM for acetic acid and between 4.55 and 9.45 mM for sorbic acid as measured among 38 strains (Stratford et al. 2013). This resistance was not due to degradation of sorbic acid, but a result of the presence of a low number of cells with a lowered cytoplasmic pH by 0.4–0.8 pH unit. These cells show reduced (approx. ten-fold) intracellular accumulation of weak organic acids in the cell in medium with a pH of 4.0. These resistant cells are very rare in exponential cultures and increase 8000-fold in stationary cultures (Stratford et al. 2014). Furthermore, this yeast does not show catabolite repression and acetic acid and glucose can be metabolized together, resulting in less accumulation of counter-anions of the acid.

The ZbACS2 gene plays an important role in acetate catabolism (Rodrigues et al. 2012), while permeability and structure of the cell wall and the plasma membrane are modulated upon weak organic acid and salt stress (Přibylova et al. 2007, reviewed by Palma et al. 2018). Remarkable are unique major changes in the lipidome; when exposed to acetic acids include a strong increase of complex sphingolipids suggested to result in reduced permeability of the plasma membrane (Lindberg et al. 2013). A transcription factor in *Z. bailii* homologous to *S. cerevisiae* Haa1 (ZbHaa1) is bifunctional, in the response and tolerance to both acetic acid and copper stresses (Palma et al. 2017). The accumulated protons are removed from the cytoplasm by the action of the Pma1 H+-ATPase (Palma et al. 2018).

Zygosaccharomyces strains have notable salt tolerance, albeit less compared to *D. hansenii* (Lages et al. 1999). With nitrogen supplementation salt tolerance had markedly increased (Solieri et al. 2016) with growth in the presence of 2 M NaCl (μ was 0.65/h compared to 1.35/h without salt). *Z. rouxii* expresses a proton-ATPase in combination with two Na+/H+ antiporter genes to remove sodium ions from the cell (Watanabe et al. 1991, 1995).

A *S. cerevisiae* strain that was very sensitive to salt stress was made more osmotolerant with the antiporter genes (ZrSod2 and ZrSod22) from *Z. rouxii* (Iwaki et al. 1998). When the antiporter was deleted from *Z. rouxii* the organism could not grow on medium with high salt concentration, but was still able to develop on very high concentrations of sugar (Watanabe et al. 1995). Further, ZrNha1 was identified, an antiporter that is thought to be indispensable to maintain potassium homeostasis (Pribylova et al. 2008). Glycerol accumulation and retention are important to handle salt stress and vary between strains and hybrids of the yeast (Solieri et al. 2016).

3.6.2 Xerophilic Fungi

The presence of water is an absolute requirement for fungal growth. Strictly speaking, all halophiles and osmophiles are xerophiles—fungi that develop at low water activity. Both water activity and relative humidity quantify the availability of water (chemical potential) for a growing cell and are interchangeable. However, one has to realize that water content in a matrix, such as food, is correlated to higher fungal growth even at the same water activity (van Laarhoven et al. 2015). Water availability can be restricted as a result of high concentrations of solutes in food matrix, but also when the relative humidity in air is low when fungi grow on surfaces from food. These are conditions that also may prevail during large-scale storage of cereals.

Fungi of the section Aspergillus subgenus *Aspergillus* including *A. chevalieri, A. proliferans, A. restrictus,* and *A. penicillioides* (Sklenář et al. 2017) and also *Xerochrysium dermatitidis, Xerochrysium xerophilum,* and the halophilic *Wallemia ichthyophaga* are extreme xerophilic fungi. The very small genome (9.8 Mb) of *W. sebi* was sequenced (Padamsee et al. 2012) and revealed adaptations to osmotic stress. The genome of *X. bisporus* with a size of 24.8 Mb and over 10.000 genes (Leong et al. 2015) contains a very low number of clusters for the production of secondary metabolites, which suggest that competition in its unique niche is not of importance. Further, genes involved in stress signaling pathways are hardly differentially expressed after a shift in water activity from 0.89 (which is optimal) toward 0.68 highlighting the adaption of the fungus to these conditions. The fungus produces and secretes copious amounts of glycerol.

A number of studies highlight the role of intra- and extracellular solutes on survival and growth at low water activities (Williams and Hallsworth 2009; Chin et al. 2010) including growth of xerophilic fungi on different media containing solutes with varying degrees of chaotropic activity. These solutes disrupt or destabilize macromolecular interactions and disorder cellular structures, while kosmotropic solutes stabilize or order these interactions, although these phenomena (Ball and Hallsworth 2015) are far from simple and non-linear. Chaotropic solutes include glycerol (only at very high concentrations), magnesium chloride, and fructose and kosmotropic solutes are ammonium sulfate, trehalose, and sucrose.

Chin et al. (2010) realized growth of *A. penicillioides* at a water activity of 0.647 on a neutral medium compared to 0.653 for *X. bisporus* in 7.6 M glycerol. (see also Stevenson et al. 2017a). Chin et al. (2010) evaluate the influence of chaotropy on growth at low temperatures and observe that *Eurotium herbariorum* (now renamed as *Aspergillus glaucus*; Samson et al. 2014) grows much better in fructose (chaotropic) at 1.7 °C and best on sucrose at 30 °C. The extreme halotolerant yeast *Mrakia frigida* even can grow slowly at −5 ° C in 1.1 M glycerol, which is not possible in 0.73 M sucrose (kosmotropic). Spores of *X. bisporus* from glycerol containing cultures survive freezing treatments better compared to conidia from sucrose containing medium that are better survivors after heat- and high-pressure treatments. *Aspergillus wentii* accumulated low-Mr polyols to ≥100 mg/g dry weight (de Lima et al. 2015). Polyols with lower molecular weight such as glycerol, erythritol, and arabitol were more effective for osmotic adjustment. Mannitol and trehalose have too high water-activity values for saturated solutions to be effective; i.e. 0.978 and 0.970 (25 °C).

Psychrophilic growth is very relevant for food spoilage in cold environments, many crops and processed food products are lost during cold storage. Five *Penicillium* and *Cladosporium* species were isolated from the blades of a ventilator in a refrigerator that must have been developed at temperatures near or even below 0 °C (Dijksterhuis 2019). A *Penicillium hirsutum* strain after inoculation on lily bulbs had fully developed after a 6-month storage at temperatures slightly below 0 °C. Nguyen Van Long et al. (2021) modeled cardinal growth temperatures for 29 strains of *Penicillium roqueforti* and predicted a T_{min} between −8.8 and −35.4 °C, but this was based on extrapolation of data above 0 °C. *P. expansum* germinates at 0.1 °C after lag times of on average 800 hours (Gougouli and Koutsoumanis 2013). The yeast *Mrakia* (Yurkov et al. 2020) is reported from frozen food and also spoils cooled juices. The extreme halotolerant yeast *Mrakia frigida* even grows slowly at −5 ° C in 1.1 M glycerol, which is not possible in 0.73 M sucrose (cosmotropic) (Chin et al. 2010).

Xerophily and psychrophily might have similarities as xerophilic fungi are isolated from high altitude Himalayan soil, where very low temperatures are common (Petrovič et al. 2000).

3.6.3 Fungal Survival Structures

Fungal spores are survival structures that enable fungi to disperse in both space and time (the reviews of Wyatt et al. 2013; Dijksterhuis 2017, 2019 provide much more information on this). Fungal spores are extremely variable, the way they are formed and their shape were leading for identification of fungal genera and species, before the era of DNA-based identification. Spore-forming structures bear different names as sporangiophores in case of *Mucuromycotina* (Dijksterhuis and Samson 2006a, b) or conidiophores when they produce mitotic asexual spores (in the case of the large group of Ascomycetes). Ascospores are sexual spores formed in ascomata (Chitarra and Dijksterhuis 2007; Magan et al. 2012; Wyatt et al. 2013). Some of the latter belong to the most stress-resistant eukaryotic cells described to date (Dijksterhuis 2007). Fungal genera that are important for food spoilage as *Aspergillus*, *Fusarium*, *Paecilomyces* , and *Penicillium* also are avid sporeformers (Berbee et al. 1995).

Hyaline conidia are dispersed by water splashes and mostly pigmented airborne conidia are more stress resistant. Survival (of high UV, drought, heat and cold stress) and dormancy of the latter spores are improved as spores in air are in a collapsed dehydrated state (Segers et al. 2023). Interestingly, aerial or water dispersion is correlated to membrane composition (ergosterol level) or cytoplasmic parameters (viscosity of the cytoplasm) of the spores (Van Leeuwen et al. 2008, 2010). Extremely heat-resistant ascospores of *Talaromyces macrosporus* and *Aspergillus fischeri* exhibit very high cytoplasmic viscosity in liquid solution (Dijksterhuis et al. 2007; Wyatt et al. 2015a).

In fungal conidia of the species *Aspergillus oryzae*, *A. nidulans*, *A. niger* and *Penicillium rubens* both the compatible solutes mannitol, a polyol, and trehalose, a disaccharide, are accumulated (Horikoshi and Ikeda 1966; D'Enfert and Fontaine 1997; Bekker et al. 2012; Van Leeuwen et al. 2013a, b; Seekles et al. 2023). Trehalose is dominant in conidia of the worldwide contaminant *Pae. variotii* (Van den Brule et al. 2020a; b) with lower amounts of arabitol, while *Penicillium roqueforti* accumulates trehalose and arabitol (Punt et al. 2020). Trehalose accumulation protects and stabilizes cell components (Wiemken 1990) including membranes (Crowe et al. 1984) and proteins (Hottiger et al. 1994). Higher levels of trehalose are clearly correlated with increased heat resistance (Punt et al. 2020; Van den Brule et al. 2020a, b; Seekles et al. 2023).

The history of the spore-forming colony affects survival properties of conidia. Hallsworth and Magan (1994, 1996) showed that different environmental conditions during cultivation affect the solute composition inside conidia in the case of insect pathogenic fungi.

Optimal conditions might result in stronger spores, which have higher shelf lives if they are used for biological control of insect pests. The principles found in the entomopathogenic species may also apply for fungal species occurring on food, with the difference that now conidia maybe have to be eradicated as effective as possible.

Conidia of *Aspergillus fumigatus* formed at higher temperatures show higher heat resistance, accompanied with higher trehalose levels. Surprisingly, conidia formed at a lower temperature are more melanized and UV resistant (Hagiwara et al. 2017). Similar trends are described for *P. roqueforti, Pae. variotii*, and *A. niger*. (Punt et al. 2020; Van den Brule et al. Unpublished results; Seekles et al. 2023). Conidia of *P. rubens* that are formed at lower water activity (0.86, by the addition of NaCl to medium) attract more water as their counterparts formed at 0.99 water activity (Van Laarhoven et al. 2017). Variation in compatible solutes of conidia is also observed during maturation of conidia (Teertstra et al. 2017).

In *A. niger* trehalose levels increased in time between 2- and 8-day-old cultures while mannitol decreased and still showed some metabolic activity, albeit much lower as growing hyphae (Novodvorska et al. 2016; Teertstra et al. 2017). Very early formed conidia of *A. niger* hardly contained compatible solutes and are very heat sensitive (Seekles et al. 2023). Conidia from a ΔmpdA, ΔtpsABC strain contained no measurable trehalose and very little mannitol and were similarly heat sensitive (approx. ten-fold lower heat resistance compared to fully matured conidia). Transcriptomics and proteomics indicated two putative heat shock proteins being upregulated at cultivation of the fungus at higher temperature. However, conidia of knock-out strains lacking these putative heat shock proteins did not show a reduced heat resistance.

Fungal colonization of food is often initiated with the deposition of (single) conidia on the product with the prerequisite that the dormant state is effectively broken and germination can occur. The transition from a dormant conidium toward a vegetative growing fungal hyphae includes changes of the cell wall (Tiedt 1993; Fontaine et al. 2010), breakdown of compatible sugars (including trehalose and mannitol) (D'Enfert et al. 1999; Fillinger et al. 2001; Van Leeuwen et al. 2013b), reorganization of the transcriptome including major mRNA breakdown, and strong down- and upregulation of specific gene categories (Van Leeuwen et al. 2013a). Water alone added to spores directly results in the formation of polyribosomes (*Neurospora crassa*, Bonnen and Brambl 1983).

Isotropic growth (swelling) of conidia of *A. niger* occurs after incubation in distilled water (Morozova et al. 2002) and conidia of *A. oryzae* germinate on water agar (Sakamoto et al. 2009). Nutrients as phosphate, amino acids, glucose, and their combinations reactivated dried sporangiospores of the tempeh fungus *Rhizopus oligosporus* to a different extent, but rich media were far more effective in this (Thanh and Nout 2004; Thanh et al. 2005). Small amounts of glucose or salts in water resulted in very low <1%) germination of *A. niger* conidia, which increased gradually in combinations of glucose and salts. Similarly, the amino acids proline and alanine combined with phosphate and magnesium chloride resulted in germination of a large part of the conidia. Other amino acids do hardly activate conidia to germination (Ijadpanahsaravi et al. 2021). Contrastingly other fungi as *Cladosporium halotolerans* and *P. rubens* produce conidia that form hyphae in ultrapure water (Segers et al. 2017), while, *A. nidulans* specifically need carbon sources as glucose for germination (D'enfert and Fontaine 1997; Osherov and May 2000). The latter authors treated conidia with various inhibitors and only the protein synthesis inhibitor cycloheximide prevented isotropic growth in contrast to inhibitors of other cell processes.

Germination of conidia is also affected by volatile compounds such as 1-octen-3-ol. This compound blocks germination of conidia and is measured in the headspace above dense suspensions of conidia of *Penicillium paneum* (Chitarra et al. 2004, 2005) and also occurs in *A. nidulans* (Herrero-Garcia et al. 2011). Spores of *A. flavus* do not produce the component (Miyamoto et al. 2014). 1-Octen-3-ol is also produced by *A. niger* (Karlshøj et al. 2007). It is hypothesized that 1-octen-3-ol acts as a fungal self-inhibitor that prevents premature germination of conidia on conidiophores.

The compound had a profound influence on protein expression patterns (Chitarra et al. 2005) and blocked isotropic growth. Volatiles also activate germination of conidia in case of *Penicillium digitatum* which causes post-harvest citrus rot. Conidia, when provoked with volatiles from wounded oranges showed enhanced germination (Eckert and Ratnayake 1994).

A whole suite of primary and secondary models for spore germination is published during recent years.

Primary models for the inactivation of fungal spores are reviewed (Dijksterhuis et al. 2012) and can be used for different sporicidal conditions including heat, drying, or vapor treatments (Dao and Dantigny 2009; Dao et al. 2010).

The asymmetric model proposed by Dantigny et al. (2011) addresses the often skewed nature of germination curves in which the increase of the number of germinating spores in time is modeled. The lag phase, which is the time before germ tube formation and the subsequent elongation of germ tubes, depends on the presence of stressors, such as antimicrobial compounds like red cabbage seed extract (Dagnas et al. 2015) or lower temperature and water activity (Dagnas et al. 2017).

The 2015 paper summarizes the different interpretations of "lag phase," which is occasionally used to describe the first visibility of a fungus on a product, very well in the first figure.

The lag phase can expand up to weeks, while the individual behavior of germinating spores becomes more unpredictable indicated by a steep increase in the standard deviations of the individual germination times (appearance of germ tubes and colonies originating from one spore) (Dagnas et al. 2017). Thus, the length of the lag phase can be taken as a sign of environmental stress, but care has to be taken (Hamill et al. 2020). After a prolonged lag phase outgrowth of hyphae can be surprisingly fast adding to the notion of "sudden" spoilage of a product after a long storage time. However, with even more sub-optimal conditions also germ tube and colony development slow down (Stevenson et al. 2017b).

When food products are contaminated by airborne spores, the variation in germination between individual spores is leading as inoculation occurs by short chains or even single cells. That conidia are heterogeneous in stress resistance and germination behavior (Gougouli and Koutsoumanis 2013; Teertstra et al. 2017; Ijadpanahsaravi et al. 2021; Wang et al. 2021, Seekles et al. 2023) is well documented and this complicates modeling. The aspect that part of a spore population does simply not germinate (which is explained as bet-hedging) under adverse conditions adds further to this challenge.

3.6.4 Heat Resistance

Fungal survival structures are generally more heat resistant compared to vegetative cells and conidia. Sclerotia, chlamydospores, and ascospores survive heat treatments between 55 and 95 °C. Yeast ascospores isolated from soft drinks and fruit products had D_{60} values that were 25–350 times higher than those of the corresponding vegetative cells (Put and De Jong 1980). Many aspects of spore survival are summarized by Wyatt et al. (2013). Recent measurements at our laboratory show that heat resistance of ascospores above 70 °C (see Figs. 3.5 and 3.6) is a common trait that occurs within the order *Eurotiales*: *Aspergillus*, *Hamigera*, *Xeromyces*, *Paecilomyces*, *Penicillium*, *Talaromyces*, and *Thermoascus* and also described for other fungal genera (e.g., *Neurospora* and *Daldinia* species). Ascospores of the fungus *Talaromyces macrosporus* survive at 85 °C for 100 min (Dijksterhuis and Teunissen 2004), which is similar to some bacterial spores (e.g., *Bacillus subtilis*). Ascospores of *A. spinosus* and *Pea. variotii* show similar heat resistance (Houbraken et al. 2008; Wyatt et al. 2015a). These spores have a constitutive dormancy, meaning that even in rich nutrient media no germination occurs. Constitutive dormancy of ascospores includes a metabolic block, a barrier to the penetration of nutrients or the production of a self-inhibitor (as defined by Sussman and Halvorson 1966) or as a result of a specific physical state of the cytoplasm, such as a glassy state.

The heat resistance of ascospores in food products generally increases with the sugar concentration of the surrounding medium (Splittstoesser and Splittstoesser 1977; Beuchat 1988a; King and Whitehand 1990). Additional factors, such as pH and the presence of organic acids (as those present in fruits or used for preservation), counteract the heat resistance of ascospores, but only at low pH values (lower than 4).

Benzoic and sorbic acid affect heat resistance of *T. flavus* and *A. fischeri* (Beuchat 1988b; Rajashekhara et al. 1998). A combination of different factors may lead to some unpredictable variations in heat resistance. For instance, *A. fischeri* exhibited a far higher heat resistance in 0.1 M phosphate buffer (pH 7.0) than in grape jellies with large amounts of cane sugar (pH 3.1–3.3, Beuchat and Kuhn 1997) and *P. niveus*, *P. fulvus,* and *A. fischeri* were approx. twice as heat resistant in tomato juice (pH of 4.2) as in phosphate buffer (pH 7.0, Kotzekidou 1997).

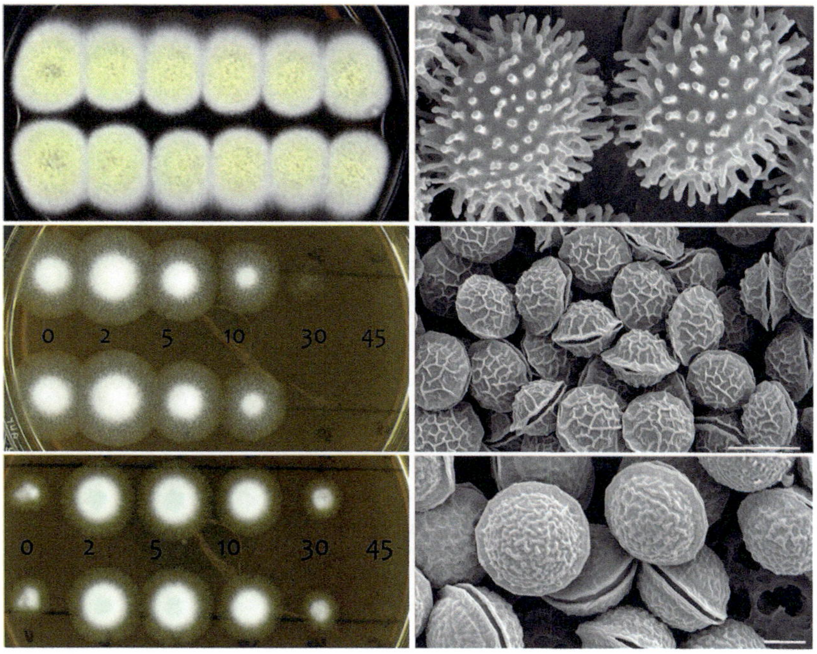

Fig. 3.5 Heat-resistant fungi survive 75°- 85 °C for periods that are longer than conventional pasteurization times. Left panel; ca. 5000 ascospores were inoculated in a small droplet on agar after a heat treatment for 0, 2, 5, 10, 30, and 45 min (from left to right). Top (*Talaromyces macrosporus*, middle *Aspergillus fischeri* and bottom *Aspergillus hiratsukae,* inoculated at 85 °C). The right panel shows images of the ascospores as taken with low temperature scanning electron microscopy. Bars are 1, 5, and 2 μm from top to bottom (Pictures made by Alex Verkennis and Jan Dijksterhuis)

Further, factors inside the ascospore are important for heat resistance, for instance as a result of the age of the culture in case of *A. fischeri*, *T. flavus,* and *P. niveus* (Conner and Beuchat 1987; Beuchat 1988a; Casella et al. 1990) or the growth temperature of the spore-generating colony (Conner and Beuchat 1987; King and Whitehand 1990).

Harvested and washed ascospores of *T. macrosporus* still showed maturation (i.e., increase of heat resistance in time, Dijksterhuis and Teunissen 2004) when stored in liquid at 30 ° C. This phenomenon did not occur at 10 ° C suggesting a temperature-dependent acquisition of resistance. King and Whitehand (1990) report higher heat resistance of *T macrosporus* on solid medium and also the type of medium used is important (Beuchat 1988a). Finally, heat resistance varies with the fungal isolate used (Bayne and Michener 1979; Beuchat 1986; King and Whitehand 1990). A combination of heat treatment and high pressure was effective in a quicker inactivation of ascospores in strawberry puree (Timmermans et al. 2020).

Ascospores often need a robust physical signal such as heat or high pressure for breaking of dormancy, where the number of viable counts after treatment is increased by several log cycles (e.g., *Aspergillus glaucus* at 60 °C, Splittstoesser et al. 1989). Further, high pressure treatments (6000 Bar) that are used for non-thermal "pasteurization" of a number of food products can activate ascospores of *T. macrosporus* to germinate (Reyns et al. 2003; Dijksterhuis and Teunissen 2004; Timmermans et al. 2020) in which physical disruption of the thick outer cell wall may play a role. For ascospores of *T. macrosporus* activation is observed at 80 °C and, at 85 °C, activation is followed by killing (Beuchat 1986). Further, after a short heat treatment of *T. macrosporus* ascospores (5 min, 85 °C) synchronized germination occurs. This is dubbed heat

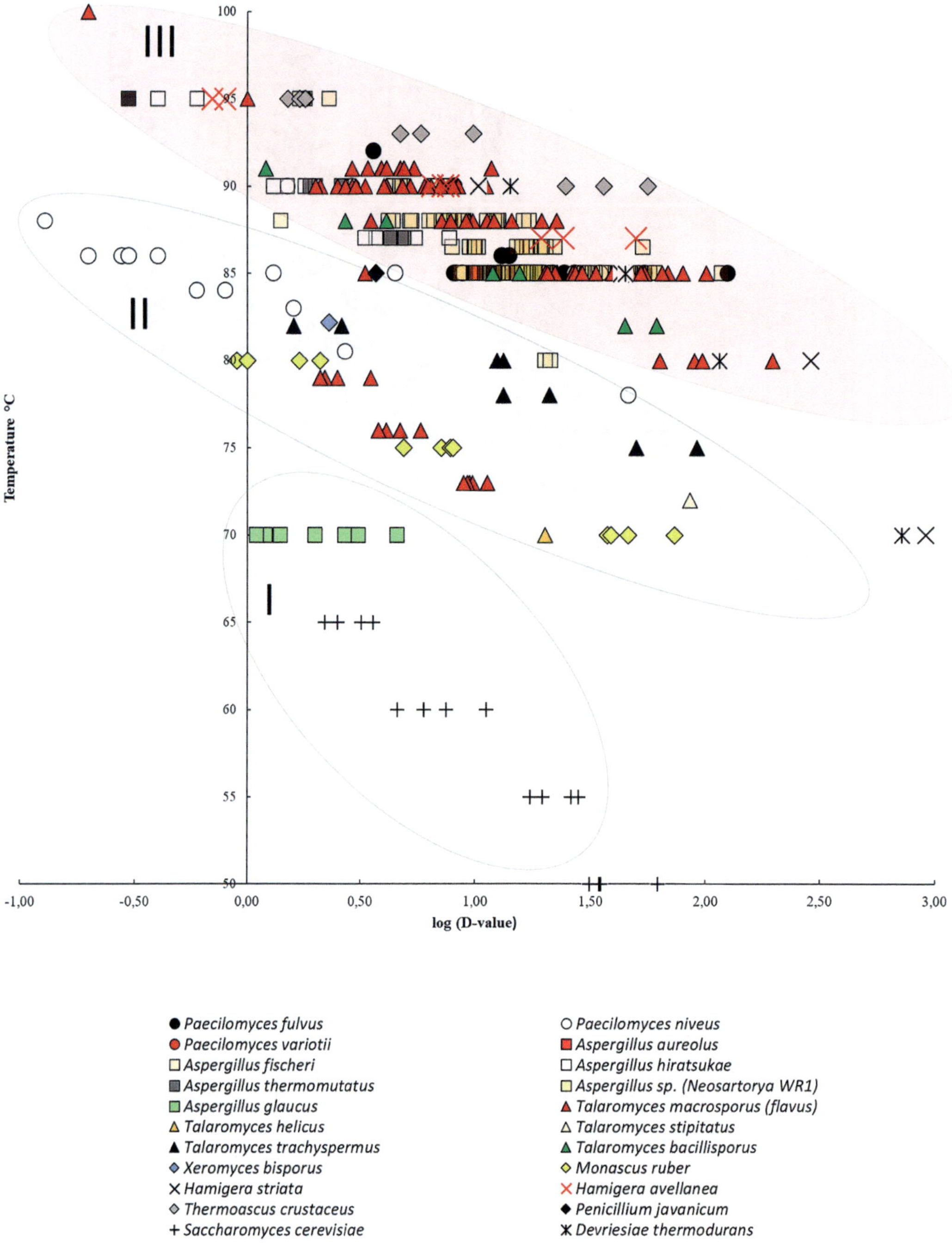

Fig. 3.6 Graphical representation of 463 D-values of 24 fungal species from 33 studies over a period of 39 years. The log of the D-value in minutes is plotted against the temperature. B.) Three degrees of heat resistance can be discerned; Group I consists out of ascospores with relatively low resistance (10 min at 60 °C). A second group (II) consists out of ascospores with heat resistance of several minutes at 80 °C to 85 °C. Group three exists out of very resistant ascospores with resistance of ten or more minutes at 90 °C. Revised from Dijksterhuis et al. (2019). The *Devriesiae* species might not produce ascospores, but extremely heat-resistant chlamydospores (Kikoku et al. 2008)

activation and can result in fungal outgrowth in products of food industry after pasteurization. ESR (Electron Spin Resonance) measurements show that the cytoplasm of the ascospores is dense, but that no glassy state occurs during heating in liquid (Dijksterhuis et al. 2007; Wyatt et al. 2015b).

This extreme dormancy may also serve a very long shelf-life of the ascospores; they can be still viable for up to 17 years in the case of *T. flavus* (Nagtzaam and Bollen 1994). At lower temperatures, activation fails and only low numbers of germinated spores are observed. Remarkably, the speed of activation increases with higher temperatures with *T. macrosporus* (Kikoku 2003; Dijksterhuis et al. 2021). Unlike *T. macrosporus* and *T. flavus*, with *A. fischeri* the dormant state can be broken by a drying treatment of 18 h at 40 °C (Beuchat 1992). Heating at 50% r.h. (dry heat treatment) at 95 ° C (for 30 or 60 min) activated *N. fischeri* ascospores, but the temperature of the wetting or recovery buffer was crucial for the viable count obtained (Gomez et al. 1989, 1993).

Drying or freezing of ascospores does not result in a change of the state of being activated, which suggests that once occurring, this state is permanent (Dijksterhuis and Samson 2006a; b). This indicates that irreversible changes occur during breaking of dormancy. In *T. macrosporus* a small molecular weight protein is released upon heat activation, suggesting that dormancy is conveyed via an impermeability of the outer cell wall (Dijksterhuis et al. 2021). Cryoplaning electron microscopy shows that rupture of the thick outer cell wall of these spores is necessary for germination (Dijksterhuis et al. 2007).

Compatible solutes accumulate to higher concentrations as observed in conidia during maturation of ascospores. In *T. macrosporus* and *T. flavus*, trehalose is the sole solute accumulating up to over 1 M. In *Aspergillus fischeri* and *A. spinosus* other and novel fungal compatible solutes are accumulated including three trehalose-based oligosaccharides (TOS) containing a α-1,1 trehalose core and one, two, or three α-1,6-glucose moieties linked to it defining isobemisiose, neosartose, and fischerose (Wyatt et al. 2015a). *T. macrosporus* has higher liquid heat resistance as *A. fischeri*, but lower resistance against dry heat, which could indicate that mixtures of different compatible solutes result in different biophysics of the cytoplasm.

Maturation and germination of ascospores are accompanied with increase or decrease in the concentration compatible solutes, which correlate precisely with microviscosity in the cytoplasm as well as a decrease in heat resistance (Dijksterhuis et al. 2002, 2007; Wyatt et al. 2015b). Ascospores showed changes in the inner cell-wall region during aging (Conner et al. 1987). Wyatt et al. (2015b) described that the composition of the mixture of compatible solutes changed during later stages of maturation with comparatively more trehalose and TOS and less mannitol after 50 days of cultivation.

3.6.5 Food Preservatives

Food preservatives are added to food products in order to prevent outgrowth of fungi. The food additive sorbic acid is widely used in food industry as a preservative of low pH sugar containing products (Stratford et al. 2012).

The minimally inhibitory concentration of sorbic acid is dependent on the density of the inoculum of conidia in case of *A. niger* (Plumridge et al. 2004). Sorbic acid delays conidial germination, but at 3 hours of germination conidia start to degrade the preservative (Plumridge et al. 2010) and resume development.

This is the result of the activity of a phenylacrylic acid decarboxylase encoded by the *padA1* gene and a putative 2-hydroxybenzoic acid decarboxylase encoded by *ohbA1* (Plumridge et al. 2008). One of the characteristic degradation products of sorbic acid is 1,3-pentadiene that has a kerosene like smell. Several osmophilic yeast including *Z. rouxii* and *D. hansenii* are able to perform degradation of sorbate, as well as several *Penicillium* species and *Trichoderma* (Cheng et al. 1999; Casa et al. 2004; Pinches and Apps 2007). Yeast cells also strongly upregulate the expression of transporters in answer on weak acid stress, which is imposed on the cell by sorbic acid, including Pdr12 that specifically removes the anions from the cells (Piper et al. 2001).

Propionic acid is also widely used, is potent against growing hyphae of different fungal species, and also inactivates conidia, albeit dependent on the fungal species (Dijksterhuis et al. 2019). With increasing molecular weight of the organic acid molecules, the M.I.C. drops to a level that the cytoplasm cannot be acidified enough to prevent growth. This suggests other cell-inhibiting mechanisms for the larger organic acids as sorbic acid (Stratford et al. 2009).

Another preservative, natamycin, is used for the protection of cheese and sausages against fungal development (Stark 2007). This polyene antibiotic binds to ergosterol, which is most available at growing tips of germinating spores and vegetative hyphae (van Leeuwen et al. 2008, 2010) and blocks active fungal growth. Natamycin is active at very low concentration (micromolars) and affects many species of fungi (Streekstra et al. 2016). Other polyenes as amphotericin B and nystatin form complexes that lead to leakage at the plasma membrane, but natamycin does not permeabilize cells (Te Welscher et al. 2008; Van Leeuwen et al. 2009), but acts directly on different aspects of cellular physiology and membrane trafficking and fusion. It inhibits endocytosis in germinating conidia of *P. discolor* (Van Leeuwen et al. 2009) and fusion of prevacuolar compartments in *S. cerevisiae* (Te Welsher et al. 2010). Natamycin inhibits transport of amino acids and sugar into cells in a reversible manner via a hitherto unknown mechanism (Te Welsher et al. 2012). In natamycin-containing solutions, development of the spores was halted at or before isotropic growth (Van Leeuwen et al. 2013b).

A micro-array study on treated conidia of *A. niger* showed that 8-hour natamycin-treated cells showed certain similarities with dormant conidia compared to the controls that had formed germ tubes. These included the elevated presence of transcripts of protective proteins (heat shock proteins, dehydrins, LEA-like proteins) and genes involved in glyoxylate cycle, fermentation, glycerol, trehalose, and mannitol synthesis. Conidia of *A. niger* and *P. discolor* are able to survive 20 hours in a concentration of natamycin of ten times the minimal inhibitory concentration and germinate at high percentages after removal of the antifungal compound (Van Leeuwen et al. 2010). Mann and Beuchat (2008) suggest a combinatory use of reduced concentrations of organic acids and natamycin to control those fungi that degrade sorbic acid.

Novel food preserving techniques include the application of the preservatives sorbate and benzoate on the surface of fruits and vegetables and maybe also to processed food by the use of edible coatings (Valencia-Chamorro et al. 2010; Mehyar et al. 2011). In addition, biocontrol agents as the yeast *Pichia anomala* are an interesting option that counteracts spoilage fungi in case of high moisture feed grain under airtight conditions (Petersson et al. 1999). Davies et al. (2021) summarize a large number of very recent approaches to counteract microbial rot on crops and food products including pulsed light, UV treatment, electrolyzed and ozonated water, essential oils, and biocontrol among them. Very recently even RNA interference techniques are considered (de Oliveira Filho et al. 2022).

3.6.6 Heterogeneity

With respect to risk assessment and modeling of eradication of fungi to prevent food spoilage it is important to use more as one strain of model fungi. The heat resistance of ascospores does vary among strains of *Saccharomyces diastaticus* that spoils beer and beer-related beverages (as Radler) (Suiker et al. 2021). Conidia of *A. niger*, *P. roqueforti*, and *Pae. variotii* show extensive and consistent variation in heat resistance when originating from different isolates. Between twenty strains of each fungus, the extent of variation in heat resistance was of a similar magnitude, indicating that variation of stress resistance between fungal isolates is an important bet-hedging strategy (van den Brule et al. 2022). For all three species, inactivation of the number of spores after a heat treatment could vary up to 5 log-values. *Pae. variotii* strains showed variations between isolates, but also within the population of conidia harvested from one culture of the fungus. Heat-resistant strains had a subpopulation of larger conidia (Van den Brule et al. 2020a; b). Marked variation was also observed for sorbic acid tolerance between strains of *P. roqueforti* (Punt et al. 2022) and to a lesser extent between *A. niger* strains (Seekles et al. 2022).

References

Adler L, Gustafsson L (1980) Polyhydric alcohol production and intracellular amino acid pool in relation to halotolerance of the yeast *Debaryomyces hansenii*. Arch Microbiol 124:123–130

Adler L, Blomberg A, Nilsson A (1985) Glycerol metabolism and osmoregulation in the salt-tolerant yeast *Debaryomyces hansenii*. J Bacteriol 162:300–306

Alkan N, Fluhr R, Sherman A, Prusky D (2008) Role of ammonia secretion and pH modulation on pathogenicity of *Colletotrichum coccodes* on tomato fruit. Mol Plant Microbe Interact 21:1058–1066

Alkan N, Davydov O, Sagi M, Fluhr R, Prusky D (2009) Ammonium secretion by *Colletotrichum coccodes* activates host NADPH oxidase activity enhancing host cell death and fungal virulence in tomato fruits. Mol Plant Microbe Interact 22:1484–1491

Alkan N, Fluhr R, Prusky D (2012) Ammonium secretion during *Colletotrichum coccodes* infection modulates salicylic and jasmonic acid pathways of ripe and unripe tomato fruit. Mol Plant Microbe Interact 25:85–96. https://doi.org/10.1094/MPMI-01-11-0020

Alkan N, Friedlander G, Ment D, Prusky D, Fluhr R (2015) Simultaneous transcriptome analysis of *Colletotrichum gloeosporioides* and tomato fruit pathosystem reveals novel fungal pathogenicity and fruit defense strategies. New Phytol 205:801–815. https://doi.org/10.1111/nph.13087

Almagro A, Prista C, Benito B, Loureiro-Dias MC, Ramos J (2001) Cloning and expression of two genes coding for sodium pumps in the salt-tolerant yeast *Debaryomyces hansenii*. J Bacteriol 183:3251–3255

Amselem J, Cuomo CA, Van Kan J, Viaud M, Benito EP, Couloux A, Coutinho PM, De Vries RP, Dyer PS, Fillinger S, Fournier E, Gout L, Hahn M, Kohn L, Lapalu N, Plummer KM, Pradier J-M, Quévillon SA, Simon A, Ten Have A, Tudzynski B, Wincker P, Andrew M, Anthouard V, Beever RE, Beffa R, Benoit I, Bouzid O, Brault B, Chen Z, Choquer M, Collémare J, Cotton P, Danchin EG, Da Silva C, Gautier A, Giraud C, Giraud T, Gonzalez C, Grossetete S, Güldener U, Henrissat B, Howlett B, Kodira C, Kretschmer M, Lappartient A, Leroch M, Levis C, Mauceli E, Neuvéglise C, Oeser B, Pearson M, Poulain J, Poussereau N, Quesneville H, Rascle C, Schumacher J, Ségurens B, Sexton A, Silva E, Sirven C, Soanes DM, Talbot NJ, Templeton M, Yandava C, Yarden O, Zeng Q, Rollins J, Lebrun M-H, Dickman M (2011) Genomic analysis of the necrotrophic fungal pathogens *Sclerotinia sclerotiorum* and *Botrytis cinerea*. PLoS Genet 7:e1002230

Andrade MJ, Córdoba JJ, Casado EM, Córdoba MG, Rodríguez M (2010) Effect of selected strains of *Debaryomyces hansenii* on the volatile compound production of dry fermented sausage "salchichón". Meat Sci 85:256–264

Ardi R, Kobiler I, Keen NT, Prusky D (1998) Involvement of epicatechin biosynthesis in the resistance of avocado fruits to postharvest decay. Physiol Mol Plant Pathol 53:269–285

Ball P, Hallsworth JE (2015) Water structure and chaotropicity: their uses, abuses and biological implications. Phys Chem Chem Phys 17:8297–8305

Barbosa-Cànovas GV, Pothakamury U, Palou E, Swanson B (1998) High hydrostatic pressure food processing. In: Barbosa-Cànovas GV, Pothakamury U, Palou E, Swanson B (eds) Nonthermal presevation of foods. Marcel Dekker, New York, pp 9–52

Bartholomew HP, Bradshaw M, Jurick WM II, Fonseca JM (2021) The good, the bad, and the ugly: mycotoxin production during postharvest decay and their influence on Tritrophic host–pathogen–microbe interactions. Front Microbiol 12:611881. https://doi.org/10.3389/fmicb.2021.611881

Bartholomew HP, Bradshaw MJ, Macarisin O, Gaskins VL, Fonseca JM, Jurick WM 2nd (2022a) More than a virulence factor: Patulin is a non-host-specific toxin that inhibits postharvest phytopathogens and requires efflux for *Penicillium* tolerance. Phytopathology 112:1165–1174. https://doi.org/10.1094/PHYTO-09-21-0371-R

Bartholomew HP, Lichtner FJ, Bradshaw M, Gaskins VL, Fonseca JM, Bennett JW, Jurick WM II. (2022b) Comparative *Penicillium* spp. transcriptomics: conserved pathways and processes revealed in ungerminated conidia and during postharvest apple fruit decay. Microorganisms 10:2414. https://doi.org/10.3390/microorganisms10122414

Bayne HG, Michener HD (1979) Heat resistance of *Byssochlamys* ascospores. Appl Environ Microbiol 37:449–453

Barad S, Espeso EA, Sherman A, Prusky D (2016) Ammonia activates pacC and patulin accumulation in an acidic environment during apple colonization by Penicillium expansum. Mol Plant Pathol 17:727–740. https://doi.org/10.1111/mpp.12327. Epub 2015 Dec 3

Bechinger C, Giebel K-F, Schnell M, Leiderer P, Deising HB, Bastneyer M (1999) Optical measurement of invasive forces exerted by appressoria of a plant pathogenic fungus. Science 285:1896–1899

Bekker M, Huinink HP, Adan OC, Samson RA, Wyatt T, Dijksterhuis J (2012) Production of an extracellular matrix as an isotropic growth phase of *Penicillium rubens* on gypsum. Appl Environ

Microbiol 78:6930–6937. https://doi.org/10.1128/AEM.01506-12

Beno-Moualem D, Prusky D (2000) Early events during quiescent infection development by *Colletotrichum gloeosporioides* in unripe avocado fruits. Phytopathology 90:553–559

Berbee ML, Yoshimura A, Sugiyama J, Taylor JW (1995) Is *Penicillium* monophyletic? An evaluation of phylogeny in the family *Trichocomaceae* from 18S, 5.8S and ITS ribosomal DNA sequence data. Mycologia 87:210–222

Beuchat LR (1986) Extraordinary heat resistance of *Talaromyces flavus* and *Neosartorya fischeri* ascospores in fruit products. J Food Sci 51:1506–1510

Beuchat LR (1988a) Thermal tolerance of *Talaromyces flavus* ascospores as affected by growth medium and temperature, age and sugar content in the inactivation medium. Trans Br Mycol Soc 90:359–374

Beuchat LR (1988b) Influence of organic acids on heat resistance characteristics of *Talaromyces flavus* ascospores. Int J Food Microbiol 6:97–105

Beuchat LR (1992) Survival of *Neosartorya fischeri* and *Talaromyces flavus* ascospores in fruit powders. Lett Appl Microbiol 14:238–240

Beuchat LR, Kuhn GD (1997) Thermal sensitivity of *Neosartorya fischeri* ascospores in regular and reduced-sugar grape jelly. J Food Prot 60:1577–1579

Bi K, Liang Y, Mengiste T, Sharon A (2023) Killing softly: a roadmap of *Botrytis cinerea* pathogenicity. Trends Plant Sci 28:211–222. https://doi.org/10.1016/j.tplants.2022.08.024

Bonnen A, Brambl R (1983) Germination physiology of *Neurospora crassa* conidia. Exp Mycol 7:197–207

Borner H (1963) Untersuchungen uber die Bildung antiphytotischer und antimicrobieller Substanzen durch Mikroorganism im Bodem und ihre mögliche Bedeutung für die Bodenmüdigkeit beim Apfels *Pirus malus* L. Phytopathol Z 49:1–28

Butz P, Funtenberger S, Haberditzl T, Tausher B (1996) High pressure inactivation of *Byssochlamys nivea* ascospores and other heat resistant moulds. LebensmWiss Technol 29:404–410

Cai S, Snyder AB (2019) Machinery Mold *(Galactomyces geotrichum*) survival following thermal and hydrostatic pressure processing. J Food Prot 82:1034–1038. https://doi.org/10.4315/0362-028X.JFP-18-568

Caldeira C, De Laurentiis V, Corrado S, Van Holsteijn F, Sala S (2019) Quantification of food waste per product group along the food supply chain in the European Union: a mass flow analysis. Resour Conserv Recycl 149:479–488

Cantu D, Vicente AR, Greve LC, Dewey FM, Bennett AB, Labavitch JM, Powell ALT (2008) The intersection between cell wall disassembly ripening, and fruit susceptibility to *Botrytis cinerea*. Proc Natl Acad Sci USA 105:859–864

Casa E, De Ancos B, Valderrama MJ, Cano P, Peinado JM (2004) Pentadiene production from potassium sorbate by osmotolerant yeasts. Int J Food Microbiol 94:93–96

Casella MLA, Matasci F, Schmidt-Lorenz W (1990) Influence of age, growth medium, and temperature on heat resistance of *Byssochlamys nivea* ascospores. Lebensm-Wiss Technol 23:404–411

Chen AJ, Hubka V, Frisvad JC, Visagie CM, Houbraken J, Meijer M, Varga J, Demirel R, Jurjević Ž, Kubátová A, Sklenář F, Zhou YG, Samson RA (2017) Polyphasic taxonomy of *Aspergillus* section *Aspergillus* (formerly *Eurotium*), and its occurrence in indoor environments and food. Stud Mycol 88:37–135. https://doi.org/10.1016/j.simyco.2017.07.001

Cheng L, Moghraby J, Piper PW (1999) Weak organic acid treatment causes a trehalose accumulation in low-pH cultures of *Saccharomyces cerevisiae*, not displayed by the more preservative-resistant *Zygosaccharomyces bailii*. FEMS Microbiol Lett 170:89–95

Chin JP, Megaw J, Magill CL, Nowotarski K, Williams JP, Bhaganna P, Linton M, Patterson MF, Underwood GJC, Mswaka AY, Hallsworth JE (2010) Solutes determine the temperature windows for microbial survival and growth. Proc Natl Acad Sci USA 107:7835–7840

Chitarra GS, Dijksterhuis J (2007) The germinating spore as a contaminating vehicle. In: Dijksterhuis J, Samson RA (eds) Food mycology: a multifaceted approach to fungi and food. Taylor and Francis, Boca Raton, FL, pp 83–100

Chitarra GS, Abee T, Rombouts FM, Posthumus MA, Dijksterhuis J (2004) Germination of *Penicillium paneum* conidia is regulated by a volatile self-inhibitor. Appl Env Microbiol 70:2823–2829

Chitarra GS, Abee T, Rombouts FM, Dijksterhuis J (2005) 1-Octen-3-ol has mild effects on membrane permeability, respiration and intracellular pH, but blocks germination and changes the protein composition of *Penicillium paneum* conidia. FEMS Microbiol Ecol 54:67–75

Choquer M, Fournier E, Kunz C, Levis C, Pradier J-M, Simon A, Viaud M (2007) *Botrytis cinerea* virulence factors: new insights into a necrotrophic and polyphageous –pathogen. FEMS Microbiol 277:1–10

Ciofini A, Negrini F, Baroncelli R, Baraldi E (2022) Management of post-harvest anthracnose: current approaches and future perspectives. Plants (Basel) 15:1856. https://doi.org/10.3390/plants11141856

Clipson NJW, Jennings DH (1992) *Dendryphiella salina* and *Debaryomyces hansenii:* models for ecophysical adaptation to salinity by fungi that grow in the sea. Can J Bot 70:2097–2105

Conner DE, Beuchat LR (1987) Efficacy of media for promoting ascospore formation by *Neosartory fischeri*, and the influence of age and culture temperature on heat resistance of ascospores. Food Microbiol 4:229–238

Conner DE, Beuchat LR, Chang CJ (1987) Age-related changes in ultrastructure and chemical composition associated with changes in heat resistance of *Neosartorya fischeri* ascospores. Trans Br Mycol Soc 89:539–550

Crippa M, Solazzo E, Guizzardi D (2021) Food systems are responsible for a third of global anthropogenic GHG emissions. Nat Food 2:198–209

Crouch J, O'Connell R, Gan P, Buiate E, Torres MF et al (2014) Genomics of *Colletotrichum*. In: Dean RA,

Lichens-Park A, Kole C (eds) Genomics of plant-associated fungi: monocot pathogens. Springer-Verlag. ISBN 978-3-662-44053-7. hal-02798712

Crowe JH, Crowe LM, Chapman D (1984) Preservation of membranes in anhydrobiotic organisms: the role of trehalose. Science 223:701–703

D'Enfert C, Fontaine T (1997) Molecular characterization of the *Aspergillus nidulans* treA gene encoding an acid trehalase required for growth on trehalose. Mol Microbiol 24:203–216

D'Enfert C, Bonini BM, PDA Z, Fontaine T, Da Silva AM, Terenzi HF (1999) Neutral trehalase catalyse intracellular trehalose breakdown in the filamentous fungi *Aspergillus nidulans* and *Neurospora crassa*. Mol Microbiol 32:471–483

Dagnas S, Gougouli OB, Koutsoumanis KP, Membré J-M (2015) Modeling red cabbage seed extract effect on *Penicillium corylophilum*: relationship between germination time, individual and population lag time. Int J Food Microbiol 211:86–94. https://doi.org/10.1016/j.ijfoodmicro.2015.07.008

Dagnas S, Gougouli M, Onno B, Koutsoumanis KP, Membré J-M (2017) Quantifying the effect of water activity and storage temperature on single spore lag times of three moulds isolated from spoiled bakery products. Int J Food Microbiol 240:75–84. https://doi.org/10.1016/j.ijfoodmicro.2016.06.013

Dakal TC, Solieri L, Giudici P (2014) Adaptive response and tolerance to sugar and salt stress in the food yeast *Zygosaccharomyces rouxii*. Int J Food Microbiol 185:140–157. https://doi.org/10.1016/j.ijfoodmicro.2014.05.01

Dantigny P, Nanguy S, Judet-Correia D, Bensoussan M (2011) A new model for germination of fungi. Int J Food Microbiol 146:176–181. https://doi.org/10.1016/j.ijfoodmicro.2011.02.022

Dao T, Dantigny P (2009) Preparation of fungal spores impacts their susceptibility to inactivation by ethanol vapours. Int J Food Microbiol 135:268–273

Dao T, Dejardin J, Bensoussan M, Dantigny P (2010) Use of the Weibull model to describe the inactivation of dry-harvested conidia of different *Penicillium* species by ethanol vapours. J Appl Microbiol 109:408–414

Davies CR, Wohlgemuth F, Young T, Violet J, Dickinson M, Sanders JW, Vallieres C, Avery SV (2021) Evolving challenges and strategies for fungal control in the food supply chain. Fungal Biol Rev 36:15–26. https://doi.org/10.1016/j.fbr.2021.01.003

de Lima AF, Stevenson A, Baxter E, Gillion JL, Hejazi F, Hayes S, Morrison IE, Prior BA, McGenity TJ, Rangel DE, Magan N, Timmis KN, Hallsworth JE (2015) Concomitant osmotic and chaotropicity-induced stresses in *Aspergillus wentii*: compatible solutes determine the biotic window. Curr Genet 61:457–477. https://doi.org/10.1007/s00294-015-0496-8

de Oliveira Filho JG, da Cruz Silva G, de Paula Gomes M, de Sousa TL, Ferreira MD, Egea MB (2022) External application of RNA interference (RNAi): an innovative tool for controlling fungi during food storage. Curr Opin Food Sci 47:100872. https://doi.org/10.1016/j.cofs.2022.100872

De Silva DD, Crous PW, Ades PK, Hyde KD, Taylor PWJ (2017) Life styles of *Colletotrichum* species and implications for plant biosecurity. Fungal Biol Rev 31:155–168. https://doi.org/10.1016/j.fbr.2017.05.001

De Vries RP (2000) Accessory enzymes from *Aspergillus* involved in xylan and pectin degradation. PhD thesis, Wageningen University, 192pp

De Vries RP, Visser J (2001) Enzymes from *Aspergilli* involved in the degradation of plant cell wall polysaccharides. Microbiol Mol Biol Rev 65:497–522

Dean R, Van Kan JAL, Pretorius ZA, Hammond-Kosack KE, Di Pietro A, Spanu PD, Rudd JJ, Dickman M, Kahmann R, Ellis J, Foster GD (2012) The top 10 fungal pathogens in molecular plant pathology. Mol Plant Pathol 13:414–430. https://doi.org/10.1111/j.1364-3703.2011.00783.x

Dijksterhuis J (2007) Heat-resistant ascospores. In: Dijksterhuis J, Samson RA (eds) Food mycology: a multifaceted approach to fungi and food. Taylor and Francis, Boca Raton, FL, pp 101–117

Dijksterhuis J (2017) The fungal spore and food spoilage. Curr Opin Food Sci 17:68–74. https://doi.org/10.1016/j.cofs.2017.10.006

Dijksterhuis J (2019) Fungal spores: highly variable and stress-resistant vehicles for distribution and spoilage. Food Microbiol 81:2–11. https://doi.org/10.1016/j.fm.2018.11.006

Dijksterhuis J, Samson RA (2006a) Zygomycetes. In: de Blackburn CW (ed) Food spoilage microorganisms. Woodhead Publishing Ltd., Cambridge, pp 415–436

Dijksterhuis J, Samson RA (2006b) Activation of ascospores by novel food preservation techniques. Adv Exp Med Biol 571:247–260

Dijksterhuis J, Samson RA (2007) Food mycology: a multifaceted approach to fungi and food. Taylor and Francis, Boca Raton, FL, p 403

Dijksterhuis J, Teunissen PGM (2004) Dormant ascospores of *Talaromyces macrosporus* are activated to germinate after treatment with ultra high pressure. J Appl Microbiol 96:162–169

Dijksterhuis J, Van Driel K, Sanders M, Houbraken J, Molenaar D, Samson RA, Kets EPW (2002) Glucose efflux and cell ejection during germination of heat activated ascospores of *Talaromyces macrosporus*. Arch Microbiol 78:1–7

Dijksterhuis J, Nijsse J, Hoekstra FA, Golovina EA (2007) High viscosity and anisotropy characterize the cytoplasm of dormant stress-resistant spores. Eukaryot Cell 6:157–170

Dijksterhuis J, Rodriquez de Massaguer P, Da Silva P, Dantigny P (2012) Primary models for inactivation of fungal spores. In: Dantigny P, Panagou EZ (eds) Predictive mycology. Nova Science Publishers, New York

Dijksterhuis J, Meijer M, Van Doorn T, Houbraken J, Bruinenberg P (2019) The preservative propionic acid differentially affects survival of conidia and germ tubes of feed spoilage fungi. Int J Food Microbiol 306:108258. https://doi.org/10.1016/j.ijfoodmicro.2019.108258

Dijksterhuis J, Wyatt T, Hanssen M, Golovina E, Hoekstra F, Lugones L (2021) Abundant small protein ICARUS inside the cell wall of stress-resistant Ascospores of *Talaromyces macrosporus* suggests a novel mechanism of constitutive dormancy. J Fungi (Basel) 17:216. https://doi.org/10.3390/jof7030216

Droby S, Eick A, Macarisin D, Cohen L, Rafael G, Stange R, McColum G, Dudai N, Nasser A, Wisniewski M, Shapira R (2008) Role of citrus volatiles in host recognition, germination and growth of *Penicillium digitatum* and *Penicillium italicum*. Postharvest Biol Technol 49:386–396. https://doi.org/10.1016/j.postharvbio.2008.01.016

Eckert JW, Ratnayake M (1994) Role of volatile compounds from wounded oranges in induction of germination of *Penicillium digitatum* conidia. Phytopathology 84:746–750

Eliskases-Lechner F, Guéguen M, Panoff JM (2022) *Geotrichum candidum*. In: McSweeney PLH, McNamara JP (eds) Encyclopedia of dairy sciences, 3rd edn. Academic Press, pp 561–568, ISBN 9780128187678. https://doi.org/10.1016/B978-0-08-100596-5.22976-4

Eshel D, Miyara I, Ailing T, Dinoor A, Prusky D (2002) pH regulates endoglucanase expression and virulence of *Alternaria alternata* in persimmon fruit. MPMI 15:774–779

FAO (2018) The future of food and agriculture–alternative pathways to 2050. Rome. 224 pp. Licence: CC BY-NC-SA 3.0 IGO

FAO (2019) The State of Food and Agriculture 2019. Moving forward on food loss and waste reduction. Rome. Licence: CC BY-NC-SA 3.0 IGO

Fillinger S, Chaveroce MK, Van Dijck P, De Vries RP, Ruijter G, Thevelein J, D'Enfert C (2001) Trehalose is required for the acquisition of tolerance to a variety of stresses in the filamentous fungus *Aspergillus nidulans*. Microbiology (Reading) 147:1851–1862

Filtenborg O, Frisvad JC, Thrane U (1996) Moulds in food spoilage. Int J Food Microbiol 33:85–102. https://doi.org/10.1016/0168-1605(96)01153-1

Filtenborg O, Frisvad JC, Samson RA (2004) Specific association of fungi to foods and influence of physical environmental factors. In: Samson RA, Hoekstra ES, Frisvad JC, Filtenborg O (eds) Introduction to food and airborne fungi, 7th edn. Centraalbureau voor Schimmelcultures, Utrecht, pp 306–320

Flaishman MA, Kolattukudy PE (1994) Timing of fungal invasion using host's ripening hormones as a signal. Proc Natl Acad Sci USA 91:6579–6583

Fontaine T, Beauvais A, Loussert C, Thevenard B, Fulgsang CC, Ohno N, Clavaud C, Prevost M-C, Latgé JP (2010) Cell wall α1-3glucans induce the aggregation of germinating conidia of *Aspergillus fumigatus*. Fungal Genet Biol 47:707–712

Frisvad JC, Filtenborg O (1988) Specific mycotoxin producing *Penicillium* and *Aspergillus* mycoflora of different foods. Proc Jpn Assoc Mycotoxicol Suppl 1:163–166

Frisvad JC, Filtenborg O (1993) Mycotoxin production by *Penicillium* and *Aspergillus* species associated with different foods and other substrates. In: Scudamore K (ed) Occurrence and significance of mycotoxins. Central Science Laboratory, Slough, pp 138–145

Frisvad JC, Thrane U, Samson RA (2007a) Mycotoxin producers. In: Dijksterhuis J, Samson RA (eds) Food mycology: a multifaceted approach to fungi and food. Taylor and Francis, Boca Raton, FL, pp 135–159

Frisvad JC, Andersen B, Samson RA (2007b) Association of moulds to foods. In: Dijksterhuis J, Samson RA (eds) Food mycology: a multifaceted approach to fungi and food. Taylor and Francis, Boca Raton, FL, pp 199–239

Garnier L, Valence F, Mounier J (2017) Diversity and control of spoilage fungi in dairy products: an update. Microorganisms 28:42. https://doi.org/10.3390/microorganisms5030042

Gomez MM, Busta FF, Pflug IJ (1989) Effect of the post-dry heat treatment temperature on the recovery of ascospores of *Neosartorya fisheri*. Lett Appl Microbiol 8:59–62

Gomez MM, Pflug IJ, Busta FF (1993) Factors affecting recovery of *Neosartorya fisheri* after exposure to dry heat. J Parenter Sci Technol 47:300–305

González-Hernández JC, Jiménez-Estrada M, Peña A (2005) Comparative analysis of trehalose production by *Debaryomyces hansenii* and *Saccharomyces cerevisiea* under saline stress. Extremophiles 9:7–16

Gougouli M, Koutsoumanis KP (2013) Relation between germination and mycelium growth of individual fungal spores. Int J Food Microbiol 15:231–239. https://doi.org/10.1016/j.ijfoodmicro.2012.12.006

Gourgues M, Brunet-Simon A, Lebrun M-H, Levis C (2004) The tetraspanin BcPls1 is required for appressorium-mediated penetration of *Botrytis cinerea* into host plant leaves. Mol Microbiol 51:619–629

Guetsky R, Kobiler I, Wang X, Perlman N, Gollop N, Avila-Quezada G, Hadar I, Prusky D (2005) Metabolism of the flavonoid epicatechin by laccase of *Colletotrichum gloeosporioides* and its effect on pathogenicity on avocado fruits. Phytopathology 95:1341–1348

Gunde-Cimerman N, Ramos J, Plemenitas A (2009) Halotolerant and halophilic fungi. Mycol Res 113:1231–1241

Guo H, Liu A, Wang Y, Wang T, Zhang W, Zhu P, Xu L (2020) Measuring light-induced fungal ethylene production enables non-destructive diagnosis of disease occurrence in harvested fruits. Food Chem 310:125827. https://doi.org/10.1016/j.foodchem.2019.125827

Hadas Y, Goldberg I, Pines O, Prusky D (2007) Involvement of gluconic acid and glucose oxidase in the pathogenicity of *Penicillium expansum* in apples. Phytopathology 97:384–390

Hagiwara D, Sakai K, Suzuki S, Umemura M, Nogawa T (2017) Temperature during conidiation affects stress tolerance, pigmentation, and trypacidin accumulation in the conidia of the airborne pathogen *Aspergillus fumigatus*. PLoS One 12(5):e0177050. https://doi.org/10.1371/journal.pone.0177050

Hallsworth JE, Magan N (1994) Effects of KCl concentrations on accumulation of acyclic sugar alcohols and trehalose in conidia of three entomopathogenic fungi. Lett Appl Microbiol 18:8–11

Hallsworth JE, Magan N (1996) Culture age, temperature, and pH affect the polyol and trehalose contents of fungal propagules. Appl Environ Microbiol 62:2435–2442

Hamill PG, Stevenson A, McMullan PE, Williams JP, Lewis ADR, Stevenson KE, Farnsworth KD, Khroustalyova G, Takemoto JY, Quinn JP, Rapoport A, Hallsworth JE (2020) Microbial lag phase can be indicative of, or independent from, cellular stress. Sci Rep 10:5948. https://doi.org/10.1038/s41598-020-62552-4

He P, Wang Y, Wang X, Zhang X, Tian C (2017) The mitogen-activated protein kinase CgMK1 governs synthesis, and plant infection of *Colletotrichum gloeosporioides*. Front Microbiol 8:2216. https://doi.org/10.3389/fmicb.2017.02216

Hegedus DD, Rimmer SR (2005) *Sclerotinia sclerotiorum*: When "to be or not to be" a pathogen? FEMS Microbiol Lett 251:177–184

Herrera R, Salazar A, Ramos-Moreno L, Ruiz-Roldan C, Ramos J (2017) Vacuolar control of subcellular cation distribution is a key parameter in the adaptation of *Debaryomyces hansenii* to high salt concentrations. Fungal Genet Biol 100:52–60. https://doi.org/10.1016/j.fgb.2017.02.002

Herrero-Garcia E, Garzia A, Cordobes S, Espeso EA, Ugalde U (2011) 8-carbon oxylipins inhibit germination and growth, and stimulate aerial conidiation in *Aspergillus nidulans*. Fungal Biol 115:393–400

Hoogerwerf SA, Kets EPW, Dijksterhuis J (2002) High-oxigen and high-carbon dioxide containing atmospheres inhibit growth of food-associated moulds. Lett Appl Microbiol 35:419–422

Horikoshi K, Ikeda Y (1966) Trehalase in conidia of *Aspergillus oryzae*. J Bacteriol 91:1883–1887

Hosono K (2000) Effect of nystatin on the release of glycerol from salt-stressed cells of the salt-tolerant yeast *Zygosaccharomyces rouxii*. Arch Microbiol 173:284–287

Hottiger T, DeVirgilio C, Hall MN, Boller T, Wiemken A (1994) The role of trehalose for the acquisition of thermotolerance in yeast. II. Physiological concentrations of trehalose increase the stability of proteins in vivo. Eur J Biochem 219:187–193

Houbraken JAMP, Varga J, Rico-Munoz E, Johnson S, Samson RA (2008) Sexual reproduction as the cause of heat resistance in the food spoilage fungus *Byssochlamys spectabilis* (Anamorph *Paecilomyces variotii*). Appl Environ Microbiol 74:1613–1619

Hawksworth DL, Crous PW, Redhead SA et al (2011) The Amsterdam Declaration on Fungal Nomenclature. IMA Fungus 2:105–111. https://doi.org/10.5598/imafungus.2011.02.01.14

Houbraken J, Kocsubé S, Visagie CM, Yilmaz N, Wang XC, Meijer M, Kraak B, Hubka V, Bensch K, Samson RA, Frisvad JC (2020) Classification of *Aspergillus*, *Penicillium*, *Talaromyces* and related genera (Eurotiales): an overview of families, genera, subgenera, sections, series and species. Stud Mycol 95:5–169. https://doi.org/10.1016/j.simyco.2020.05.002

Ijadpanahsaravi M, Punt M, Wösten HAB, Teertstra WR (2021) Minimal nutrient requirements for induction of germination of *Aspergillus niger* conidia. Fungal Biol 125:231–238. https://doi.org/10.1016/j.funbio.2020.11.004

Iwaki T, Higashida Y, Tsuji T, Tamai Y, Watanabe Y (1998) Characterization of a second gene (ZSOD22) of Na+/H+ antiporter from salt-tolerant yeast *Zygosaccharomyces rouxii* and functional expression of ZSOD2 and ZSOD22 in *Saccharomyces cerevisiae*. Yeast 14:1167–1174

Jennings DH, Burke RM (1990) Compatible solutes – the mycological dimension and their role as physiological buffering agents. New Phytol 116:277–283

Jiao W, Liu X, Li Y, Li B, Du Y, Zhang Z et al (2022) Organic acid, a virulence factor for pathogenic fungi, causing postharvest decay in fruits. Mol Plant Pathol 23:304–312. https://doi.org/10.1111/mpp.13159

Jurick WM 2nd, Peng H, Beard HS, Garrett WM, Lichtner FJ, Luciano-Rosario D, Macarisin O, Liu Y, Peter KA, Gaskins VL, Yang T, Mowery J, Bauchan G, Keller NP, Cooper B (2020) Blistering1 modulates *Penicillium expansum* virulence via vesicle-mediated protein secretion. Mol Cell Proteomics 19:344–361. https://doi.org/10.1074/mcp.RA119.001831

Kanashiro AM, Akiyama DY, Kupper KC, Fill TP (2020) *Penicillium italicum*: an underexplored postharvest pathogen. Front Microbiol 11:606852. https://doi.org/10.3389/fmicb.2020.606852

Karlshøj K, Nielsen PV, Larsen TO (2007) Fungal volatiles: biomarkers of good and bad food quality. In: Dijksterhuis J, Samson RA (eds) Food mycology: a multifaceted approach to fungi and food. Taylor and Francis, Boca Raton, FL, pp 279–302

Kikoku Y (2003) Heat activation characteristics of *Talaromyces ascospores*. J Food Sci 68:2331–2335

Kikoku Y, Tagashira N, Nakano H (2008) Heat resistance of fungi isolated from frozen blueberries. J Food Prot 71:2030–2035. https://doi.org/10.4315/0362-028x-71.10.2030

Kim YK, Li D, Kolattukudy PA (1998) Induction of Ca^{2+}calmodulin signalling by hard-surface contact primes *Colletotrichum gloeosporioides* conidia to germinate and form appressoria. J Bacteriol 180:5144–5150

Kim YT, Prusky D, Rollins JA (2007) An activating mutation of the *Sclerotinia sclerotiorum pac*1 gene increases oxalic acid production at low pH but decreases virulence. Mol Plant Pathol 8:611–622

King AD, Whitehand LC (1990) Alteration of *Talaromyces flavus* heat resistance by growth conditions and heating medium composition. J Food Sci 55:830–832

Kolattukudy PA, Rogers LM, Li D, Hwang CS, Flaishman MA (1995) Surface signalling in pathogenesis. Proc Natl Acad Sci USA 92:4080–4087

Konuchová M, Valík L (2021) Modelling the radial growth of *Geotrichum candidum*: effects of temperature and water activity. Microorganisms 9:532. https://doi.org/10.3390/9030532

Kotzekidou P (1997) Heat resistance of *Byssochlamys nivea*, *Byssochlamys fulva* and *Neosartorya fischeri* isolated from canned tomato paste. J Food Sci 62:410–412

Kuanyshev N, Adamo GM, Porro D, Branduardi P (2017) The spoilage yeast *Zygosaccharomyces bailii*: Foe or friend? Yeast 34:359–370. https://doi.org/10.1002/yea.3238

Lages F, Silva-Graça M, Lucas C (1999) Active glycerol uptake is a mechanism underlying halotolerance in yeasts: a study of 42 species. Microbiology (Reading) 145:2577–2585

Leandro MJ, Sychrová H, Prista C, Loureiro-Dias MC (2011) The osmotolerant fructophilic yeast *Zygosaccharomyces rouxii* employs two plasma-membrane fructose uptake systems belonging to a new family of yeast sugar transporters. Microbiology (Reading) 157:601–608. https://doi.org/10.1099/mic.0.044446-0

Leong SL, Vinnere-Pettersson RT, Hocking AD, Schnürer J (2011) The extreme xerophilic mould *Xeromyces bisporus* – growth and competition at various water activities. Int J Food Microbiol 145:57–63

Leong S-L, Lantz H, Pettersson OV, Frisvad JC, Thrane U, Heipieper HJ, Dijksterhuis J, Grabherr M, Pettersson M, Tellgren-Roth C, Schnürer J (2015) *Xeromyces bisporus* genome, transcriptome and physiology. Environ Microbiol 17:496–513. https://doi.org/10.1111/1462-2920.12596

Leyva JS, Manrique M, Prats L, Loureiro-Dias MC, Peiñado JM (1999) Regulation of fermentative CO_2 production by the food spoilage yeast *Zygosaccharomyces bailii*. Enzyme Microb Technol 24:270–275

Li B, Zong Y, Du Z, Chen Y, Zhang Z, Qin G, Zhao W, Tian S (2015) Genomic characterization reveals insights into Patulin biosynthesis and pathogenicity in *Penicillium* species. Mol Plant-Microbe Interact 28:635–647. https://doi.org/10.1094/MPMI-12-14-0398-FI

Lin SY, Okuda S, Ikeda K, Okuno T, Takano Y (2012) LAC2 encoding a secreted laccase is involved in appressorial melanization and conidial pigmentation in *Colletotrichum orbiculare*. Mol Plant-Microbe Interact 25:1552–1561. https://doi.org/10.1094/MPMI-05-12-0131-R. PMID: 22934563

Lindberg L, Santos AXS, Riezman H, Olsson L, Bettiga M (2013) Lipidomic profiling of *Saccharomyces cerevisiae* and *Zygosaccharomyces bailii* reveals critical changes in lipid composition in response to acetic acid stress. PLoS One 8:e73936. https://doi.org/10.1371/journal.pone.007393

Lucas C, Da Costa M, Van Uden N (1990) Osmoregulatory active sodium-glycerol co-transport in the halotolerant yeast *Debaryomyces hansenii*. Yeast 6:187–191

Luciano-Rosario D, Keller NP, Jurick WM 2nd (2020) *Penicillium expansum*: biology, omics, and management tools for a global postharvest pathogen causing blue mould of pome fruit. Mol Plant Pathol 21:1391–1404. https://doi.org/10.1111/mpp.12990

Lücking R, Aime MC, Robbertse B (2021) Fungal taxonomy and sequence-based nomenclature. Nat Microbiol 6:540–548. https://doi.org/10.1038/s41564-021-00888-x

Ludwig N, Löhrer M, Hempel M, Mathea S, Schliebner I, Menzel M, Kiesow A, Schaffrath U, Deising HB, Horbach R (2014) Melanin is not required for turgor generation but enhances cell-wall rigidity in appressoria of the corn pathogen *Colletotrichum graminicola*. Mol Plant-Microbe Interact 27:315–327. https://doi.org/10.1094/MPMI-09-13-0267-R. PMID: 24261846

Luxo C, Fernanda Nobre M, Da Costa MS (1993) Intracellular polyol accumulation by yeast-like fungi of the genera *Geotrichum* and *Endomyces* in response to water stress (NaCl). Can J Microbiol 39:868–873

Magan N, Aldred D, Dijksterhuis J (2012) The physiological state of fungal spores and survival dynamics. In: Dantigny P, Panagou EZ (eds) Predictive mycology. Nova Science Publishers, New York

Mann DA, Beuchat LR (2008) Combinations of antimycotics to inhibit the growth of molds capable of producing 1,3-pentadiene. Food Microbiol 25:144–153

Maor U, Barda O, Sadhasivam S, Bi Y, Levin E, Zakin V, Prusky DB, Sionov E (2021) Functional roles of LaeA, polyketide synthase, and glucose oxidase in the regulation of ochratoxin a biosynthesis and virulence in *Aspergillus carbonarius*. Mol Plant Pathol 22:117–129. https://doi.org/10.1111/mpp.13013

Martínez JL, Sychrova H, Ramos J (2011) Monovalent cations regulate expression and activity of the Hak1 potassium transporter in *Debaryomyces hansenii*. Fungal Genet Biol 48:177–184

Martorell P, Stratford M, Steels H, Fernández-Espinar MT, Querol A (2007) Physiological characterization of spoilage strains of *Zygosaccharomyces bailii* and *Zygosaccharomyces rouxii* isolated from high sugar environments. Int J Food Microbiol 114:234–242

Mehyar GF, Al-Qadiri HM, Abu-Blan HA, Swanson BG (2011) Antifungal effectiveness of potassium sorbate incorporated in edible coatings against spoilage molds of apples, cucumbers, and tomatoes during refrigerated storage. J Food Sci 76:210–217

Membre JM, Kubaczka M, Chene C (1999) Combined effects of pH and sugar on growth rate of *Zygosaccharomyces rouxii*, a bakery product spoilage yeast. Appl Environ Microbiol 65:4921–4925

Miyamoto et al (2014) Formation of 1-octen-3-ol from *Aspergillus flavus* conidia is accelerated after disruption of cells independently of Ppo oxygenases, and is not a main cause of inhibition of germination. PeerJ 2:e395. https://doi.org/10.7717/peerj.395

Miyara I, Shafran H, Kramer Haimovich H, Rollings J, Sherman A, Prusky D (2008) Multi-factor regulation of pectate lyase secretion by *Colletotrichum gloeosporioides* pathogenic on avocado fruit. Mol Pathol 9:281–291

Morita Y, Hyon GS, Hosogi N, Miyata N, Nakayashiki H, Muranaka Y, Inada N, Park P, Ikeda K (2013) Appressorium-localized NADPH oxidase B is essential for aggressiveness and pathogenicity in the host-specific, toxin-producing fungus *Alternaria alternata* Japanese pear pathotype. Mol Plant Pathol 14:365–378. https://doi.org/10.1111/mpp.12013

Morozova EV, Kozlov VP, Tereshina VM, Memorskaya AS, Feofilova EP (2002) Changes in lipid composition and carbohydrate composition of *Aspergillus niger* conidia during germination. Prikl Biokhim Mikrobiol 38:149–154

Münch S, Lingner U, Floss DS, Ludwig N, Sauer N, Deising H (2008) The hemibiotrophic lifestyle of *Colletotrichum* species. J Plant Physiol 165:41–51

Nagtzaam MPM, Bollen GJ (1994) Long shelf life of *Talaromyces flavus* in coating material of pelleted seed. Eur J Plant Pathol 100:279–282

Nguyen Van Long N, Rigalma K, Jany JL, Mounier J, Vasseur V (2021) Intraspecific variability in cardinal growth temperatures and water activities within a large diversity of *Penicillium roqueforti* strains. Food Res Int 148:110610. https://doi.org/10.1016/j.foodres.2021.110610

Niem J, Miyara I, Ettedgui Y, Reuveni M, Flaishman M, Prusky D (2007) Core rot development in red delicious apples is affected by susceptibility of the seed locule to *Alternaria alternata* colonization. Phytopathology 97:1415–1421

Novodvorska M, Stratford M, Blythe MJ, Wilson R, Beniston RG, Archer DB (2016) Metabolic activity in dormant conidia of *Aspergillus niger* and developmental changes during conidial outgrowth. Fungal Genet Biol 94:23–31. https://doi.org/10.1016/j.fgb.2016.07.002

O'Connell RJ, Thon MR, Hacquard S, Amyotte SG, Kleemann J, Torres MF, Damm U, Buiate EA, Epstein L, Alkan N, Altmuller J, Alvarado-Balderrama L, Bauser CA, Becker C, Birren BW, Chen Z, Choi J, Crouch JA, Duvick JP et al (2012) Lifestyle transitions in plant pathogenic *Colletotrichum* fungi deciphered by genome and transcriptome analyses. Nat Genet 44:1060–1065. https://doi.org/10.1038/ng.2372

Osherov N, May G (2000) Conidial germination in *Aspergillus nidulans* requires RAS signaling and protein synthesis. Genetics 155:647–656

Padamsee M, Arun Kumar TK, Riley R, Binder M, Boyd A, Calvo AM et al (2012) The genome of the xerotolerant mould *Wallemia sebi* reveals adaptations to osmotic stress and suggests cryptic sexual reproduction. Fungal Genet Biol 49:217–226

Palma M, Goffeau A, Spencer-Martins I, Baret PV (2007) A phylogenetic analysis of the sugar porters in hemiascomycetous yeasts. J Mol Microbiol Biotechnol 12:241–248

Palma M, Dias PJ, Roque FC, Luzia L, Guerreiro JF, Sá-Correia I (2017) The *Zygosaccharomyces bailii* transcription factor Haa1 is required for acetic acid and copper stress responses suggesting subfunctionalization of the ancestral bifunctional protein Haa1/Cup2. BMC Genomics 18:75. https://doi.org/10.1186/s12864-016-3443-2

Palma M, Guerreiro JF, Sá-Correia I (2018) Adaptive response and tolerance to acetic acid in *Saccharomyces cerevisiae* and *Zygosaccharomyces bailii*: a physiological genomics perspective. Front Microbiol 9:274. https://doi.org/10.3389/fmicb.2018.00274

Palou E, Lopez-Malo A, Barbosa-Canovas GV, WeltiChanes J, Davidson PM, Swanson BG (1998) Effect of oscillatory high hydrostatic pressure treatments on *Byssochlamys nivea* ascospores suspended in fruit juice concentrates. Lett Appl Microbiol 27:375–378

Pel HJ, de Winde JH, Archer DB, Dyer PS, Hofmann G, Schaap PJ, Turner G, de Vries RP, Albang R, Albermann K, Andersen MR, Bendtsen JD, Benen JA, van den Berg M, Breestraat S, Caddick MX, Contreras R, Cornell M, Coutinho PM, Danchin EG, Debets AJ, Dekker P, van Dijck PW, van Dijk A, Dijkhuizen L, Driessen AJ, d'Enfert C, Geysens S, Goosen C, Groot GS, de Groot PW, Guillemette T, Henrissat B, Herweijer M, van den Hombergh JP, van den Hondel CAMJJ, van der Heijden RT, van der Kaaij RM, Klis FM, Kools HJ, Kubicek CP, van Kuyk PA, Lauber J, Lu X, van der Maarel MJ, Meulenberg R, Menke H, Mortimer MA, Nielsen J, Oliver SG, Olsthoorn M, Pál K, van Peij NNME, Ram AFJ, Rinas U, Roubos JA, Sagt CMJ, Schmoll M, Sun J, Ussery D, Varga J, Vervecken W, van de Vondervoort PJ, Wedler H, Wösten HAB, Zeng AP, van Ooyen AJ, Visser J, Stam H (2007) Genome sequencing and analysis of the versatile cell factory *Aspergillus niger* CBS 513.88. Nat Biotechnol 25:221–231

Petersson S, Jonsson N, Schnürer J (1999) *Pichia anomala* as a biocontrol agent during storage of high-moisture feed grain under airtight conditions. Postharvest Biol Technol 15:175–184

Petrovič U, Gunde-Zimmerman N, Zalar P (2000) Xerotolerant mycobiota from high altitude Annapurna soil, Nepal. FEMS Microbiol Lett 182:339–342

Pinches SE, Apps P (2007) Production in food of 1,3-pentadiene and styrene by *Trichoderma* species. Int J Food Sci 116:182–185

Piper P, Ortiz Calderon C, Hatzixanthis K, Mollapour M (2001) Weak acid adaptation: the stress response that confers yeasts with resistance to organic acid food preservatives. Microbiology (Reading) 147:2635–2642

Pitt JI, Hocking AD (2009) Fungi and food spoilage, 3rd edn. Springer, Dordrecht, p 519

Plumridge A, Hesse SJA, Watson AJ, Lowe KC, Stratford M, Archer DB (2004) The weak-acid preservative, sorbic acid, inhibits conidial germination and mycelial growth of *Aspergillus niger* through intracellular acidification. Appl Environ Microbiol 70:3506–3511

Plumridge A, Stratford M, Lowe KC, Archer DB (2008) The weak-acid preservative, sorbic acid, is decarboxylated and detoxified by a phenylacrylic acid decarboxylase, PadA1, in the spoilage mould *Aspergillus niger*. Appl Environ Microbiol 74:550–552

Plumridge A, Melin P, Stratford M, Novodvorska M, Shunburne L, Dyer PS, Roubos JA, Menke H, Stark J, Stam H, Archer DB (2010) The decarboxylation of the weak-acid preservative, sorbic acid, is encoded by linked genes in *Aspergillus* spp. Fungal Genet Biol 47:683–692

Podila GK, Rogers LM, Kolattukudy PE (1993) Chemical signals from avocado wax trigger ger9mination and appressorium formation in *Colletotrichum gloeosporioides*. Plant Physiol 103:267–272

Porquier A, Morgant G, Moraga J, Dalmais B, Luyten I, Simon A, Pradier J-M, Amselem J, González Collado I, Viaud M (2016) The botrydial biosynthetic gene cluster of *Botrytis cinerea* displays a bipartite genomic structure and is positively regulated by the putative Zn(II)2Cys6 transcription factor BcBot6. Fungal Genet Biol 96:33–46. https://doi.org/10.1016/j.fgb.2016.10.003

Přibylova L, Farkas V, Slaninova I, De Montigny J, Sychrova H (2007) Differences in osmotolerant and cell-wall properties of two *Zygosaccharomyces rouxii* strains. Folia Microbiol 52:241–245

Pribylova L, Papouskova K, Sychrova H (2008) The salt tolerant yeast *Zygosaccharomyces rouxii* possesses two plasma-membrane Na^+/H^+ -antiporters (*Zr*Nha1p and *Zr*Sod2-22p) playing different roles in cation homeostasis and cell physiology. Fung Genet Biol 45:1439–1447

Prista C, Loureiro-Dias C, Montiel V, García R, Ramos J (2005) Mechanisms underlying the halotolerant way of *Debaryomyces hansenii*. FEMS Yeast Res 5:693–701

Prista C, Michán C, Miranda IM, Ramos J (2016) The halotolerant *Debaryomyces hansenii*, the Cinderella of non-conventional yeasts. Yeast 33:523–533. https://doi.org/10.1002/yea.3177

Prusky D, Kolattukudy PE (2007) Cross-talk between host and fungus in postharvest situations and its effect on symptom development. In: Dijksterhuis J, Samson RA (eds) Food mycology: a multifaceted approach to fungi and food. Taylor and Francis, Boca Raton, FL, pp 3–25

Prusky D, Lichter A (2007) Activation of quiescent infections by postharvest pathogens during transition from the biotrophic to the necrotrophic stage. FEMS Microbiol Lett 268:1–8

Prusky D, McEvoy JL, Saftner R, Conway WS, Jones R (2004) Relationship between host acidification and virulence of *Penicillium* spp. on apple and citrus fruit. Phytopathology 94:44–51

Prusky D, Alkan N, Mengiste T, Fluhr R (2013) Quiescent and necrotrophic lifestyle choice during postharvest disease development. Annu Rev Phytopathol 51:155–176. https://doi.org/10.1146/annurev-phyto-082712-102349

Punt M, Van den Brule T, Teertstra WR, Dijksterhuis J, Den Besten HMW, Ohm RA, Wösten HAB (2020) Impact of maturation and growth temperature on cell-size distribution, heat-resistance, compatible solute composition and transcription profiles of *Penicillium roqueforti* conidia. Food Res Int 136:109287. https://doi.org/10.1016/j.foodres.2020.109287

Punt M, Seekles SJ, van Dam JL, de Adelhart TC, Martina RR, Houbraken J, Ram AFJ, Wösten HAB, Ohm RA (2022) High sorbic acid resistance of *Penicillium roqueforti* is mediated by the SORBUS gene cluster. PLoS Genet 18:e1010086. https://doi.org/10.1371/journal.pgen.1010086

Put HMC, De Jong J (1980) The heat resistance of selected yeasts causing spoilage of canned soft drinks and fruit products. In: Skinner FA, Passmore SM, Davenport RR (eds) Biology and activities of yeasts. Soc Appl Bacteriol Symp Series No 9. Academic Press, London

Rajashekhara E, Suresh ER, Ethiraj S (1998) Thermal death rate of ascospores of *Neosartorya fischeri* ATCC 200957 in the presence of organic acids and preservatives in fruit juices. J Food Prot 61:1358–1362

Reyns KMFA, Verbeke EAV, Michiels CEW (2003) Activation and inactivation of *Talaromyces macrosporus* ascospores by high hydrostatic pressure. J Food Prot 66:1035–1042

Rico-Munoz E, Samson RA, Houbraken J (2019) Mould spoilage of foods and beverages: using the right methodology. Food Microbiol 81:51–62. https://doi.org/10.1016/j.fm.2018.03.016

Rodrigues F, Sousa MJ, Ludovico P, Santos H, Corte-Real M, Leao C (2012) The fate of acetic acid during glucose co-metabolism by the spoilage yeast *Zygosaccharomyces bailii*. PLoS One 7:e52402

Rodriguez F, Côrte-Real M, Leão C, Van Dijken JP, Pronk JT (2001) Oxygen requirements of the food spoilage yeast *Zygosaccharomyces bailii* in synthetic and complex media. Appl Environ Microbiol 67:2123–2128

Ryder LS, Cruz-Mireles N, Molinari C, Eisermann I, Eseola AB, Talbot NJ (2022) The appressorium at a glance. J Cell Sci 135(14):jcs259857. https://doi.org/10.1242/jcs.259857. Epub 2022 Jul 20. PMID: 35856284

Sakamoto K, Iwashita K, Yamada O, Kobayashi K, Mizuno A, Akita O, Mikami S, Shimoi H, Gomi K (2009) *Aspergillus oryzae* atfA controls conidial germination and stress tolerance. Fungal Genet Biol 46:887–897

Samson RA, Frisvad JC (2004) Polyphasoc taxonomy of *Penicillium* subgenus *Penicillium*. A guide to identification of food and air-borne terverticillate Penicillia and their mycotoxins. Stud Mycol 49:1–174

Samson RA, Houbraken J (2011) Phylogenetic and taxonomic studies on the genera *Penicillium* and *Talaromyces*. Stud Mycol 70. 183 pp

Samson RA, Varga J (2007) *Aspergillus* systematic in the genomic era. Stud Mycol 59:1–206

Samson RA, Visagie CM, Houbraken J, Hong SB, Hubka V, Klaassen CH, Perrone G, Seifert KA, Susca A, Tanney JB, Varga J, Kocsubé S, Szigeti G, Yaguchi T, Frisvad JC (2014) Phylogeny, identification and nomenclature of the genus *Aspergillus*. Stud Mycol 78:141–173. https://doi.org/10.1016/j.simyco.2014.07.004

Samson RA, Houbraken J, Thrane U, Frisvad JC, Andersen B (2019) Food and indoor fungi, 2nd edn. Westerdijk Fungal Biodiversity Institute/Centraalbureau voor Schimmelcultures, Utrecht. 481 pp

Santos JLPD, Samapundo S, Biyikli A, Van Impe J, Akkermans S, Höfte M, Abatih EN, Sant'Ana AS, Devlieghere F (2018) Occurrence, distribution and contamination levels of heat-resistant moulds throughout the processing of pasteurized high-acid fruit products. Int J Food Microbiol 281:72–81. https://doi.org/10.1016/j.ijfoodmicro.2018.05.019

Scholte RPM, Samson RA, Dijksterhuis J (2004) Spoilage fungi in the industrial processing of food. In: Samson RA, Hoekstra ES, Frisvad JC, Filtenborg O (eds) Introduction to food- and airborne fungi, 7th edn. Centraalbureau voor Schimmelcultures, Utrecht, pp 339–356

Schoch CL, Seifert KA, Huhndorf S et al (2012) Nuclear ribosomal internal transcribed spacer (ITS) region as a universal DNA barcode marker for Fungi. Proc Natl Acad Sci USA 109(16):6241–6246

Seekles SJ, van Dam J, Arentshorst M, Ram AFJ (2022) Natural variation and the role of Zn2Cys6 transcription factors SdrA, WarA and WarB in sorbic acid resistance of *Aspergillus niger*. Microorganisms 10:221. https://doi.org/10.3390/microorganisms10020221

Seekles SJ, van den Brule T, Punt M, Dijksterhuis J, Arentshorst M, Ijadpanahsaravi M, Roseboom W, Meuken G, Ongenae V, Zwerus J, Ohm RA, Kramer G, Wösten HAB, de Winde JH, Ram AFJ (2023) Compatible solutes determine the heat resistance of conidia. Fungal Biol Biotechnol 13:21. https://doi.org/10.1186/s40694-023-00168-9

Segers FJJ, Van Laarhoven KA, Wösten HAB, Dijksterhuis J (2017) Growth of indoor fungi on gypsum. J Appl Microbiol 123:429–435. https://doi.org/10.1111/jam.13487

Segers FJJ, Dijksterhuis J, Giesbers M, Debets AJM (2023) Natural folding of airborne fungal spores: a mechanism for dispersal and long-term survival? Fungal Biolog Rev 44:100292. https://doi.org/10.1016/j.fbr.2022.10.005

Silva CJ, Van Den Abeele C, Ortega-Salazar I, Papin V, Adaskaveg JA, Wang D, Casteel CL, Seymour GB, Blanco-Ulate B (2021) Host susceptibility factors render ripe tomato fruit vulnerable to fungal disease despite active immune responses. J Exp Bot 72:2696–2709

Silva CJ, Adaskaveg JA, Mesquida-Pesci SD, Ortega-Salazar IB, Pattathil S, Zhang L, Hahn MG, van Kan JAL, Cantu D, Powell ALT, Blanco-Ulate B (2023) *Botrytis cinerea* infection accelerates ripening and cell wall disassembly to promote disease in tomato fruit. Plant Physiol 191:575–590. https://doi.org/10.1093/plphys/kiac408

Sklenář F, Jurjević Ž, Zalar P, Frisvad JC, Visagie CM, Kolařík M, Houbraken J, Chen AJ, Yilmaz N, Seifert KA, Coton M, Déniel F, Gunde-Cimerman N, Samson RA, Peterson SW, Hubka V (2017) Phylogeny of xerophilic aspergilli (subgenus *Aspergillus*) and taxonomic revision of section *Restricti*. Stud Mycol 88:161–236. https://doi.org/10.1016/j.simyco.2017.09.002

Smelt JPPM (1998) Recent advances in the microbiology of high pressure processing. Trends Food Sci Technol 9:152–158

Smid EJ, Hendriks L, Boerrigter HAM, Gorris LGM (1996) Surface disinfection of tomatoes using the natural plant compound trans-cinnamaldehyde. Postharvest Biol Technol 9:343–350

Snowdon AL (1990) A colour atlas of post-harvest diseases and disorders of fruits and vegetables, General introduction and fruits, vol 1. Wolfe Scientific Ltd, London

Snyder AB, Worobo RW (2018) Fungal spoilage in food processing. J Food Prot 81:1035–1040. https://doi.org/10.4315/0362-028X.JFP-18-031. PMID: 29768029

Snyder AB, Churey JJ, Worobo RW (2019) Association of fungal genera from spoiled processed foods with physicochemical food properties and processing conditions. Food Microbiol 83:211–218. https://doi.org/10.1016/j.fm.2019.05.012

Solieri L, Vezzani V, Cassanelli S, Dakal TC, Pazzini J, Giudici P (2016) Differential hypersaline stress response in *Zygosaccharomyces rouxii* complex yeasts: a physiological and transcriptional study. FEMS Yeast Res 16(6):fow063. https://doi.org/10.1093/femsyr/fow063

Splittstoesser DF, Splittstoesser CM (1977) Ascospores of *Byssochlamys fulva* compared with those of a heat resistant *Aspergillus*. J Food Sci 4:685–688

Splittstoesser DF, Lammers JM, Downing DL, Churey JJ (1989) Heat resistance of *Eurotium herbariorum*, a xerophilic mold. J Food Sci 54:683–685

Stark J (2007) Cheese and fermented sausages. In: Dijksterhuis J, Samson RA (eds) Food mycology: a multifaceted approach to fungi and food. Taylor and Francis, Boca Raton, FL, pp 319–331

Steels H, James SA, Roberts IN, Stratford M (2000) Sorbic acid resistance: the inoculum effect. Yeast 16:1173–1183

Stevenson A, Cray JA, Williams JP, Santos R, Sahay R, Neuenkirchen N, McClure CD, Grant IR, Houghton JD, Quinn JP, Timson DJ, Patil SV, Singhal RS, Antón J, Dijksterhuis J, Hocking AD, Lievens B, Rangel DE, Voytek MA, Gunde-Cimerman N, Oren A, Timmis KN, McGenity TJ, Hallsworth JE (2015) Is there a common water-activity limit for the three domains of life? ISME J 9:1333–31351. https://doi.org/10.1038/ismej.2014.219

Stevenson A, Hamill PG, Medina Á, Kminek G, Rummel JD, Dijksterhuis J, Timson DJ, Magan N, Leong SL, Hallsworth JE (2017a) Glycerol enhances fungal germination at the water-activity limit for life. Environ Microbiol 19:947–967. https://doi.org/10.1111/1462-2920.13530

Stevenson A, Hamill PG, Dijksterhuis J, Hallsworth JE (2017b) Water-, pH- and temperature relations of germination for the extreme xerophiles *Xeromyces bisporus* (FRR 0025), *Aspergillus penicil-*

lioides (JH06THJ) and *Eurotium halophilicum* (FRR 2471). Microb Biotechnol 10:330–340. https://doi.org/10.1111/1751-7915.12406

Stratford M, Plumridge A, Nebe-von-Caron G, Archer DB (2009) Inhibition of spoilage mould conidia by acetic acid and sorbic acid involves different modes of action, requiring modification of the classical weak-acid theory. Int J Food Microbiol 136:37–43. https://doi.org/10.1016/j.ijfoodmicro.2009.09.025

Stratford M, Plumridge A, Pleasants MW, Novodvorska M, Baker-Glenn CAG, Pattenden G, Archer DB (2012) Mapping the structural requirements of inducers and substrates for decarboxylation of weak acid preservatives by the food spoilage mould *Aspergillus niger*. Int J Food Microbiol 157:375–383

Stratford M, Steels H, Nebe-von-Caron G, Novodvorska M, Hayer K, Archer DB (2013) Extreme resistance to weak-acid preservatives in the spoilage yeast *Zygosaccharomyces bailii*. Int J Food Microbiol 166:126–134

Stratford M, Steels H, Nebe-von-Caron G, Avery SV, Novodvorska M, Archer DB (2014) Population heterogeneity and dynamics in starter culture and lag phase adaptation of the spoilage yeast *Zygosaccharomyces bailii* to weak acid preservatives. Int J Food Microbiol 181:40–47

Streekstra H, Verkennis AE, Jacobs R, Dekker A, Stark J, Dijksterhuis J (2016) Fungal strains and the development of tolerance against natamycin. Int J Food Microbiol 238:15–22. https://doi.org/10.1016/j.ijfoodmicro.2016.08.006

Suiker IM, Arkesteijn GJA, Zeegers PJ, Wösten HAB (2021) Presence of *Saccharomyces cerevisiae* subsp. *diastaticus* in industry and nature and spoilage capacity of its vegetative cells and ascospores. Int J Food Microbiol 347:109173. https://doi.org/10.1016/j.ijfoodmicro.2021.109173

Sussman S, Halvorson HO (1966) Spores, their dormancy and germination. Harper and Row, New York

Suzuki T, Shinogi T, Narusaka Y et al (2003) Infection behavior of *Alternaria alternata* Japanese pear pathotype and localization of 1,3-β-D-glucan in compatible and incompatible interactions between the pathogen and host plants. J Gen Plant Pathol 69:91–100. https://doi.org/10.1007/s10327-002-0001-3

Talhinhas P, Baroncelli R (2021) *Colletotrichum* species and complexes: geographic distribution, host range and conservation status. Fungal Divers 110:109–198. https://doi.org/10.1007/s13225-021-00491-9

Tao N-G, Chen Y, Wu Y, Wang X, Li L, Zhu A (2018) The terpene limonene induced the green mold of citrus fruit through regulation of reactive oxygen species (ROS) homeostasis in *Penicillium digitatum* spores. Food Chem 277:414–422. https://doi.org/10.1016/j.foodchem.2018.10.142

Te Welscher YM, ten Napel HH, Balagué MM, Souza CM, Riezman H, de Kruijff B, Breukink E (2008) Natamycin blocks fungal growth by binding specifically to ergosterol without permeabilizing the membrane. J Biol Chem 283:6393–6401

Te Welsher YM, Jones L, Van Leeuwen MR, Dijksterhuis J, De Kruijff B, Eitzen G, Breukink EJ (2010) Natamycin inhibits vacuole fusion at the priming phase via a specific interaction with ergosterol. Antimicrob Agents Chemother 54:2618–2625

Te Welsher YM, Van Leeuwen MR, De Kruijff B, Dijksterhuis J, Breukink EJ (2012) Polyene antibiotic that inhibits membrane transport proteins. Proc Natl Acad Sci USA 109:11156–11159

Teertstra WR, Tegelaar M, Dijksterhuis J, Golovina EA, Ohm RA, Wösten HAB (2017) Maturation of conidia on conidiophores of Aspergillus niger. Fungal Genet Biol 98:61–70. https://doi.org/10.1016/j.fgb.2016.12.005

Thanh NV, Nout MJ (2004) Dormancy, activation and viability of *Rhizopus oligosporus* sporangiospores. Int J Food Microbiol 92:171–179

Thanh NV, Rombouts FM, Nout MJ (2005) Effect of individual amino acids and glucose on activation and germination of *Rhizopus oligosporus* sporangiospores in tempe starter. J Appl Microbiol 99:1204–1214

Thomma BPHJ (2003) *Alternaria* spp.: from general saprophyte to specific parasite. Mol Plant Pathol 4:225–236

Tiedt LR (1993) An electron microscope study of conidiogenesis and wall formation of conidia of *Aspergillus niger*. Mycolog Res 12:1459–1462

Timmermans R, Hayrapetyan H, Vollebregt M, Dijksterhuis J (2020) Comparing thermal inactivation to a combined process of moderate heat and high pressure: Effect on ascospores in strawberry puree. Int J Food Microbiol 16(325):108629. https://doi.org/10.1016/j.ijfoodmicro.2020.108629

Tournas V (1994) Heat-resistant fungi of importance to the food and beverage industry. Crit Rev Microbiol 20:243–263

Valencia-Chamorro SA, Pérez-Gago MB, Del Río MA, Palou L (2010) Effect of antifungal hydroxypropyl methylcellulose-lipid edible composite coatings on *Penicillium* decay development and postharvest quality of cold-stored "Ortanique" mandarins. J Food Sci 75:418–426

Valle M, Nguyen Van Long N, Jany J-L, Bregier T, Pawtowski A, Barbier G, Rigalma K, Vasseur V, Huchet V, Coroller L (2022) Impact of water activity on the radial growth of fungi in a dairy environment. Food Res Int 157:111247. https://doi.org/10.1016/j.foodres.2022.111247

Valle M, Nguyen Van Long N, Jany JL, Koullen L, Couvert O, Huchet V, Coroller L (2023) Impact of carbon dioxide on the radial growth of fungi isolated from dairy environment. Food Microbiol 115:104324. https://doi.org/10.1016/j.fm.2023.104324

Van den Brule T, Punt M, Teertstra W, Houbraken J, Wösten H, Dijksterhuis J (2020a) The most heat-resistant conidia observed to date are formed by distinct strains of *Paecilomyces variotii*. Environ Microbiol 22:986–999. https://doi.org/10.1111/1462-2920.14791

Van den Brule T, Lee CLS, Houbraken J, Haas PJ, Wösten H, Dijksterhuis J (2020b) Conidial heat resistance of various strains of the food spoilage fungus *Paecilomyces variotii* correlates with mean spore size, spore shape and size distribution. Food Res Int 137:109514. https://doi.org/10.1016/j.foodres.2020.109514

Van den Brule T, Punt M, Seekles SJ, Segers FJJ, Houbraken J, Hazeleger WC, Ram AFJ, Wösten HAB, Zwietering MH, Dijksterhuis J, den Besten HMW (2022) Intraspecific variability in heat resistance of fungal conidia. Food Res Int 156:111302. https://doi.org/10.1016/j.foodres.2022.111302

Van Kan JAL (2006) Licensed to kill: the lifestyle of a necrotrophic plant pathogen. Trends Plant Sci 11:247–253

Van Laarhoven KA, Huinink HP, Segers FJ, Dijksterhuis J, Adan OC (2015) Separate effects of moisture content and water activity on the hyphal extension of *Penicillium rubens* on porous media. Environ Microbiol 17:5089–5099. https://doi.org/10.1111/1462-2920.13012

Van Laarhoven KA, Peeters LHM, Bekker M, Huinink HP, Adan OCG (2017) Conidia of *Penicillium rubens* formed at low water activities can attract more water. MicrobiologyOpen 6:e526. https://doi.org/10.1002/mbo3.526

Van Leeuwen MR, De Boer W, Smant W, Dijksterhuis J (2008) Filipin is a reliable in situ marker of ergosterol in the plasma membrane of germinating conidia (spores) of *Penicillium discolor* and stains intensively at the site of germ tube formation. J Microbiol Meth 74:64–73

Van Leeuwen MR, Golovina EA, Dijksterhuis J (2009) The polyene antimycotics nystatin and filipin disrupt the plasma membrane, whereas natamycin inhibits endocytosis in germinating conidia of *Penicillium discolor*. J Appl Microbiol 106:1908–1918

Van Leeuwen MR, Van Doorn T, Golovina EA, Stark J, Dijksterhuis J (2010) Water- and air-distributed conidia exhibit differences in sterol content and cytoplasmic microviscosity. Appl Env Microbiol 67:366–369

Van Leeuwen MR, Krijgsheld P, Bleichrodt R, Menke H, Stam H, Stark J, Wösten HA, Dijksterhuis J (2013a) Germination of conidia of *Aspergillus niger* is accompanied by major changes in RNA profiles. Stud Mycol 15(74):59–70. https://doi.org/10.3114/sim0009

Van Leeuwen MR, Krijgsheld P, Wyatt TT, Golovina EA, Menke H, Dekker A, Stark J, Stam H, Bleichrodt R, Wösten HA, Dijksterhuis J (2013b) The effect of natamycin on the transcriptome of conidia of *Aspergillus niger*. Stud Mycol 74:71–85. https://doi.org/10.3114/sim0013

Vinnere-Pettersson O, Leong SL, Lantz H, Rice T, Dijksterhuis J, Houbraken J, Samson RA, Schnürer J (2011) Phylogeny and intraspecific variation of the extreme xerophile, *Xeromyces bisporus*. Fungal Biolog 115:1100–1111

Visagie CM, Houbraken J, Frisvad JC, Hong SB, Klaassen CH, Perrone G, Seifert KA, Varga J, Yaguchi T, Samson RA (2014) Identification and nomenclature of the genus *Penicillium*. Stud Mycol. 78:343–371. https://doi.org/10.1016/j.simyco.2014.09.001

Vivek-Ananth RP, Mohanraj K, Vandanashree M et al (2018) Comparative systems analysis of the secretome of the opportunistic pathogen *Aspergillus fumigatus* and other *Aspergillus species*. Sci Rep 8:6617. https://doi.org/10.1038/s41598-018-25016-4

Wang CL, Shaw BD (2016) F-actin localization dynamics during appressorium formation in *Colletotrichum graminicola*. Mycologia 108:506–514. https://doi.org/10.3852/15-068

Wang X, Beno-Moualem D, Kobiler I, Leikin-Frenkel A, Lichter A, Prusky D (2004) Expression of Δ^{12} fatty acid desaturase during the induced accumulation of the antifungal dienen in avocado fruits. Mol Pant Pathol 5:575–585

Wang H, Hu Z, Long F, Guo C, Niu C, Yuan Y, Yue T (2016) Combined effect of sugar content and pH on the growth of a wild strain of Zygosaccharomyces rouxii and time for spoilage in concentrated apple juice. Food Control 59:2980305. https://doi.org/10.1016/j.foodcont.2015.05.040

Wang F, Sethiya P, Hu X, Guo S, Chen Y, Li A, Tan K, Wong KH (2021) Transcription in fungal conidia before dormancy produces phenotypically variable conidia that maximize survival in different environments. Nat Microbiol 6:1066–1081. https://doi.org/10.1038/s41564-021-00922-y

Watanabe Y, Yamaguchi M, Sanemitsu Y, Tarnai Y (1991) Characterisation of plasma membrane protonATPase from salt-tolerant yeast *Zygosaccharomyces rouxii* cells. Yeast 7:599–606

Watanabe Y, Miwa S, Tarnai Y (1995) Characterization of Na+/H(+)-antiporter gene closely relate salt-tolerance of yeast *Zygosaccaromyces rouxii*. Yeast 11:829–838

Westerdijk J (1949) The concept "Association" in mycology. Antonie van Leeuwenhoek 15:187–189. https://doi.org/10.1007/BF02062645

Wiemken A (1990) Trehalose in yeast, stress protectant rather than reserve carbohydrate. Antonie van Leeuwenhoek 68:209–217

Wildman JD, Clark PB (1947) Some examples of the occurrence of machinery slime in caning factories. J Ass Off Anal Chern 30:582–585

Williams JP, Hallsworth JE (2009) Limits of life in hostile environments: no barriers to biosphere function? Env Microbiol 11:3292–3308

Williams B, Kabbage M, Kim H-J, Britt R, Dickman MB (2011) Tipping the balance: *Sclerotinia sclerotiorum* secreted oxalic acid suppresses host defences by manipulating the host redox environment. PLOS Pathogens 7:e1002107

Wingfield MJ, De Beer ZW, Slippers B, Wingfield BD, Groenewald JZ, Lombard L, Crous PW (2012) One fungus, one name promotes progressive plant pathology. Mol Plant Pathol 13(6):604–613. https://doi.org/10.1111/j.1364-3703.2011.00768.x

Wyatt TT, Wösten HAB, Dijksterhuis J (2013) Fungal spores for dispersion in space and time. In: Sariaslani S, Gadd GM (eds) Advances in applied microbiology, vol 85. Academic Press, pp 43–91. ISBN 9780124076723. https://doi.org/10.1016/B978-0-12-407672-3.00002-2

Wyatt TT, Golovina EA, van Leeuwen R, Hallsworth JE, Wösten HA, Dijksterhuis J (2015a) A decrease in bulk water and mannitol and accumulation of trehalose and trehalose-based oligosaccharides define a two-stage maturation process towards extreme stress resistance in ascospores of *Neosartorya fischeri* (*Aspergillus fischeri*). Environ Microbiol 17:383–394. https://doi.org/10.1111/1462-2920.12557

Wyatt TT, van Leeuwen MR, Golovina EA, Hoekstra FA, Kuenstner EJ, Palumbo EA, Snyder NL, Visagie C, Verkennis A, Hallsworth JE, Wösten HA, Dijksterhuis J (2015b) Functionality and prevalence of trehalose-based oligosaccharides as novel compatible solutes in ascosporesof *Neosartorya fischeri* (*Aspergillus fischeri*) and other fungi. Environ Microbiol 17:395–411. https://doi.org/10.1111/1462-2920.12558

Yilmaz N, Visagie CM, Houbraken J, Frisvad JC, Samson RA (2014) Polyphasic taxonomy of the genus *Talaromyces*. Stud Mycol 78:175–341. https://doi.org/10.1016/j.simyco.2014.08.001

Yurkov AM, Sannino C, Turchetti B (2020) *Mrakia fibulata* sp. nov., a psychrotolerant yeast from temperate and cold habitats. Antonie van Leeuwenhoek 113:499–510. https://doi.org/10.1007/s10482-019-01359-4

Zhang Y, An B, Wan W, Zhan B, He C, Luo H (2022) Actin-bundling protein fimbrin regulates pathogenicity via organizing F-actin dynamics during appressorium development in *Colletotrichum gloeosporioides*. Mol Plant Pathol 23:1472–1486. https://doi.org/10.1111/mpp.13242

Ergot Alkaloids

4

Kyle A. Davis, Corinne M. Hazel, Abigail M. Jones, Samantha J. Fabian, and Daniel G. Panaccione

Abstract

Ergot alkaloids are specialized metabolites of agricultural and medical importance produced by at least two different lineages of fungi. Fungi from different lineages share early steps of the biosynthetic pathway, and functional genetic studies and genomic analyses indicate a common set of clustered genes encodes these early pathway steps. Different taxa follow different terminal branches of the pathway to produce profiles of ergot alkaloids that vary by lineage. In addition to the pathway end-product, the profile of ergot alkaloids observed in a given fungus often includes pathway intermediates that accumulate to high levels and sometimes products of pathway spurs. Ergot alkaloid-producing fungi fill a range of environmental niches and studies with mutants containing altered ergot alkaloid pathways indicate these specialized metabolites contribute to resource acquisition and resource preservation in the fungi that produce them.

Keywords

Ergot alkaloids · Lysergic acid · Clavines · Ergopeptines · *Aspergillus* · *Claviceps* · *Metarhizium* · *Periglandula* · *Epichloë*

4.1 Introduction

Ergot alkaloids are a class of tryptophan-derived mycotoxins that have affected human health and agricultural productivity in both positive and negative ways. This complex group of specialized metabolites is produced by a diverse range of fungi including several species in the orders the Eurotiales and Hypocreales (Wallwey and Li 2011; Robinson and Panaccione 2015; Florea et al. 2017; Tasker and Wipf 2021; Panaccione 2023). These fungi occupy a wide variety of ecological niches such as soil saprotrophs, plant endophytes, pathogens of plants or insects, and opportunistic pathogens of humans and other mammals (Schardl et al. 2006, Gerhards et al. 2014, Young et al. 2015, Leadmon et al. 2020, Panaccione 2023). Ergot alkaloids share a common biosynthetic origin, with all characterized pathway branches being derived from a set of biosynthetic reactions encoded by a cluster of five shared genes (Robinson and Panaccione 2015;

This chapter is an updated version of 'Ergot Alkaloids' published in Vol. 10: Industrial Applications, 2011, ISBN 978-3-642-11457-1, https://link.springer.com/book/10.1007/978-3-642-11458-8.

K. A. Davis · C. M. Hazel · A. M. Jones · S. J. Fabian · D. G. Panaccione (✉)
West Virginia University, Division of Plant and Soil Sciences, Morgantown, WV, USA
e-mail: danpan@wvu.edu

S. Grüttner et al. (eds.), *Agricultural and Industrial Applications*, The Mycota 16,
https://doi.org/10.1007/978-3-031-81904-9_4

Florea et al. 2017; Tasker and Wipf 2021; Panaccione 2023). After these early, shared steps, pathways diverge in different fungi to yield the diverse structures observed in different ergot alkaloid producers. Additional diversity of structures results from inefficiency in the respective pathways such that intermediates and spur products accumulate to relatively high levels (Panaccione 2005). Thus, the various ergot alkaloids observed in a fungus at a given time may include pathway end-products, pathway intermediates, and products of spur pathways of the main ergot alkaloid pathway of the particular fungus.

Ergot alkaloids have affinities for receptors of the monoamine neurotransmitters 5-hydroxytryptamine (5-HT, also called serotonin), dopamine, noradrenaline, and adrenaline (Pertz 1996; Pertz and Eich 1999; Tfelt-Hansen and Koehler 2008; Zajdel et al. 2015). Because of their structural diversity, some ergot alkaloids act as agonists at these receptors, whereas others are antagonists. The interactions are further complicated by the fact that receptors for each of these neurotransmitters occur as different subtypes, and affinities and activities for different alkaloids at different subtypes vary. As a result, ergot alkaloids affect a variety of systems and often have undesirable side-effects.

Historically, ergot alkaloids associated with the ergot fungus *Claviceps purpurea* have achieved notoriety because of inadvertent human poisoning in populations that relied heavily upon rye bread for a large portion of their calories (Matossian 1989; Haarmann et al. 2009, Florea et al. 2017). The ability of some ergot alkaloids to cause psychotropic and narcotic effects has been utilized by Native American cultures (Schultes and Hofmann 1973) and, more recently, recreational drug users (Hofmann 1980). Clinically, ergot alkaloids have been used to treat migraines, Parkinson's disease, prolactinomas, and to stimulate uterine contractions and reduce bleeding during child birth (Schade et al. 2007; Zanettini et al. 2007; Tfelt-Hansen and Koehler 2008; Smakosz et al. 2021). Most ergot alkaloids, such as lysergic acid, possess vasoconstrictive properties while dihydrogenated ergot alkaloids, which are rare in nature, tend to possess vasorelaxant properties. Semisynthetic derivatives of both classes have also been developed as pharmaceuticals to treat a wide variety of neurological diseases and disorders (Roberto et al. 2015; Lee et al. 2019; Montastruc et al. 2017; Stiles et al. 2021). However, their toxicity and frequent side-effects have limited their utility as drugs. In agriculture, ergot alkaloids have directly and negatively affected plant quality through their antimammalian activities and indirectly and positively affected plant productivity as a result of their activities against invertebrate and vertebrate herbivores (Caradus et al. 2022; Saikkonen et al. 2016; Young et al. 2013; Schardl et al. 2013). All these effects of ergot alkaloids are discussed in detail below.

The main objectives of this review are to illustrate the diversity of alkaloids in the ergot family with respect to structure and activity and to discuss the potential roles for ergot alkaloids in the recently expanded catalogue of fungi that produce these compounds and occupy a wide range of ecological niches.

4.2 Diversification of Structure and Activity in the Ergot Alkaloid Pathway

Individual ergot alkaloid producers often contain a diverse set of ergot alkaloids, and even greater diversity may be observed among producers from different taxa. The output of the ergot pathway in some of these fungi is not a single toxin but rather a complex profile of alkaloids consisting of the pathway end-product, one or more intermediates that accumulate to relatively high levels, and in some cases the products of pathway spurs from which some portion of an intermediate is modified to an alternate product (Panaccione 2005, 2023). In this section, the profile of ergot alkaloids observed in representative ergot alkaloid producers is described in the context of where those ergot alkaloids occur along the pathway. The ergot alkaloid pathway also is considered as a set of taxon-specific branches that arise from a common shared root.

Ergot alkaloids represent a continuum of structures of increasing complexity as they are produced along a long, branching pathway. Traditionally, they have been grouped into several structural classes: clavine alkaloids, simple amides of lysergic acid, and ergopeptines. Clavines are generally among the simpler ergot alkaloids observed accumulating in fungi. In some fungi, such as *Aspergillus fumigatus* (synonym *Neosartorya fumigata*) and *Claviceps fusiformis*, clavines are the ultimate products of the ergot alkaloid pathway. In other fungi (e.g., *C. purpurea* and several *Epichloë* spp.), clavines accumulate to varying levels but also serve as intermediates on the pathway to lysergic acid, which is subsequently incorporated into lysergic acid amides and non-ribosomally synthesized peptides (ergopeptines).

A global scheme of the ergot alkaloid pathway and its branches in different groups of fungi is represented in Fig. 4.1. Comparisons of ergot alkaloid profiles of different fungi and the ergot

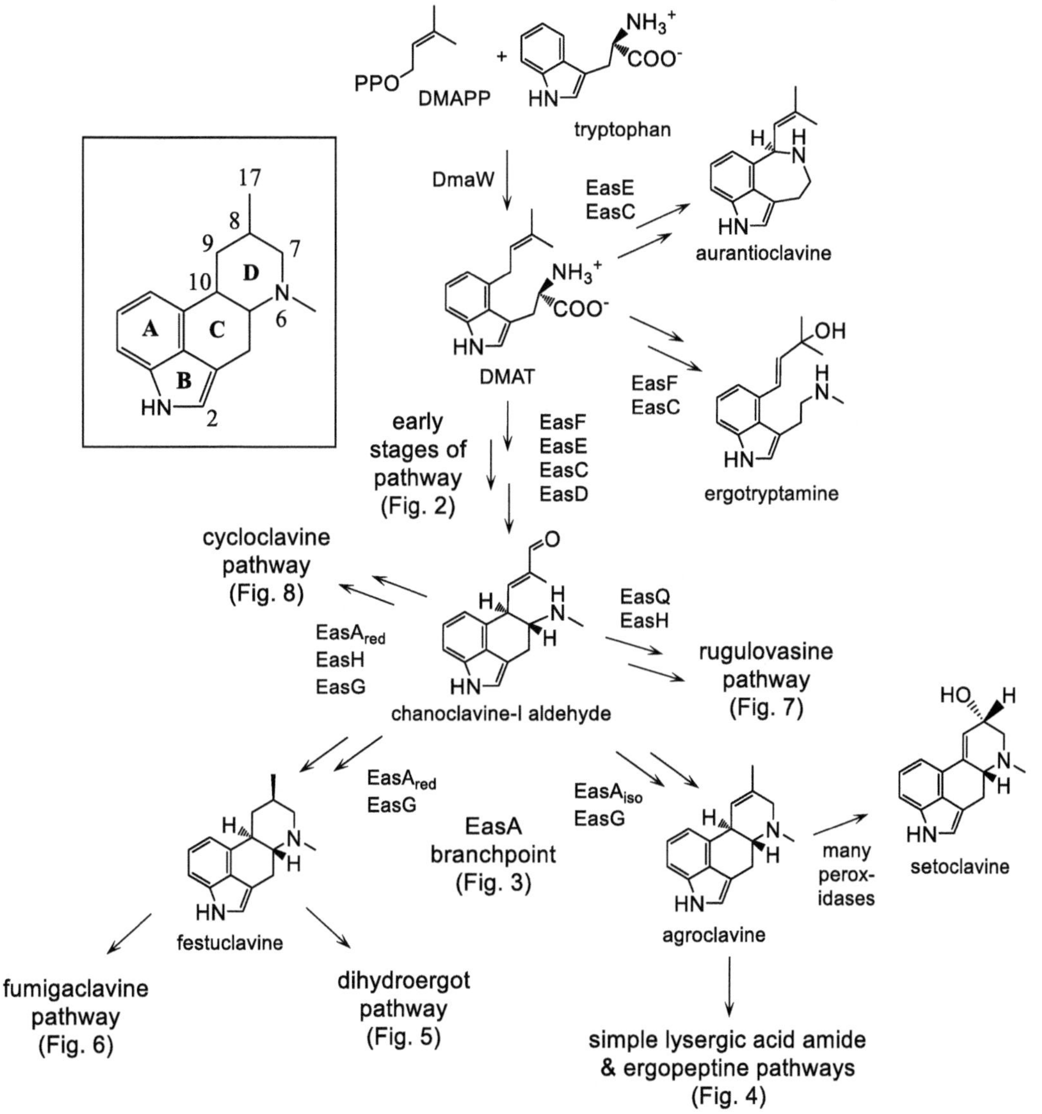

Fig. 4.1 Pathways showing the enzymatic steps for the biosynthesis of different ergot alkaloids. Multiple arrows indicate steps omitted for brevity. *DMAPP* dimethylallylpyrophosphate, *DMAT* dimethylallyltryptophan, *red* reductase, *iso* isomerase. Inset: ring- and position-labeling; for simplicity, position numbering is shown only for atoms referred to in the text

alkaloid biosynthesis gene clusters in these fungi support the hypothesis of a shared early portion of the pathway followed by different terminal branches in different ergot alkaloid-producing fungi. Experiments have also shown that strains may be engineered to produce products from branches found in other lineages as well as entirely novel ergot alkaloids predicted by enzyme activity (e.g., Robinson and Panaccione 2014; Davis et al. 2020).

4.2.1 Early Pathway Steps Shared Among Producers

Ergot alkaloid biosynthesis is encoded by a set of ergot alkaloid synthesis (*eas*) genes clustered together within a biosynthetic gene cluster (Tudzynski et al. 1999; Coyle and Panaccione 2005; Unsöld and Li 2005; Schardl et al. 2013; Panaccione 2023). Studies on the early steps of the ergot alkaloid pathway have been reviewed in detail recently (e.g., Florea et al. 2017; Tasker and Wipf 2021; Panaccione 2023) (Fig. 4.2). In this chapter we will follow the *eas* gene-naming convention proposed by Schardl et al. (2006); synonyms in published works by other authors are listed in Table 4.1. Five genes (*dmaW*, *easF*, *easE*, *easC*, and *easD*) are required for biosynthesis of chanoclavine-I aldehyde, the precursor to several other branches of the ergot alkaloid pathway (Fig. 4.2). The first pathway step is carried out by 4-γ,γ-dimethylallyltryptophan synthase encoded by *dmaW*, which utilizes dimethylallylpyrophosphate to prenylate tryptophan at position 4 (Tsai et al. 1995), generating 4-dimethylallyltryptophan (DMAT). DMAT is then methylated at the primary amine by the methyltransferase encoded by *easF* (Rigbers and Li 2008), producing *N*-methyl-4-dimethylallyltryptophan. The oxidoreductase and catalase products of *easE* and *easC*, respectively, then carry out decarboxylative closure of third (or C) ring of the ergoline nucleus (Fig. 4.2; also refer to Fig. 4.1, inset), generating chanoclavine-I (Lorenz et al. 2010; Goetz et al. 2011; Yao et al. 2019). Closure of the C ring involves an unexpected oxidative cyclization by EasC, an enzyme

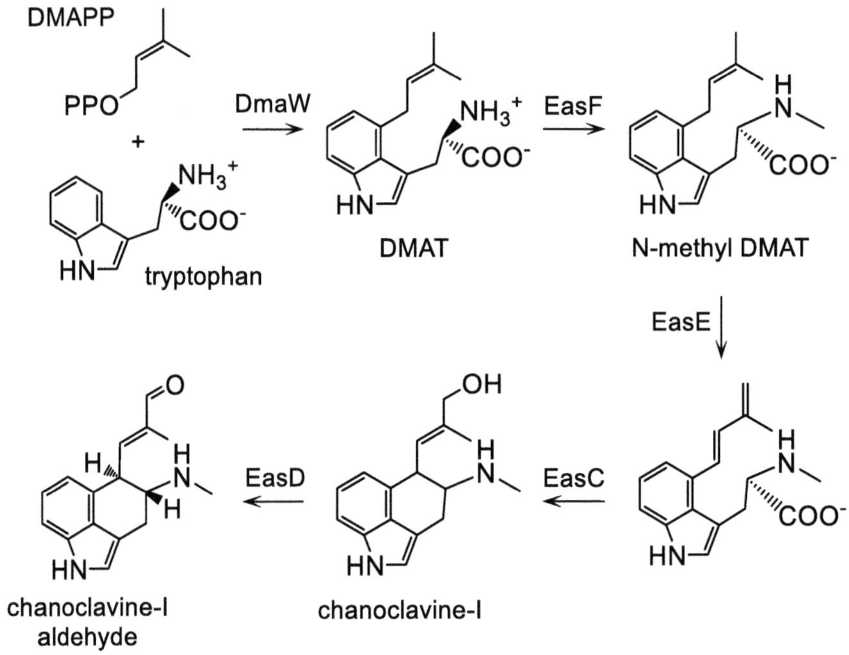

Fig. 4.2 Early shared steps of the ergot alkaloid pathway. *DMAPP* dimethylallylpyrophosphate, *DMAT* dimethylallyltryptophan

Table 4.1 Genes found in ergot alkaloid clusters and synonyms among species[a]

Gene name	Function or proposed function	*Aspergillus fumigatus*	*Claviceps purpurea*
dmaW	Dimethylallyltryptophan synthase	*fgaPT2*	*cpd1*
easA	Oxidoreductase	*fgaOx3*	*easA*
easC	Catalase	*fgaCat*	*easC*
easD	Oxidoreductase	*fgaDH*	*easD*
easE	Oxidoreductase	*fgaOx1*	*ccsA*
easF	*N*-methyltransferase	*fgaMT*	*easF*
easG	Oxidoreductase	*fgaFS*	*easG*
easH[b]	Dioxygenase	*easH*$^{\Psi}$	*easH*
easK	FAD monooxygenase	*fgaP450-1*	–
easL	Reverse prenyl transferase	*fgaPT1*	–
easM	P450 monooxygenase	*fgaP450-2*	–
easN	*O*-acetyltransferase	*fgaAT*	–
cloA	P450 monooxygenase	–	*cloA*
lpsA	Lysergyl peptide synthetase 1	–	*Cpps1* and *Cpps4*
lpsB	Lysergyl peptide synthetase 2	–	*Cpps2*
lpsC	Lysergyl peptide synthetase 3	–	*Cpps3*
easO	FAD-containing monooxygenase	–	–
easP	α/β hydrolase fold protein	–	–
easQ	Aldehyde dehydrogenase	–	–

[a]Superscripted psi (Ψ) indicates a pseudogene
[b]Enzymes encoded by *easH* vary in function among pathway branches

that bioinformatically is classified as a catalase (Yao et al. 2019). The primary alcohol of chanoclavine-I is then oxidized to an aldehyde by the short-chain alcohol dehydrogenase product of *easD*, generating chanoclavine-I aldehyde (Wallwey et al. 2010a).

Chanoclavine-I aldehyde serves as the major branchpoint to several other ergot alkaloid classes including different clavines, lysergic acid amides, and ergopeptines, as well as their dihydrogenated forms (Fig. 4.1). A major distinction among branches of the ergot alkaloid pathway involving a saturated fourth (or D) ring (e.g., the fumigaclavine and the dihydroergot alkaloids) compared to those involving an unsaturated D ring (i.e., the lysergic acid derivatives) is whether the version of *easA* present in an organism's *eas* gene cluster encodes a reductase or isomerase version of EasA (Fig. 4.3) (Cheng et al. 2010a; Cheng et al. 2010b; Coyle et al. 2010). In both cases, the product of *easA* catalyzes closure of the D ring of the ergoline ring system, generating an iminium ion. Subsequent reduction of the N6,C7 bond through the reductase encoded by *easG* completes ring closure (Wallwey et al. 2010b; Matuschek et al. 2011), generating either agroclavine (via the product of the isomerase allele of *easA*) or festuclavine (via the reductase allele *easA*) (Coyle et al. 2010; Cheng et al. 2010b). Agroclavine serves as precursor to lysergic acid (Fig. 4.4), whereas festuclavine is the precursor to the dihydrolysergic acid (DHLA) (Fig. 4.5) and fumigaclavine (Fig. 4.6) branches. The isomerase allele, such as that found in the fungus *Epichloë festucae* (formerly called *Neotyphodium lolii*), encodes an enzyme that temporarily reduces the C8,C9 double bond of chanoclavine-I aldehyde to allow for isomerization before the C8,C9 double bond is restored. This isomerization brings the aldehyde and amine groups into proximity, facilitating ring closure. The reductase allele of *easA* found in *A. fumigatus* encodes an enzyme that fully reduces the double bond, allowing for rotation of the aldehyde functional group such that it can react with the secondary amine. In either case, the resulting iminium ion is subsequently reduced by the product of *easG* (Wallwey et al. 2010a; Matuschek et al. 2011).

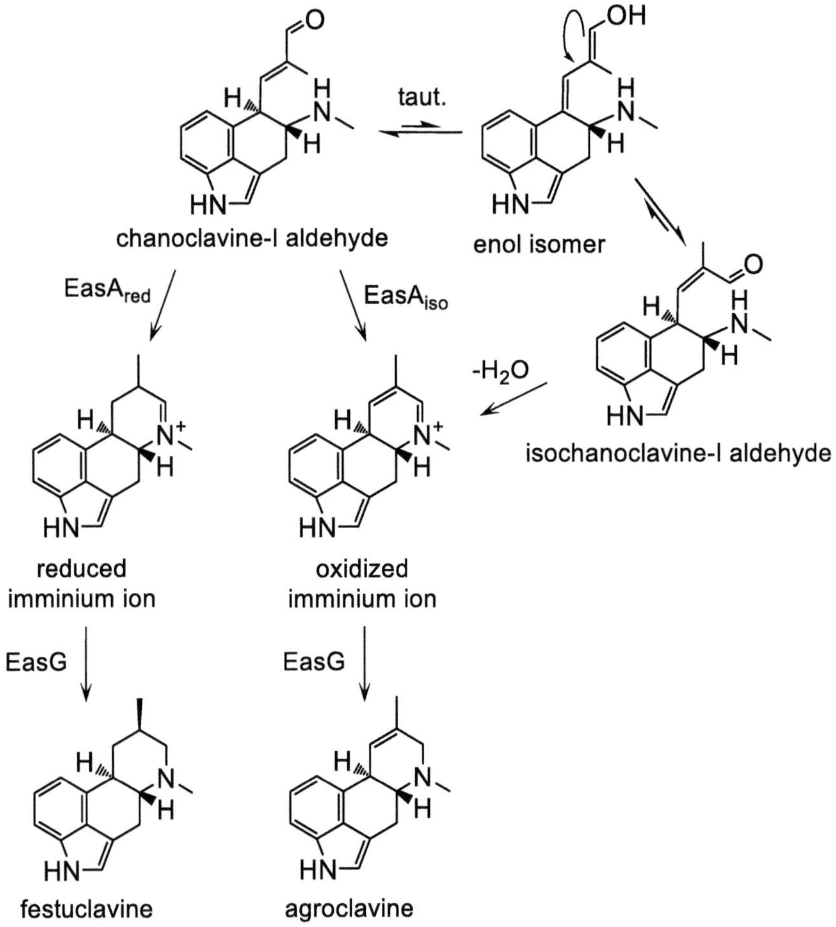

Fig. 4.3 Pathway showing the divergence of the ergot alkaloid pathway branches at EasA. The noncatalyzed, tautomerization-based ring closure mechanism leading to low levels of agroclavine is also shown. *red* reductase, *iso* isomerase, *taut* tautomerization

In a well-functioning pathway, expression of these early genes appears to be coordinated such that intermediates from the early, shared portions do not accumulate to levels that would allow them to be detected by high-performance liquid chromatography (HPLC). The earliest intermediate detected in a non-modified fungus is N-Me-DMAT in *Penicillium camemberti* (Fabian et al. 2018), which requires only *dmaW* and *easF* to encode its biosynthesis. Two spur products derived from the early stages of the pathway, ergotryptamine and aurantioclavine, occasionally accumulate to detectable levels in certain fungi (Ryan et al. 2013, 2015; Lin et al. 2015). Studies expressing combinations of early pathway steps from *A. fumigatus* in *A. nidulans*, a non-ergot alkaloid producer, revealed that the expression of *dmaW*, *easF*, and *easC* resulted in formation of ergotryptamine (Ryan et al. 2015). The authors noted that the oxygen of ergotryptamine is likely added by peroxidase activity common to many fungi and that ergotryptamine is equivalent to the provisional 6,7-secolysergine described by Panaccione et al. (2003) in *Epichloë*-infected

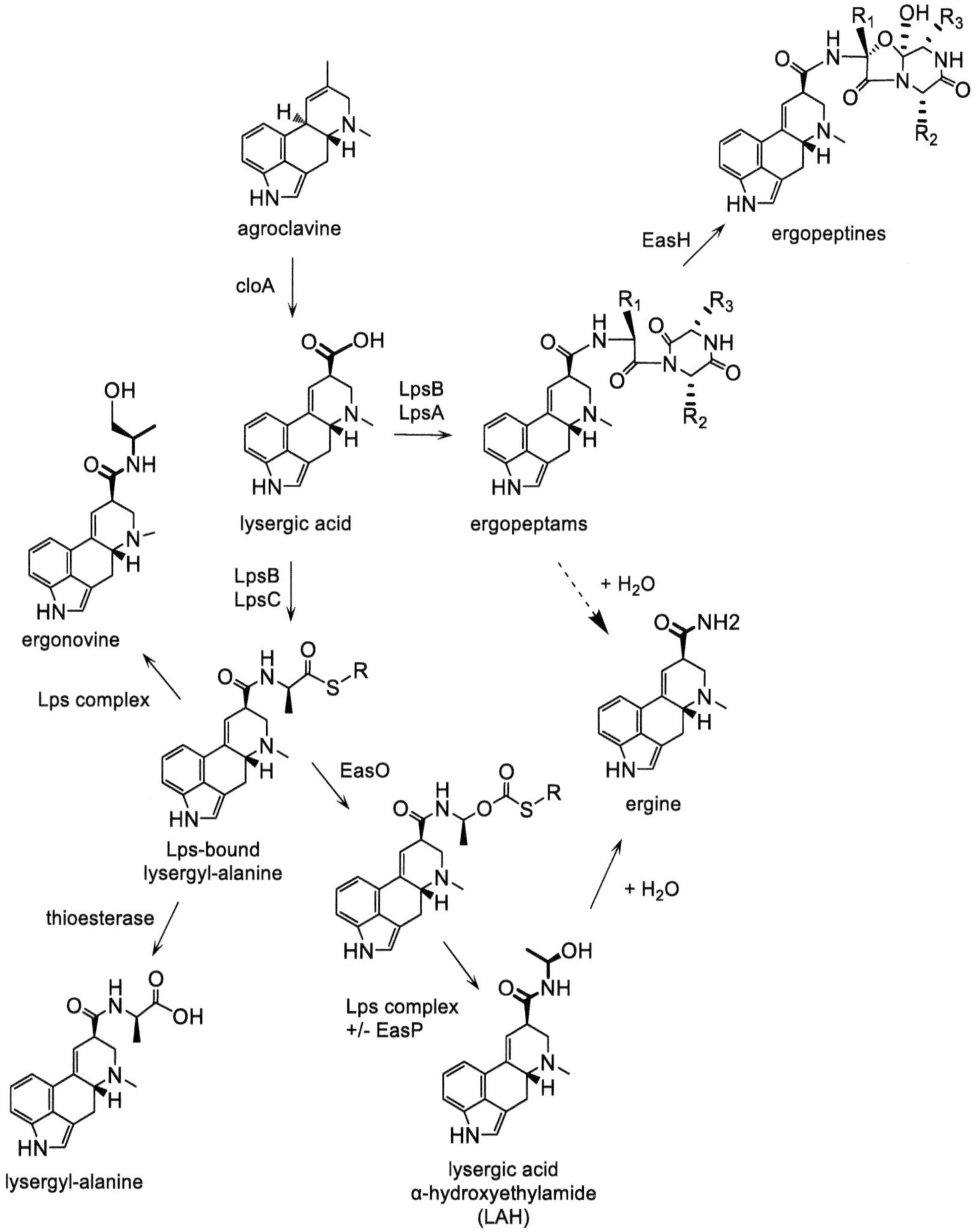

Fig. 4.4 Pathways to simple lysergic acid amides, ergopeptines, and the hydrolysis product ergine. Not all possible spur products or alternate end-products are shown. EasP is marked by ± as it contributes to, but is not required for, production of LAH. Dashed arrow represents a hypothetical step. Ergopeptines differ by amino acid side chains at positions R1, R2, and R3 and are outlined in Uhlig and Petersen (2008) and Schardl et al. (2006)

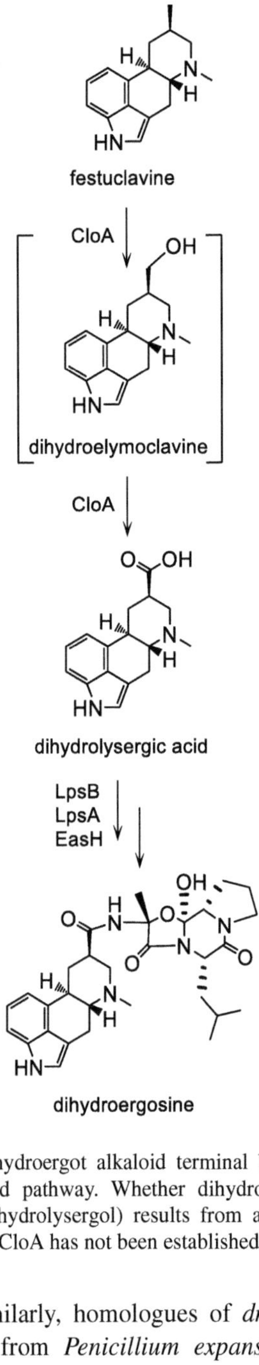

Fig. 4.5 Dihydroergot alkaloid terminal branch of the ergot alkaloid pathway. Whether dihydroelymoclavine (synonym dihydrolysergol) results from a two-electron oxidation by CloA has not been established

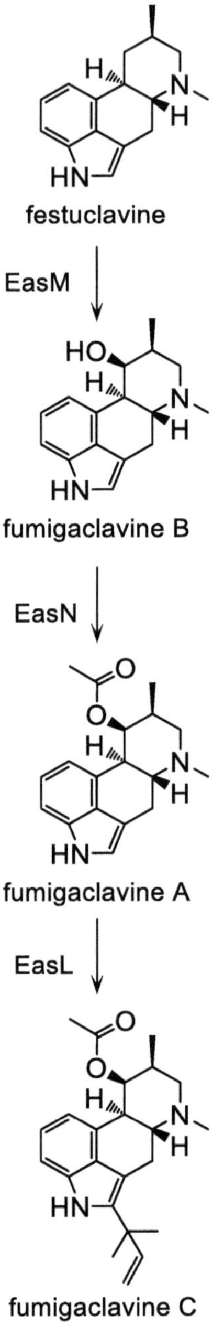

Fig. 4.6 Fumigaclavine branch of the ergot alkaloid pathway. *Aspergillus fumigatus* completes the illustrated branch, whereas *P. commune*, *P. roqueforti* (which makes products with different stereochemistry around carbons 8 and 9), and some isolates of *A. fumigatus* stop at fumigaclavine A

plants. Similarly, homologues of *dmaW*, *easE*, and *easC* from *Penicillium expansum* control production of aurantioclavine (Lin et al. 2015). The combination of *dmaW*, *easF*, *easE*, and *easC* is sufficient to encode synthesis of chanoclavine-I (Ryan et al. 2013; Nielsen et al. 2014). Chanoclavine-I is observed as the pathway end-product of *Epichloë canadensis* (Schardl et al. 2013) and frequently is observed as accumulating to varying degrees in fungi that produce more complex ergot alkaloids from any of several other branches of the pathway (Panaccione 2023).

Agroclavine is readily oxidized to the spur product setoclavine and its isomer by peroxidase activities common to fungi and bacteria (Coyle et al. 2010; Béliveau and Ramstad 1967; Panaccione et al. 2003).

4.2.2 Branches to Different Families of Ergot Alkaloids

4.2.2.1 Lysergic Acid-Derived Branches

In the typical clavicipitaceous pathway (observed in several *Claviceps*, *Epichloë*, and *Metarhizium* species), agroclavine is the key intermediate that is further oxidized and modified to yield different ergot alkaloid profiles in different producers (Fig. 4.4). Agroclavine is oxidized at position 17 (Fig. 4.1, inset) in two successive steps to yield lysergic acid. Robinson and Panaccione (2014) proposed that the enzyme responsible for this six-electron oxidation, CloA, acts with high processivity because no elymoclavine (the agroclavine derivative with a primary alcohol at position 17) was detected when the *cloA* gene of *Epichloë hybrida* (originally called *E. typhina* × *E. festucae* isolate Lp1; Schardl et al. 1994) was expressed in a strain of *A. fumigatus* engineered to accumulate agroclavine as substrate for CloA. In *C. fusiformis*, elymoclavine is the most complex ergot alkaloid observed (Floss et al. 1968). The idea that elymoclavine observed in *C. fusiformis* and some other *Claviceps* species arises from a two-electron oxidation of agroclavine by CloA is not supported by two observations: (1) the allele of *cloA* present in the *C. fusiformis* ergot gene cluster is transcribed in culture but does not encode an enzyme that catalyzes this reaction (Lorenz et al. 2007); and (2) a strain of *C. purpurea* in which *cloA* was knocked out still accumulated elymoclavine (Haarmann et al. 2006). Elymoclavine observed in these contexts must, thus, arise from some mechanism other than a two-electron oxidation by CloA.

Once lysergic acid is formed, it typically is converted into any of several different amides in a relatively efficient manor, such that free lysergic acid intermediate is not frequently or abundantly detected. In some cases, an individual fungus will produce more than one lysergic acid amide. These amides can be classified into two groups, simple amides and ergopeptines, depending on their complexity. The ergopeptines are a family of non-ribosomally synthesized peptides containing lysergic acid and three amino acids. With rare exception the second amino acid is L-proline, and variation at the first and third amino acids produces a family of different ergopeptines. The ergopeptines are formed through the interaction of two peptide synthetases (Riederer et al. 1996; Walzel et al. 1997). The two-component peptide synthetase is unique among fungal peptide synthetases, which are usually comprised of a single multifunctional polypeptide (Walton et al. 2004). The three amino acids are recognized and incorporated into the ergopeptine by lysergyl peptide synthetase 1 (LPS1), a multifunctional non-ribosomal peptide synthetase, whereas lysergic acid is recognized and activated for incorporation into the ergopeptine by an enzyme named lysergyl peptide synthetase 2 (LPS2; Riederer et al. 1996; Walzel et al. 1997). Functional analysis of the role of LPS1 in ergopeptine formation in *E. hybrida* was demonstrated by gene knockout (Panaccione et al. 2001). Genetic evidence for the role of LPS2 was provided in *C. purpurea* by Correia et al. (2003) and in *E. festucae* by Fleetwood et al. (2007). A version of the dioxygenase EasH is then responsible for hydroxylating the α-carbon of the amino acid adjacent to D-lysergic acid, driving cyclolization of the α-carbon with the terminal lactam carbonyl group generated by LPS1/LPS2 (Havemann et al. 2014). Some strains of *C. purpurea* isolated from wild Norwegian grasses have been found to produce ergopeptams, the penultimate step in the ergopeptine pathway, as their primary end-products and thus presumably lack a functional EasH (Uhlig and Petersen 2008).

Simple amides of lysergic acid include ergine (also called lysergic acid amide, LSA, or lysergamide), lysergyl-alanine, ergonovine (also called ergometrine, ergobasine), and lysergic acid α-hydroxyethylamide (LAH). The existence of a separate enzyme for lysergic acid activation (LPS2) allows for an elegant combinatorial system in which LPS2 combines with LPS1 to form ergopeptines (as described above) and with a second mono-modular peptide synthetase (LPS3) capable of activating alanine (and encoded by ergot cluster gene *lpsC*; Haarmann et al. 2005; Ortel and Keller 2009) to form the simple amide ergonovine (Ortel and Keller 2009). The alanine-activating peptide synthetase LPS3 contains a reductase domain in its carboxy-terminal region to convert the lysergyl-alanine intermediate into ergonovine. Biosynthesis of LAH in *Metarhizium brunneum* has been characterized more recently (Davis et al. 2020; Steen et al. 2021; Britton et al. 2022). This process relies upon the activity of a Baeyer-Villiger monooxygenase encoded by *easO*, which acts on lysergyl-alanine bound to LPS3 (Steen et al. 2021). A model involving the reductase domain of LPS3 acting on the product of EasO to liberate LAH has been proposed (Steen et al. 2021; Panaccione 2023). Another gene, *easP*, encodes an α/β hydrolase fold protein that was found to contribute to, but is not required for, LAH production in *M. brunneum* (Britton et al. 2022). Jones et al. (2021) recently observed LAH accumulation in three species of *Aspergillus*. In phylogenetic analyses, all seven genes required to produce lysergic acid in the *Aspergillus* species clustered tightly with those of the Clavicipitaceous lysergic acid amide producers. The genes required to derive LAH from lysergic acid, however, appeared to have evolved from different ancestral genes, indicating that the ability to put lysergic acid into LAH evolved separately in the two lineages.

Ergine is an interesting lysergic acid amide because in most lysergic acid amide producers, it appears to arise spontaneously by hydrolysis of LAH (Kleinerova and Kybal 1973; Flieger et al. 1982; Leadmon et al. 2020). The side chain of LAH is unusual in that it contains a hemiaminal group that shares its nitrogen with the amide group common to all lysergic acid amides (Fig. 4.4). Hydrolysis of the hemiaminal yields ergine, whereas hydrolysis of the amide would yield lysergic acid. Both of these reactions occur at some rate in LAH-accumulating fungi, but hydrolysis of the hemiaminal appears to occur more readily because ergine frequently is detected at higher concentrations in extracts of LAH-producing fungi (Table 4.2). One fungus that appears to produce ergine at a rate much faster than others is *A. leporis* (Jones et al. 2021). The differences in rates of conversion of LAH to ergine in *A. leporis* compared to other LAH producers such as *A. homomorphus* or *M. brunneum* indicate catalysis of the reaction in *A. leporis*.

The endophyte *E. hybrida* lacks a homologue of LPS3 and does not produce ergonovine or LAH but still accumulates ergine as well as another lysergic acid amide, lysergyl-alanine (Panaccione et al. 2003). Panaccione et al. (2003) hypothesized that these particular simple amides form as breakdown products during the synthesis of ergopeptines, because a strain of *E. hybrida* in which the peptide synthetase LPS1 was knocked out accumulated no ergine or lysergyl-alanine. The lysergyl-peptide lactam product of LPS1/LPS2 and its oxidized derivative (Fig. 4.4) are potential stages that might yield the simple amides in *Epichloë* spp.

4.2.2.2 Dihydroergot Alkaloid Branch

The sorghum ergot pathogen *C. africana* produces a family of dihydroergot alkaloids including dihydroergosine, dihydroergotamine, and dihydrolysergol in which the D ring is completely saturated (Fig. 4.5) (Mantle and Waight, 1968; Shimshoni et al. 2017). In *C. africana*, a series of oxidations analogous to those described in *C. purpurea* occur at position 17 (Fig. 4.1, inset),

Table 4.2 Proportion of lysergic acid amide (from among LAH, ergine, or lysergic acid) in ergine or lysergic acid in different fungi

Species	Proportion in ergine	Proportion in lysergic acid	*P* value for difference
A. hancockii	23 ± 3	1 ± 0.05	<0.0001
A. homomorphus	2 ± 1	1 ± 0.02	0.03
M. brunneum	2 ± 1	1 ± 0.03	<0.0001
Periglandula sp.	51 ± 17	17 ± 2	<0.0001

but with festuclavine as opposed to agroclavine as the substrate (Barrow et al. 1974). The resulting product, DHLA, is then incorporated into dihydroergopeptines. The maize ergot fungus *C. gigantea* has been reported to accumulate festuclavine and dihydroelymoclavine in certain specimens (Agurell and Ramstad 1965) and festuclavine, dihydroelymoclavine, and DHLA in others (Bragg et al. 2017). When CloA from *C. africana* was expressed in a strain of *A. fumigatus* previously engineered to produce festuclavine as substrate, DHLA, but not dihydrolysergol, was detected (Arnold and Panaccione 2017). A trace amount of dihydrolysergol was detected, in addition to DHLA, when the *C. gigantea* allele of *cloA* was expressed in that same *A. fumigatus* background (Bragg et al. 2017). These observations indicate a rapid turnover of festuclavine to DHLA.

The most parsimonious explanation of the origin of the dihydroergot alkaloids observed in *C. africana* and *C. gigantea* is that it involves a pathway that incorporates enzymes of the typical clavicipitaceous branch but with festuclavine serving as the entry point into that pathway. *Claviceps africana* and *C. gigantea* contain reductase alleles of *easA* to provide them with festuclavine and specialized alleles of *cloA* that encode products capable of oxidizing festuclavine to DHLA (Arnold and Panaccione 2017; Bragg et al. 2017). Versions of CloA from lysergic acid-producing fungi were not capable of converting festuclavine to DHLA, though the festuclavine-oxidizing version of CloA from *C. africana* (Arnold and Panaccione 2017) and *C. gigantea* (Bragg et al. 2017) accepted agroclavine as well as festuclavine as substrates. Lysergyl peptide synthetases appeared to have more relaxed substrate specificity, as the LPS systems of lysergic acid amide producers *C. purpurea* (Riederer et al. 1996; Ortel and Keller 2009) and *M. brunneum* (Davis et al. 2020) accepted DHLA as substrate. In fact, LPS1 of *C. purpurea* has the same K_m for DHLA, which contains an unsaturated D ring, as it does for its natural C8,C9-unsaturated substrate lysergic acid (Riederer et al. 1996). Davis et al. (2020) engineered pathways to DHLA and the dihydro form of LAH (named dihydroLAH) in *M. brunneum* by utilizing a combination of CRISPR/Cas9 gene editing and heterologous expression. The *easA* allele from *A. fumigatus* was fused to a synthetic, intron-free version of *cloA* based on sequences from *C. africana* (Arnold and Panaccione 2017) and both were expressed in *M. brunneum* while silencing its *easA* allele. As a result, the pathway was redirected to production of dihydroLAH, a completely novel ergot alkaloid. Silencing of the *lpsB* locus during addition of *easA* and *cloA* caused the typical pathway branch to stop at DHLA, which would have otherwise served as substrate for the typical downstream enzymes.

4.2.2.3 Fumigaclavine Branch

Festuclavine represents the key intermediate in the Eurotiales-associated branches of the ergot alkaloid pathway (Fig. 4.6). Festuclavine differs from agroclavine, the key intermediate in the typical clavicipitaceous pathway, in having a fully saturated D ring. In *A. fumigatus* and certain *Penicillium* spp., festuclavine is hydroxylated at position 9 (Fig. 4.1, inset) to form fumigaclavine B (Bilovol and Panaccione 2016), and the hydroxyl group oxygen is then subsequently acetylated to form fumigaclavine A (Liu et al. 2009a, b) (Fig. 4.6). In *A. fumigatus* only, reverse prenylation of fumigaclavine A at position 2 results in the formation of fumigaclavine C (Unsöld and Li 2006) (Fig. 4.6). Fumigaclavines may serve as another source of pharmaceuticals as fumigaclavine C possesses anti-atherogenic, anti-inflammatory, antitumor properties (Wu et al. 2005; Zhao et al. 2010; Li et al. 2013; Du et al. 2011). Positions 9 and 2 are not the sites of activity for any known enzyme in the pathway branches for clavicipitaceous fungi. In the clavicipitaceous fungi, oxidations occur at position 17 and, in some cases, position 10 (Fig. 4.1, inset).

Penicillium commune and several other *Aspergillus* and *Penicillium* species that produce ergot alkaloids appear to have their pathways stopped at fumigaclavine A as they lack a functional copy of the reverse prenyl transferase FgaPT1 (Robinson and Panaccione 2012; Martín et al. 2017). Interestingly, the same set of genes also is found in *Penicillium roqueforti*, which produces isofumigaclavines A and B (stereoisomers of fumigaclavines A and B), but the genes are split into two clusters encoded by separate areas of the genome (Fernández-Bodega et al., 2017).

4.2.2.4 Rugulovasine Branch

Rugulovasines A and B are a stereoisomeric pair of clavines originally isolated from several *Penicillium* species including *P. rugulosum* and *P. islandicum* (Abe et al. 1969; Dorner et al. 1980) and exhibit hypotensive and toxic properties in mammals (Nagaoka et al. 1972). An aldehyde dehydrogenase, encoded by *easQ*, and a version of the dioxygenase encoded by *easH* have been genetically associated with roles in the formation of rugulovasines in *Penicillium biforme* (Fabian et al. 2018). Biosynthesis of rugulovasines in *Penicillium* species is hypothesized to begin with the formation of chanoclavine-I aldehyde and its spontaneous isomerization to isochanoclavine-I aldehyde (Fabian et al. 2018). Chanoclavine-I aldehyde can undergo keto-enol tautomerization to allow for nonenzymatic rotation around the alpha carbon that results from the shifting of carbon–carbon double bonds (Coyle et al. 2010) (Fig. 4.3). This rotation forms isochanoclavine-I aldehyde and allows for spontaneous formation of the Schiff base that results in the iminium ion that is usually reduced by the reductase encoded by *easG*. When EasQ is present without EasA or EasG, isochanoclavine-I aldehyde is hypothesized to be converted into isochanoclavine-I acid instead (Fabian et al. 2018). This product may then serve as substrate for EasH to act upon carbon 10 (Fig. 4.1, inset) to form 10-hydroxychanoclavine-I acid (Fig. 4.7). Ultimately, the formation of carboxyl and hydroxyl groups within proximity to each other would allow for spontaneous dehydration, forming a furan ring and generating rugulovasine A (Fig. 4.7). Rugulovasine A has been demonstrated to interconvert with rugulovasine B in polar solvents (Cole et al. 1976; Rebek et al. 1984).

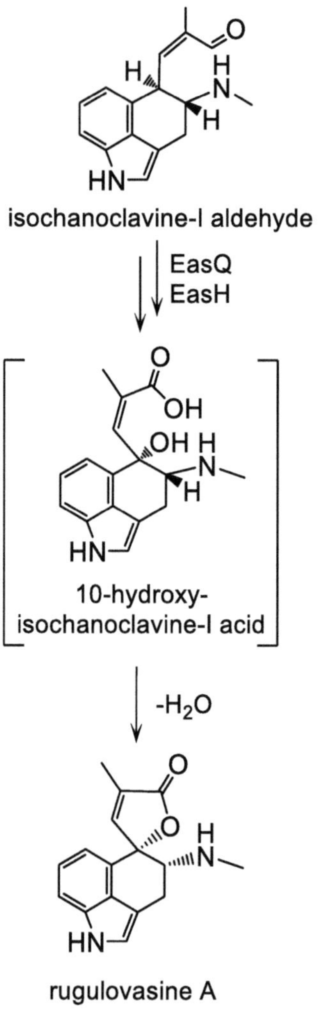

Fig. 4.7 Rugulovasine branch of the ergot alkaloid pathway found in *Penicillium biforme*. The individual products of EasQ and EasH activity are not known, so their activities are suggested as one bracketed step

4.2.2.5 Cycloclavine Branch

Cycloclavine is a unique ergot alkaloid containing a cyclopropyl ring moiety (Fig. 4.8). This ergot alkaloid was originally isolated from seeds of the morning glory *Ipomoea hildebrandtii* (Stauffacher et al. 1969), which were later shown to be symbiotic with an ergot alkaloid-producing species of *Periglandula* (Beaulieu et al. 2015). The same compound was later found in the fungus *Aspergillus japonicus* (Furuta et al. 1982).

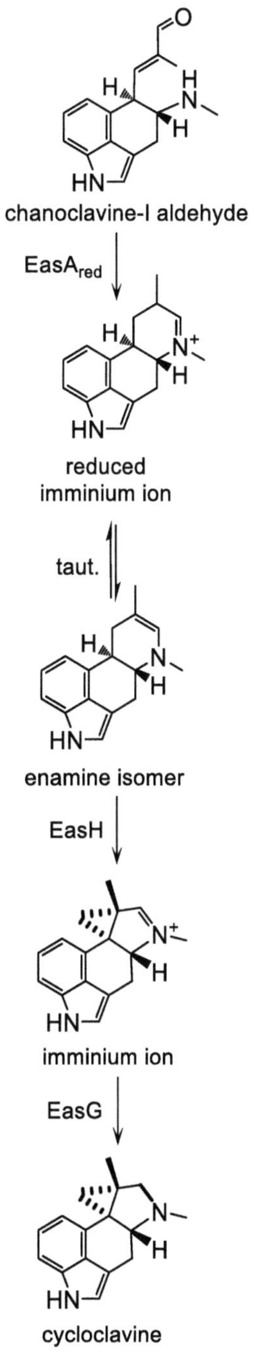

Fig. 4.8 Cycloclavine branch of the ergot alkaloid pathway found in specific *Periglandula* species-infected morning glories and *A. japonicus*. *red* reductase, *taut* tautomerization

Jakubczyk et al. (2015, 2016) determined that cycloclavine production in *A. japonicus* is dependent upon a version of EasH acting as a dioxygenase in combination with the reductase version of EasA and EasG to catalyze formation of a cyclopropyl ring moiety attached to ring D of the clavine ring system (Fig. 4.8). Formation of cycloclavine is hypothesized to begin with imine–enamine tautomerization of the reduced iminium ion formed by the reductase version of EasA (Jakubczyk et al. 2015; Jakubczyk et al. 2016). Jakubczyk et al. (2016) hypothesized that cyclopropyl ring formation is driven by radical cyclization following oxidative activity of EasH at carbon 10 (Fig. 4.8; also refer to Fig. 4.1, inset), but alternative explanations involving either hydroxylation or halogenation of this carbon by EasH are also presented as potential models. The version of EasH utilized by *A. japonicus* also displays a degree of promiscuity as it is capable of catalyzing asymmetric hydroxylation of other clavines such as elymoclavine and festuclavine (An et al. 2022). This result is consistent with the diversity of roles for EasH observed across species.

4.3 Potential Roles for Ergot Alkaloids in Fungi Occupying Different Ecological Niches

Ergot alkaloid-producing fungi have a long history of association with agricultural and medical problems. Two lineages of ergot alkaloid-producing fungi have been studied extensively: (1) members of the family Clavicipitaceae (order Hypocreales) such as *Claviceps*, *Epichloë*, *Periglandula*, and *Metarhizium* species and (2) several members of the Eurotiales including the common saprotroph and opportunistic human pathogen *A. fumigatus* and several related *Aspergillus* and *Penicillium* species. The ergot alkaloid producers display trophic strategies that range from soil saprotrophs (*Aspergillus* and *Penicillium* spp.), to plant parasites (*Claviceps* spp.), to mutualistic plant symbionts (*Epichloë* and *Periglandula* spp.). Other species even display diversity among their trophic strategies as with *Metarhizium* species, which can live as both mutualistic root symbionts or entomopathogens.

4.3.1 Ergot Alkaloid Producers in the Clavicipitaceae

4.3.1.1 *Claviceps* Species

Historically, the most widely studied ergot fungus is *C. purpurea*, a pathogen of rye (*Secale cereale*) and other wild and cultivated grasses. Like other ergot pathogens in the genus *Claviceps*, *C. purpurea* infects flowers of grasses and can replace the developing seed with a hardened fungal overwintering structure known in mycological terms as a sclerotium (plural sclerotia) or more casually as an ergot. The fungus has a life cycle with an active infective stage restricted to flowers of grasses and an overwintering stage in its sclerotial form (reviewed in detail by Luttrell 1981; Tudzynski and Scheffer 2004; Haarmann et al. 2009). The initial infection results from ascospores that arise from perithecia embedded in stalked stroma, which develop in the spring on ergot sclerotia that had been produced during the previous year. Ascospores are actively discharged from perithecia and infect through open flowers. The fungus colonizes the floret and often produces a sugary, sticky fluid called honeydew during the early stage of infection, sometimes referred to as the sphacelial stage. Honeydew contains large numbers of conidia (the asexual, clonal spores of the fungus) that may serve as inoculum for other flowers when introduced to those flowers by insects, rain splash, or simply dripping down the infected grasses spikelet. Later in the season, as the uninfected flowers develop into seeds, the fungus in the infected florets forms the dense, typically darkly pigmented sclerotium, in which the fungus will rest idly through the fall and winter. The sclerotium is made up of densely packed hyphae of the fungus and is packed with oils and relatively high concentrations of ergot alkaloids. The ergot alkaloids most notoriously associated with *C. purpurea* are the ergopeptines, including ergotamine, ergosine, and ergocristine (Haarmann et al. 2009; Gerhards et al. 2014) as well as the lysergic acid amide ergonovine, in some isolates (Ortel and Keller 2009).

The contamination of rye and other grain crops with ergot alkaloid-containing sclerotia of *C. purpurea* caused gangrenous and convulsive toxicoses known as ergotism or St. Anthony's fire. These toxicoses plagued people who relied upon rye for a large portion of their calories until diets of affected populations became diversified and screening of grain to remove ergot sclerotia became common practice (Matossian 1989). The ability of different ergot alkaloids to act as agonists or antagonists at different monoamine neurotransmitter receptors results in deleterious effects on nervous, circulatory, reproductive, and immune systems (Pertz 1996; Gröger and Floss 1998; Pertz and Eich 1999).

Claviceps purpurea typically is considered a plant pathogen, because it reduces yields from individual plants by replacing grain with ergot sclerotia. Moreover, the presence of the fungus harmed the health of those consuming infected grain. From an ecological perspective, however, the argument has been made that *C. purpurea* protects its host plant from herbivory and thus on a field scale may improve fitness of infected populations (Lev-Yadun and Halpern 2007). Data from Wali et al. (2013) with red fescue (*Festuca rubrum*) subjected to herbivory by sheep indicated an increase is seed set, due to the protective effect of the pathogen, is plausible when large herbivores are present. These observations prompted Wali et al. (2013) to speculate that the general lack of genes conferring resistance to ergot in grasses is consistent with an overall net benefit to grasses of ergot infection.

Other species of *Claviceps* also accumulate ergot alkaloids in sclerotia that are produced in place of infected grain. These other *Claviceps* species have not been studied as intensively as *C. purpurea*, presumably because the grasses they infect are not as widely consumed by humans (e.g., *C. africana* on sorghum, *Sorghum bicolor*, which accumulates dihydroergot alkaloids) or because the diseases have limited geographical distribution (e.g., ergot of maize, *Zea mays*, caused by *C. gigantea*, which is restricted to regions of Mexico and accumulates dihydroergot alkaloids). *Claviceps fusiformis* (which infects pearl millet, *Pennisetum glaucum*, and accumulates elymoclavine), *C. paspali* (which occurs on *Paspalum* species and produces LAH and ergine), *C. gigantea* (which causes ergot of maize), and one of the three *Claviceps* species (*C. africana*) that infects sorghum each produces a characteristic profile of ergot alkaloids that differs from that produced by *C. purpurea* but has a life cycle similar to that described for *C. purpurea*.

4.3.1.2 *Epichloë* Species

From an agricultural perspective, *Epichloë* species may be the most important of the ergot alkaloid-producing fungi. These fungi are typically found in mutualistic symbioses with grasses including important species of forage and turf grasses, where the fungi grow intercellularly and nondestructively, thus earning the name "endophyte." The symbioses have been reviewed extensively (e.g., Caradus et al. 2022; Saikkonen et al. 2016; Young et al. 2013; Schardl et al. 2013; Tanaka et al. 2012). The ergot alkaloid most commonly associated with *Epichloë* species is the ergopeptine ergovaline; however, a range of ergot alkaloids including ergotryptamine, chanoclavine-I, and ergine also have been reported (Tor-Agbidye et al. 2001; Panaccione et al. 2003; Ryan et al. 2015; Schardl et al. 2013; Young et al. 2015).

In the asexual phase, these fungi are not pathogenic, but rather live symbiotically in aboveground parts of the plant. Mycelium of the fungus generally proliferates between cells in the pseudostem of the grass and to a lesser extent in the leaf blades. Eventually, mycelium of the fungi colonizes developing seeds and is transmitted vertically to the next generation seedling through infected seeds. In those species that also have a sexual state, sexual reproduction by the fungus results in eruption of the fungus through the host cuticle, damaging host tissue and preventing sexual reproduction by the host (Schardl et al. 2004). Thus, when forming sexual structures, *Epichloë* species cross the line from mutualist to antagonist.

The *Epichloë* species that produce ergot alkaloids primarily accumulate those alkaloids in their hyphae rather than in any specialized structure. Hyphae growing in different tissue of their host plant all appear to accumulate ergot alkaloids but the profile of alkaloids produced is not consistent among tissues. The endophytes have significant effects on their host plant and on other organisms that interact with their host plant as herbivores, parasites, or competitors. Grass-endophyte interactions that are of particular relevance to agriculture include the interactions between tall fescue (*Lolium arundinaceum*, synonym *Festuca arundinacea*) and *Epichloë coenophiala*, and between perennial ryegrass (*Lolium perenne*) and its endophytic fungi *Epichloë festucae* or in some cases *E. hybrida* (Schardl et al. 1994; Campbell et al. 2017). Numerous other *Epichloë* species have been described from other wild and cultivated grasses (Saikkonen et al. 2016; Schardl et al. 2013).

From an agronomic perspective, endophytic fungi are beneficial because they increase host fitness and stress tolerance. However, endophytes are associated with a variety of animal toxicoses. Symptoms include the following: weight loss or poor weight gain, low fertility, reduced lactation, inability to control body temperature, low immunity, and gangrene of the limbs and tail (Schardl and Phillips 1997; Porter and Thompson 1992; Gadberry et al. 2003). Many of the symptoms of the endophyte-associated toxicoses are typical of the symptoms associated with human poisoning by ergot alkaloids. Several carefully performed experiments have shown that ergot alkaloids produced by the endophytes are major factors in the toxicoses associated with grazing endophyte-infected grasses (e.g., Hill et al. 1994; Filipov et al. 1998; Tor-Agbidye et al. 2001; Gadberry et al. 2003; Parish et al. 2003a, b).

Roles for ergot alkaloids in endophytes have been studied from an agricultural perspective. Significance of ergot alkaloids in an agricultural setting may provide a fair reflection of their ecological role(s). Since fitness of the fungus is intimately tied to success and reproduction of its host plant, ecological benefit to *Epichloë* species for producing ergot alkaloids includes protecting their host from vertebrate and invertebrate herbivory. Indeed, effects of *Epichloë*-derived ergot alkaloid on insect herbivores have been documented in studies in which the ergot alkaloid pathway was truncated at different points by gene knockouts. Potter et al. (2008) found that a strain of *E. hybrida* producing lysergic acid derivatives in perennial ryegrass deterred herbivory by, and reduced survival of, black cutworm (*Agrotis ipsilon*) compared to strains producing only clavine ergot alkaloids. Similarly, Hudson et al. (2021), comparing genetically defined strains of *E. festucae* in perennial ryegrass, observed an increase in protection from African black beetle (*Heteronychus arator*) when lysergic acid derivatives were part of the endophyte's ergot alkaloid profile. These results are consistent with previous studies in which African black beetles were fed diets spiked with purified ergot alkaloids (Ball et al., 1997). Data from these in vitro studies indicated ergopeptines and the lysergic acid amide ergonovine served as major feeding deterrents for African black beetle.

Studies with endophyte-grass symbiota varying in chemistry due to natural differences in endophyte chemotype generally support the findings observed with the near-isogenic knockout lines. In comparing populations of sleepy grass (*Achnatherum robustum*) that were symbiotic with *Epichloë* endophytes with different ergot alkaloid profiles, Shymanovich et al. (2015) found a population containing lysergic acid amides was significantly more resistant to aphids than was a population containing an endophyte producing chanoclavine-I. When these same symbiota were tested with fall armyworm (*Spodoptera frugipeda*), however, the population containing lysergic acid amides did not outperform the other (Shymanovich and Faeth 2018). Additional studies have been conducted with two species of dung beetles (*Onthophagus taurus* and *Digitonthophagus gazella*) and varieties of tall fescue infected with ergot alkaloid-containing *E. coenophiala* compared to varieties infected with an ergot alkaloid-nonproducing *Epichloë* endophyte. In these studies, Shymanovich et al. (2020) observed a preference for oviposition in dung obtained from cattle grazing pastures comprised of the ergot alkaloid-deficient endophyte as well as better larval survival on the ergot alkaloid-deficient dung.

Evidence for ergot alkaloids protecting endophytes' hosts from vertebrate herbivory includes field studies with beef cattle reared on tall fescue infected with a strain of *E. coenophiala* incapable of producing ergot alkaloids compared to control cattle reared on typical ergot alkaloid-producing *E. coenophiala* (Parish et al. 2003a). Cattle raised on ergot alkaloid-deficient grasses had higher daily weight gain, higher dry matter intake, and used less water than those reared on ergot alkaloid-containing grasses. In a similar study with ergot alkaloid-nonproducing *E. festucae* in perennial ryegrass (Parish et al. 2003b), sheep grazing grasses containing ergot alkaloid-deficient endophyte performed better in terms of resistance to heat stress and average daily weight gain than those consuming ergot alkaloid-infested grasses. Although the endophytes tested in these studies may differ in traits beyond the ergot alkaloid pathway, the results are consistent with demonstrated physiological effects of ergot alkaloids and provide evidence for ergot alkaloids as deterrents of herbivory by mammals.

In lab-based studies with genetically defined ergot alkaloid mutants of *E. hybrida* in perennial ryegrass, clavines (primarily chanoclavine-I and ergotryptamine) deterred rabbit feeding (Panaccione et al. 2006). Accumulation of lysergic acid derivatives (ergovaline and ergine), in addition to the clavines, in endophyte strains did not increase deterrence significantly during individual feedings. The ergopeptine ergovaline, however, was the only measured ergot alkaloid that significantly increased satiety of rabbits, an effect that might decrease feeding over a longer term. This study indicates differences in activities for ergot alkaloids arising at different points in the pathway and suggests that accumulation, as opposed to efficient turnover, of early pathway alkaloids can benefit the fungus and its host.

4.3.1.3 *Periglandula* Species

Another group of ergot alkaloid-producing fungi in the Clavicipitaceae—the *Periglandula* species—has been discovered only recently. The association of ergot alkaloids with certain morning glories in the family Convolvulaceae had been a puzzle for decades (Hofmann and Tscherter 1960; Stauffacher et al. 1969). The work of Leistner and colleagues first correlated the accumulation of ergot alkaloids in certain morning glories with the presence of previously uncharacterized fungi (Kucht et al. 2004; Steiner et al. 2006; Ahimsa-Müller et al. 2007). The genus *Periglandula* was established to accommodate these fungi (Steiner et al. 2011). Only two species have been formally described to date: *Periglandula ipomoeae*, which is symbiotic with *Ipomoea asarifolia*, and *P. turbinae* from host plant *Turbina corymbosa* (Steiner et al. 2011). The genome of *P. ipomoeae* has been sequenced with DNA prepared from hyphae scraped from the foliar surfaces of its host plant (Schardl et al. 2013). Analysis of this genome revealed the presence of an ergot alkaloid biosynthetic gene cluster that closely resembles those of other ergot alkaloid-producing members of the Clavicipitaceae. *Periglandula* species have been detected in nine additional ergot alkaloid-containing members of the Convolvulaceae by PCR and gene sequencing (Ahimsa-Müller et al. 2007; Beaulieu et al. 2015). The presence of ergot alkaloids in 23 of 79 species in one survey (Eich 2008) and in 53 of 210 species in another (Beaulieu et al. 2021) indicates the presence of *Periglandula* species symbionts in approximately 200 of the estimated 800 species of morning glories (Beaulieu et al. 2021). The lysergic acid amides LAH, ergine, and ergonovine are common and abundant in *Periglandula* species, and certain species accumulate the ergopeptines ergobalansine and ergosine (Kucht et al. 2004; Ahimsa-Müller et al. 2007; Beaulieu et al. 2015; Beaulieu et al. 2021).

Periglandula species are seed transmitted and obligately associated with their host plants (Kucht et al. 2004; Steiner et al. 2006; Ahimsa-Müller et al. 2007; Beaulieu et al. 2013). In some *Periglandula*-morning glory associations, the fungus emerges from its host to form visible hyphal colonies on the adaxial side of newly opened leaves. Ergot alkaloids have been detected in most tissues of infected plants and reach particularly high concentration in seeds (Beaulieu et al. 2013; Beaulieu et al. 2015). Species of morning glories that produce larger seeds are more likely to associate with ergot alkaloid-producing *Periglandula* species, and seeds of higher mass contain higher concentrations of ergot alkaloids than lighter seeds (Beaulieu et al. 2021).

Steiner et al. (2015) extensively characterized the associations of *P. ipomoeae* with *I. asarifolia* and *P. turbinae* with *Turbina corymbosa*. Imaging by transmission electron microscopy showed the *Periglandula* species covering and penetrating the cuticles of glandular trichomes on leaf surfaces. These plant structures release lipophilic oil components that may be used as nutrients by the fungus. Based on their ability to uptake low molecular weight compounds like 5(6)-carboxyfluorescein diacetate and Nile red stain, glandular trichomes could be where fungus-produced ergot alkaloids are moved into the plant (Steiner et al. 2015). The hyphae that surround the glandular trichomes have a higher number of globular structures that stain with Nile red compared to other hyphae. These globules are hypothesized to be analogous to lipid vesicles noted in ergot alkaloid-producing hyphae of *C. purpurea*. Highest accumulation of ergot alkaloids was observed to correlate with the highest density of *Periglandula* (using the fungal sterol ergosterol as a surrogate for fungal biomass) and trichomes per unit area. The area with the highest ergot alkaloid concentration in both plant–fungus symbioses was the leaf bud. Steiner et al. (2015) noted that the hyphae of the fungi did not penetrate the epidermal cells of the host plants; however, since the fungi are seed-transmitted, they must enter the plant at some stage and grow into the developing seed.

Olaranont et al. (2022) also studied the distribution of *P. ipomoeae* in tissues of *I. asarifolia* by using several different methods. Fungus was visible to the naked eye on the adaxial surface of leaves. Based on PCR with primers detecting *dmaW* (the first gene of the ergot alkaloid pathway), the fungus was found in young folded leaves, mature leaves, flower buds, mature flowers, young seeds, and mature seeds. Fungus was also found on these plant parts by several imaging methods: calcofluor white staining with fluorescence microscopy, paraffin sectioning followed by toluidine blue staining, and scanning electron microscopy. These imaging methods were also used to detect fungus on the plant stem, floral sepals, the outer sides of petals, and ovules. The association of the fungus with glandular trichomes was again noted. Fungal presence in flowers and on seed coats is consistent with the vertical transmission of *Periglandula* into the next generation of its host plant.

Data from several studies indicate the presence of ergot alkaloid-producing *Periglandula* species protects host plants from invertebrate pests. Kaur et al. (2018) measured the success of the potato psyllid (*Bactericera cockerelli*) on different species of morning glories and the potential role of ergot alkaloid-producing *Periglandula* species with limiting that success. Five morning glory species (*Ipomoea imperati*, *Ipomoea leptophylla*, *Ipomoea pandurata*, *Ipomoea tricolor*, and *Turbina corymbosa*) contained *Periglandula* species (detected by PCR analysis for *dmaW*) and high concentrations of ergot alkaloids (as detected by LC-MS/MS). On these morning glory species, investigators observed high mortality of psyllids, with death occurring an average of 24–48 h after egg hatch and with no psyllids surviving to adulthood. Conversely, psyllids survived and developed to adulthood on seven of the nine species of Convolvulaceae that lacked detectable ergot alkaloids. Psyllid mortality may have been due to alkaloid toxicity, feeding deterrence, or some undetermined factor associated with the presence of ergot alkaloid-producing *Periglandula* species. To address this question further and to study the efficacy of ergot alkaloids against a psyllid of great economic importance, Chen et al. (2021) investigated the effects of ergot alkaloids on the Asian citrus psyllid (*Diaphorina citri*). This insect devastates citrus trees by vectoring the prokaryotic pathogen causing huanglongbing (HLB, or citrus greening disease). Foliar application of crude extracts from *Periglandula*-infected host plants (*I. imperati*, *I. leptophylla*, *I. pandurata*, *I. tricolor*, and *Turbina corymbosa*) to *Citrus reshni* and *Citrus paradise* × *Poncirus trifoliata* increased psyllid nymph mortality over a 72-h observation period. There was greater mortality in nymphs than adults; however, adults settled (landed on foliar surfaces) approximately three times less frequently on leaves treated with crude extracts than on control leaves. Foliar application of commercially available, purified ergot alkaloids (ergonovine maleate, ergocornine, ergosine, lysergol, and agroclavine) also increased nymph mortality and lowered adult settlement. The authors state that their findings may help reduce citrus crop damage, manage HLB spread, and that ergot alkaloids are beneficial to plants as protection from herbivores (Chen et al. 2021).

Periglandula-produced ergot alkaloids also have been associated with resistance to nematode damage in *I. tricolor* (Durden et al. 2019). *Ipomoea tricolor* differs from other investigated species by accumulating very high concentrations of lysergic acid amides in its roots (Beaulieu et al. 2013; Hazel and Panaccione, unpublished data). *Ipomoea tricolor* populations differing only in the presence of the undescribed *Periglandula* species were obtained by treating a subset of the plants with fungicide. Second-generation plants from fungicide treated and untreated lines were inoculated with the root gall-forming, endoparasitic nematode *Meloidogyne incognita*. Plants lacking the *Periglandula* species and its ergot alkaloids were afflicted with approximately twice the number of nematode galls on their roots than were plants containing the *Periglandula* species (Durden et al. 2019). Interestingly, in the absence of nematodes, *Periglandula* infection reduced root dry weight by a small amount. This observation suggests a trade-off to the plant of maintaining its symbiont, which is presumably outweighed by benefits of the symbiosis. Overall, the available data indicate that ergot alkaloids contribute to the ability of *Periglandula* species to serve as effective defensive mutualists for their morning glory hosts.

4.3.1.4 *Metarhizium* Species

Fungi in the genus *Metarhizium* are commonly found in soil where they act as root symbionts and insect pathogens (St. Leger and Wang 2020). Certain fungi in the genus *Metarhizium* colonize roots, providing the plant with increased mineral nutrition along with enhanced resistance to insects (Liao et al. 2014; Krell et al. 2018; Behie and Bidochka 2014; St. Leger and Wang 2020). Plant host range varies among species of

Metarhizium, with some species associating more frequently with grasses and other species favoring trees and shrubs (Sheng et al. 2022). *Metarhizium* species are also effective insect pathogens, with the ability to enter insects by use of an appressorium and release of cuticle degrading enzymes (St. Leger and Wang 2020). Once inside, *Metarhizium* species slowly kill the insect host through deployment of specialized metabolites such as destruxin, sequestering nutrients from the insect during that process, and eventually emerge and sporulate to infect other insects (Xu et al. 2016). Some commercially available insecticides contain species of *Metarhizium* as active ingredients and are used in the control of members of different insect orders. For example, Bayer BIO 1020 and Green Guard® contain species of *Metarhizium* as active ingredients and are used in the control of insects of the Coleoptera and Orthoptera, respectively (Aw and Hue 2017; Reddy et al. 2014; Han et al. 2021). Some *Metarhizium* species, such as *M. acridum*, are only capable of infecting narrow host ranges (Wang et al. 2011; St. Leger and Wang 2020). Others, such as *M. brunneum*, are generalists capable of parasitizing a range of insect orders including Coleoptera, Thysanoptera, Hemiptera, Lepidoptera, Orthoptera, Blattodea, Hymenoptera, and Diptera (Russell et al. 2010; Castrillo et al. 2011; Dogan et al. 2017; Nishi and Sato 2017). Insect cadavers that result from fungal infection nearby colonized roots are thought to help provide nitrogen to the plant in exchange for carbon (Hu and Bidochka 2021).

Among the specialized metabolites produced by several species of *Metarhizium* are ergot alkaloids, including the pharmaceutically relevant lysergic acid amides, most abundantly LAH with lesser quantities of its hydrolysis product ergine, and occasionally small quantities of ergonovine and lysergyl-alanine (Leadmon et al. 2020; Davis et al. 2020; Steen et al. 2021). These metabolites are produced through enzymes encoded in a biosynthetic gene cluster (Gao et al. 2011; Leadmon et al. 2020). A recent study in which LAH was eliminated by a CRISPR/Cas9-based gene knockout, leading to accumulation of ergonovine and lysergyl-alanine, provided evidence that LAH increased the virulence of *M. brunneum* (Steen et al. 2021). Several species of *Metarhizium* including *M. brunneum*, *M. anisopliae*, *M. robertsii*, *M. acridum*, and *M. flavoviride* produce abundant ergot alkaloids in live insects, but ergot alkaloids have not been detected in spores of producing fungi or in inoculated plant roots (Leadmon et al. 2020). This pattern of gene expression suggests tight regulation of the *eas* gene cluster of *Metarhizium* species. This distribution of ergot alkaloids is also consistent with safe application and use of *Metarhizium* species as biological control agents, which are sometimes applied in high spore densities. Both wild-type and engineered *M. brunneum* strains secrete the majority of their ergot alkaloids, a feature that is very useful for their purification from culture (Leadmon et al. 2020, Davis et al. 2020).

4.3.1.5 Other Clavicipitaceous Fungi

Some fungi in the genus *Balansia* also produce ergot alkaloids including the ergopeptine ergobalansine (Bacon et al. 1981; Powell et al. 1990; Bacon et al. 2018). *Balansia* spp. grow intercellularly in infected warm season grasses and typically produce relatively large sclerotia, which may be associated with vegetative or reproductive tissues of the host (Bacon et al. 1986). The use of ergot alkaloid-containing sclerotia of *Balansia cyperi* in obstetrics by an Amazonian culture (Lewis and Elvin-Lewis 1990) is analogous to the use of *C. purpurea* sclerotia for similar purposes in European cultures. A member of the closely related genus *Atkinsonella* contains the first four genes of the ergot alkaloid pathway and thus has the genetic capacity to produce chanoclavine-I though this product has not yet been demonstrated from the fungus in the literature (Young et al. 2015).

4.3.2 Ergot Alkaloid Producers in the Eurotiales

4.3.2.1 *Aspergillus fumigatus*

Aspergillus fumigatus is a fungus that exists in nature primarily as a free-living saprotroph on decaying vegetation. It accumulates high concentrations of ergot alkaloids on its very small and highly buoyant asexual spores (conidia) (Panaccione 2005; Panaccione and Coyle 2005; Coyle et al. 2007). Despite its natural existence as a saprotroph, *A. fumigatus* is preadapted to cause disease in animals, particularly in individuals with compromised immune systems. In fact, *A. fumigatus* is the most common airborne fungal pathogen of humans (Latgé 1999).

The ability of *A. fumigatus* to cause disease in animals and humans may be considered an unfortunate accident, since infection of humans does not allow the fungus to complete its life cycle and thus is of no apparent ecological benefit to the fungus. Its pathogenic abilities may be due in part to its extremely small and buoyant spores, which may be inhaled deeply into the lungs, its ability to grow at 37 °C, and its ability to sometimes avoid destruction by cells of the innate immune system. In individuals whose acquired immune system has been suppressed by drugs (such as the organ transplant drug cyclosporine) or virus infection (most notably HIV), the fungus can grow invasively in the lungs resulting in invasive aspergillosis. The infection may also spread from the lungs into other organs and tissues. In immunocompromised individuals, invasive aspergillosis is almost always fatal (Latgé 1999).

Aspergillus fumigatus is also an agriculturally important fungus. It can grow in silage and in wet hay, causing spoilage of these feeds (Cole et al. 1977; Dutkiewicz et al. 1989; Melo dos Santos et al. 2002; Sanchez-Monedero and Stentiford 2003). Its thermotolerance and good saprotrophic abilities also allow the fungus to grow in compost. Proliferation of the fungus at large municipal composting facilities can be a health concern (Fischer and Dott 2003). The abundant spores that may be produced in composting facilities or agricultural conditions can cause clinical syndromes separate from invasive aspergillosis.

A role for fumigaclavines, the class of ergot alkaloids produced by *A. fumigatus*, in success of the fungus in its natural habitat or infected animals has only recently been studied. Studies with the insect *Galleria mellonella*, which serves as model for fungal pathogenesis of animals, indicate that fumigaclavine C is one of the preadaptations that help the fungus cause disease in animals. Using strains in which genes at different points in the ergot alkaloid pathway had been knocked out, Panaccione and Arnold (2017) found that strains with an intact fumigaclavine pathway, and thus accumulating fumigaclavine C, were significantly more virulent to *G. mellonella* than were strains accumulating the intermediates festuclavine or chanoclavine-I. Strains accumulating any of these ergot alkaloids were more virulent than strains lacking ergot alkaloids entirely. Earlier studies also indicated a link between fumigaclavine C and virulence in *G. mellonella* (O'Hanlon et al. 2012). These researchers identified two peptide synthetase genes, *pes1* and *pesL*, that were required for fumigaclavine C accumulation in *A. fumigatus*. This result was unexpected because in vitro studies had shown reverse the prenyl transferase encoded by *easL* to be sufficient for conversion of fumigaclavine A to fumigaclavine C (Unsöld and Li 2006). Mutants in which *pesL* was knocked out lacked fumigaclavine C and had lower virulence to *G. mellonella*; the relative virulence of *pes1* mutants varied by strain background (O'Hanlon et al. 2012). In addition to the loss of fumigaclavine C, differences in fumitremorgin concentrations and sensitivity to antifungal compounds were observed in *pes1* and *pesL* knockout strains (Reeves et al. 2006; O'Hanlon et al. 2012).

4.3.2.2 Lysergic Acid Amide-Producing Species of *Aspergillus*

Three species of *Aspergillus* were recently discovered to produce the lysergic acid amide LAH (Jones et al. 2021). This observation was surprising because lysergic acid amides had not previously been detected in fungi outside the family Clavicipitaceae and because the pathway steps required to put lysergic acid into LAH appeared to have evolved independently in the *Aspergillus* species as compared to the LAH producers in the Clavicipitaceae.

Concerning the ecology of the three LAH-producing *Aspergillus* species, *A. hancockii* has been found in soil in Australia and on dried peas in that same country (Pitt et al. 2017). *Aspergillus hancockii* has fair pathogenic potential when inoculated into larvae of the model insect *G. mellonella* (Jones and Panaccione 2023). *Aspergillus homomorphus* has been isolated from soil in Israel (Steiman et al. 1995) and from millet grain in Yemen (Hussein et al. 2017). This species had very limited pathogenic potential in insects (Jones and Panaccione 2023). The literature lacks significant other information on the ecology of these two species. More ecological information, however, has been published on the third LAH-producing *Aspergillus* species, *A. leporis*. *Aspergillus leporis* was originally found in fresh dung of the white-tailed jackrabbit (States and Christensen 1966) and later from several soil samples in the arid western USA (States and Christensen 2001). Isolates of the fungus from rhizospheres of selenium-hyperaccumulating plants growing in the western USA were tolerant of high concentrations of selenium (Wangeline et al. 2011). Association of a selenium-tolerant isolate of *A. leporis* with roots of the selenium-nonaccumulator *Stanleya elata* resulted in higher selenium accumulation in the plants' roots. Another ecological niche that *A. leporis* is at least preadapted to fill is that of insect pathogen. Based on the similarity of its ergot alkaloid profile to that of the insect pathogens in the genus *Metarhizium*, Jones and Panaccione (2023) tested the ability of *A. leporis* to infect the model insect *G. mellonella*. Larvae of the insect injected with spores of the fungus died at a relatively rapid rate. The fungus emerged from insect cadavers and completed its life cycle by conidiating. When larvae were inoculated topically, the rate of infection was lower than in injection trials but still greater than controls. Ergot alkaloids accumulated in larvae to levels comparable to those observed in larvae infected with *M. brunneum*. Elimination of LAH by knocking out a gene early in the ergot alkaloid pathway (*easD* in this case) reduced the virulence of the fungus. The association of *A. leporis* with plant rhizosphere, its LAH-rich ergot alkaloid profile, and its ability to infect insects parallel the ecology and chemistry of its distant relative in the Clavicipitaceae, *M. brunneum*.

4.3.2.3 Other Ergot Alkaloid Producers in the Eurotiales

Several other *Aspergillus* species and a few *Penicillium* species have been reported to produce ergot alkaloids (Flieger et al. 1997; Kozlovsky 1999; Boichenko et al. 2001). Among the *Penicillium* species, the production of fumigaclavine A by *P. commune* is noteworthy because this fungus is a saprotroph commonly associated with spoilage of dairy products and is the fungus most commonly associated with contamination of cheese (Lund et al. 1995, 2003). Another cheese-associated ergot alkaloid producer, *P. roqueforti*, is commonly used in the production of blue cheeses (e.g., Roquefort, Stilton, and Gorgonzola), though it produces low quantities of its mycotoxins, including isofumigaclavine A, in this environment (Scott 1981). The species *P. camemberti* is a domesticated *Penicillium* used for Camembert and Brie cheese production and is a close relative of the rugulovasine-producing *P. biforme*. Fabian et al. (2018) speculated that domestication of *P. camemberti* may have contributed to the loss of its ability to produce rugulovasines due to differences in environmental conditions and selection pressure. *Aspergillus japonicus* (synonym: *Aspergillus violaceofuscus*) is a member of the industrial workhorse group known as black Aspergilli (section *Nigri*) and has been isolated from both marine and soil environments (Varga et al. 2011). The cycloclavine produced by *A. japonicus* possesses insecticidal activities and has potential applications in agriculture (Körber et al. 2014).

4.3.3 Ergot Alkaloid Producers in Other Taxa

This review has focused on ergot alkaloids produced by members of the family Clavicipitaceae (order Hypocreales) and by several *Aspergillus* and *Penicillium* species of the order Eurotiales. Evidence of ergot alkaloids or the genetic potential for ergot alkaloid biosynthesis also has been found outside these taxa.

Several fungi in the ecologically important order Xylariales have been either demonstrated to produce ergot alkaloids or appear to have the biosynthetic capacity to produce ergot alkaloids based on genomic DNA sequence. The fungus *Dicyma* sp. (Xylariales) produces a unique ergopeptine with its tripeptide side chain derived from alanine, leucine, and phenylalanine (Vázquez et al. 2003). Fruiting bodies of *Xylaria nigripes* have been reported to contain agroclavine and three novel ergot alkaloids with unsaturations in the C ring (Hu and Li 2017). Several additional members of the Xylariales have gene clusters typical of those required for ergot alkaloid synthesis (Kuhnert et al. 2021) but have not yet been associated with specific ergot alkaloids. Based on these data, the order Xylariales is very likely to contain several additional ergot alkaloid producers.

Analyses of whole genome sequence data have led to predictions of ergot alkaloid synthesis gene clusters in several species in the Onygenales (Gerhards et al. 2015), though chemical evidence of ergot alkaloid production by these species is still pending. The order Onygenales is noteworthy for containing numerous species that are accidental or opportunistic pathogens of humans and other animals.

Novel derivatives of festuclavine called pibocin A and B also have been reported from the sea squirt *Eudistoma* sp. (Makarieva et al. 1999, 2001). The possibility that a marine fungus symbiotic with the sea squirt is the source of the compounds may be something to consider further. Altogether these observations suggest that numerous other organisms capable of producing ergot alkaloids remain to be discovered and that additional chemical analyses and bioinformatic analyses will lead to discovery of such species.

4.4 Conclusions

The results of biochemical, molecular genetic, and genomic studies have successfully been combined to answer important questions about the biosynthesis and function of ergot alkaloids. Comparative genomic analyses along with biochemical or molecular genetic characterization of individual genes indicate that ergot alkaloids produced by fungi in the Clavicipitaceae and Eurotiales have a common biosynthetic and genetic origin (e.g., Florea et al. 2017; Jones et al. 2021). After the early, shared steps, the pathways of different ergot alkaloid-producing fungi diverge to follow different terminal branches, resulting in unique profiles of ergot alkaloids. Despite the common origin, the last few steps of the lysergic acid amide branch appear to have evolved independently in the three *Aspergillus* species as compared to the ergot alkaloid producers in the Clavicipitaceae (Jones et al. 2021).

Our knowledge of the distribution of ergot alkaloids among fungi has expanded considerably due to simple bioinformatic searching approaches based on the identification of the genes involved in the pathway and the wealth of sequenced fungal genomes that are available to be searched. Our awareness of new sources of ergot alkaloids and perhaps even new branches of the ergot alkaloid pathway will likely increase from continued analysis of the constantly expanding library of novel fungal genome sequences.

Our understanding of the ecological importance that ergot alkaloids have to the fungi that produce them has increased with the discovery of ergot alkaloids in new sources and with the ability to genetically manipulate the ergot alkaloid pathway to generate near isogenic strains that may be compared. Chemical ecology experiments described above have already shown that fungi in diverse ecological niches rely upon ergot alkaloids as means to succeed in certain environmental niches (e.g., as virulence factors against insects) or, in the case of mutualistic ergot alkaloid producers, protect their hosts from vertebrate or invertebrate herbivores and thus increase the fitness of their hosts and themselves. Opportunities for investigation of the roles of ergot alkaloids in fungi should continue to grow with our ability to alter the pathway and with the discovery of ergot alkaloids in new sources.

Acknowledgments The work in the authors' laboratory was supported by grant 2R15-GM114774 from the United States National Institute of General Medical Sciences and project NC1183 from the United States Department of

Agriculture National Institute of Food and Agriculture. Abigail Jones is a Beckman Scholar supported by the Arnold and Mabel Beckman Foundation. This chapter is published with the approval of the Director of the WV Agricultural and Forestry Experiment Station.

References

Abe M, Ohmomo S, Ohashi T, Tabuchi T (1969) Isolation of chanoclavine-(I) and two new interconvertible alkaloids, rugulovasine A and B, from the cultures of *Penicillium concavo-rugulosum*. Ag Biolog Chem 33:469–471

Agurell S, Ramstad E (1965) A new ergot alkaloid from Mexican maize ergot. Acta Pharm Suecica 2:231–238

Ahimsa-Müller MA, Markert A, Hellwig S, Knoop V, Steiner U, Drewke C, Leistner E (2007) Clavicipitaceous fungi associated with ergoline alkaloid-containing Convolvulaceae. J Nat Prod 70:1955–1960

An C, Zhu F, Yao Y, Zhang K, Wang W, Zhang J, Wei G, Xia Y, Gao Q, Gao SS (2022) Beyond the cyclopropyl ring formation: fungal Aj_EasH catalyzes asymmetric hydroxylation of ergot alkaloids. Appl Microbiol Biotechnol 106:2981–2991

Arnold SL, Panaccione DG (2017) Biosynthesis of the pharmaceutically important fungal ergot alkaloid dihydrolysergic acid requires a specialized allele of *cloA*. Appl Environ Microbiol 83:e00805–e00817

Aw KMS, Hue SM (2017) Mode of infection of *Metarhizium* spp. fungus and their potential as biological control agents. J Fungi 3:30

Bacon CW, Porter JK, Robbins JD (1981) Ergot alkaloid biosynthesis by isolates of *Balansia epichloë* and *B. henningsiana*. Can J Bot 59:2534–2538

Bacon CW, Lyons PC, Porter JK, Robbins JD (1986) Ergot toxicity from endophyte-infected grasses: a review. Agron J 78:106–116

Bacon CW, Hinton DM, Mitchell TR, Palencia ER (2018) In situ ergot alkaloid detection in three *Balansia epichloe*-infected grass species. J Appl Microbiol 125:976–985

Ball OJ, Miles CO, Prestidge RA (1997) Ergopeptine alkaloids and *Neotyphodium lolii*-mediated resistance in perennial ryegrass against adult *Heteronychus arator* (Coleoptera: Scarabaeidae). J Econ Entomol 90:1382–1391

Barrow KD, Mantle PG, Quigley FR (1974) Biosynthesis of dihydroergot alkaloids. Tet Lett 16:1557–1560

Beaulieu WT, Panaccione DG, Hazekamp CS, Mckee MC, Ryan KL, Clay K (2013) Differential allocation of seed-borne ergot alkaloids during early ontogeny of morning glories (Convolvulaceae). J Chem Ecol 39:919–930

Beaulieu WT, Panaccione DG, Ryan KL, Kaonongbua W, Clay K (2015) Phylogenetic and chemotypic diversity of *Periglandula* species in eight new morning glory hosts (Convolvulaceae). Mycologia 107:667–678

Beaulieu WT, Panaccione DG, Quach QN, Smoot KL, Clay K (2021) Diversification of ergot alkaloids and heritable fungal symbionts in morning glories. Commun Biol 4:1362

Behie SW, Bidochka MJ (2014) Nutrient transfer in plant-fungal symbioses. Trends Plant Sci 19:734–740

Béliveau J, Ramstad E (1967) 8-Hydroxylation of agroclavine and elymoclavine by fungi. Llyodia 29:234–238

Bilovol Y, Panaccione DG (2016) Functional analysis of the gene controlling hydroxylation of festuclavine in the ergot alkaloid pathway of *Neosartorya fumigata*. Curr Genet 62:853–860

Boichenko LV, Boichenko DM, Vinokurova NG, Reshetilova TA, Arinbasarov MU (2001) Screening for ergot alkaloid producers among microscopic fungi by means of the polymerase chain reaction. Microbiology 70:30–310 (translated from Mikrobiologiya 70:36–364

Bragg PE, Maust MD, Panaccione DG (2017) Ergot alkaloid biosynthesis in the Maize (*Zea mays*) ergot fungus *Claviceps gigantea*. J Agric Food Chem 65:10703–10710

Britton K, Steen CR, Davis KA, Sampson JK, Panaccione DG (2022) Contribution of a novel gene to lysergic acid amide synthesis in *Metarhizium brunneum*. BMC Res Notes 15:183

Campbell MA, Tapper BA, Simpson WR, Johnson RD, Mace W, Ram A, Lukito Y, Dupont PY, Johnson LJ, Scott DB, Ganley AR (2017) *Epichloë hybrida*, sp. nov., an emerging model system for investigating fungal allopolyploidy. Mycologia 109:715–729

Caradus JR, Card SD, Finch SC, Hume DE, Johnson LJ, Mace WJ, Popay AJ (2022) Ergot alkaloids in New Zealand pastures and their impact. New Zealand J Agri Res 65:1–41

Castrillo LA, Griggs MH, Ranger CM, Reding ME, Vandenberg JD (2011) Virulence of commercial strains of *Beauveria bassiana* and *Metarhizium brunneum* (Ascomycota: Hypocreales) against adult *Xylosandrus germanus* (Coleoptera: Curculionidae) and impact on brood. Biol Control 58:121–126

Chen X-D, Kaur N, Horton DR, Cooper WR, Qureshi JA, Stelinski LL (2021) Crude extracts and alkaloids derived from *Ipomoea-Periglandula* symbiotic association cause mortality of Asian citrus psyllid *Diaphorina citri* Kuwayama (Hemiptera: Psyllidae). Insects 12:929

Cheng JZ, Coyle CM, Panaccione DG, O'Connor SE (2010a) A role for old yellow enzyme in ergot alkaloid biosynthesis. J Am Chem Soc 132:1776–1777

Cheng JZ, Coyle CM, Panaccione DG, O'Connor SE (2010b) Controlling a structural branch point in ergot alkaloid biosynthesis. J Am Chem Soc 132:12835–12837

Cole RJ, Kirksey JW, Clardy J, Eickman N, Weinreb SM, Singh P, Kim D (1976) Structures of rugulovasine-A and-B and 8-chlororugulovasine-A and -B. Tetrahedron Lett 17:3849–3852

Cole RJ, Kirksey JW, Dorner JW, Wilson DM, Johnson JC Jr, Johnson AN, Bedell DM, Springer JP, Chexal KK,

Clardy JC, Cox RH (1977) Mycotoxins produced by *Aspergillus fumigatus* species isolated from molded silage. J Agric Food Chem 25:826–830

Correia T, Grammel N, Ortel I, Keller U, Tudzynski P (2003) Molecular cloning and analysis of the ergopeptine assembly system in the ergot fungus *Claviceps purpurea*. Chem Biol 10:1281–1292

Coyle CM, Panaccione DG (2005) An ergot alkaloid biosynthesis gene and clustered hypothetical genes from *Aspergillus fumigatus*. Appl Environ Microbiol 71:3112–3118

Coyle CM, Kenaley SC, Rittenour WR, Panaccione DG (2007) Association of ergot alkaloids with conidiation in *Aspergillus fumigatus*. Mycologia 99:80–811

Coyle CM, Cheng JZ, O'Connor SE, Panaccione DG (2010) An old yellow enzyme gene controls the branch point between *Aspergillus fumigatus* and *Claviceps purpurea* ergot alkaloid pathways. Appl Environ Microbiol 76:3898–3903

Davis KA, Sampson JK, Panaccione DG (2020) Genetic reprogramming of the ergot alkaloid pathway of *Metarhizium brunneum*. Appl Environ Microbiol 86:e01251–e01220

Dogan YO, Hazir S, Yildiz A, Butt TM, Cakmak I (2017) Evaluation of entomopathogenic fungi for the control of *Tetranychus urticae* (Acari: Tetranychidae) and the effect of *Metarhizium brunneum* on the predatory mites (Acari: Phytoseiidae). Biol Control 111:6–12

Dorner JW, Cole RJ, Hill R, Wicklow D, Cox RH (1980) *Penicillium rubrum* and *Penicillium biforme*, new sources of rugulovasines A and B. Appl Environ Microbiol 40:685–687

Du RH, Li EG, Cao Y, Song YC, Tan RX (2011) Fumigaclavine C inhibits tumor necrosis factor α production via suppression of toll-like receptor 4 and nuclear factor κB activation in macrophages. Life Sci 89:235–240

Durden L, Wang D, Panaccione D, Clay K (2019) Decreased root-knot nematode gall formation in roots of the morning glory *Ipomoea tricolor* symbiotic with ergot alkaloid-producing fungal *Periglandula* sp. J Chem Ecol 45:879–887

Dutkiewicz J, Olenchock SA, Sorenson WG, Gerencser VF, May JJ, Pratt DS, Robinson VA (1989) Levels of bacteria, fungi, and endotoxin in bulk and aerosolized corn silage. Appl Environ Microbiol 55:1093–1099

Eich E (2008) Solanaceae and Convolvulaceae: secondary metabolites: biosynthesis, chemotaxonomy, biological and economic significance. Springer-Verlag, Berlin, Germany, pp 213–259

Fabian SJ, Maust MD, Panaccione DP (2018) Ergot alkaloid synthesis capacity of *Penicillium camemberti*. Appl Environ Microbiol 84:e01583–e01518

Fernández-Bodega Á, Álvarez-Álvarez R, Liras P, Martín JF (2017) Silencing of a second dimethylallyltryptophan synthase of *Penicillium roqueforti* reveals a novel clavine alkaloid gene cluster. Appl Microbiol Biotechnol 101:6111–6121

Filipov NM, Thompson FN, Hill NS, Dawe DL, Stuedemann JA, Price JC, Smith CK (1998) Vaccination against ergot alkaloids and the effect of endophyte-infected fescue seed-based diets on rabbits. J Anim Sci 76:2456–2463

Fischer G, Dott W (2003) Relevance of airborne fungi and their secondary metabolites for environmental, occupational and indoor hygiene. Arch Microbiol 179:75–82

Fleetwood DJ, Scott B, Lane GA, Tanaka A, Johnson RD (2007) A complex ergovaline gene cluster in *Epichloë* endophytes of grasses. Appl Environ Microbiol 73:2571–2579

Flieger M, Sedmera P, Vokoun J, Řičicová A, Řeháček Z (1982) Separation of four isomers of lysergic acid α-hydroxyethylamide by liquid chromatography and their spectroscopic identification. J Chromatogr 236:441–452

Flieger M, Wurst M, Shelby R (1997) Ergot alkaloids–sources, structures and analytical methods. Folia Microbiol 42:3–30

Florea S, Panaccione DG, Schardl CL (2017) Ergot alkaloids of the family Clavicipitaceae. Phytopathology 107:504–518

Floss HG, Hornemann U, Schilling N, Kelley K, Groeger D, ERge D (1968) Biosynthesis of ergot alkaloids. Evidence for two isomerizations in the isoprenoid moiety during the formation of tetracyclic ergolines. J Am Chem Soc 90:6500–6507

Furuta T, Koike M, Abe M (1982) Isolation of cycloclavine from the culture broth of *Aspergillus japonicus* SAITO. Agric Biol Chem 46:1921–1922

Gadberry MS, Denard TM, Spiers DE, Piper EL (2003) Effects of feeding ergovaline on lamb performance in a heat stress environment. J Anim Sci 81:1538–1545

Gao Q, Jin K, Ying SH, Zhang Y, Xiao G, Shang Y, Duan Z, Hu X, Xie XQ, Zhou G, Peng G (2011) Genome sequencing and comparative transcriptomics of the model entomopathogenic fungi *Metarhizium anisopliae* and *M. acridum*. PLoS Genet 7:e1001264

Gerhards N, Neubauer L, Tudzynski P, Li SM (2014) Biosynthetic pathways of ergot alkaloids. Toxins (Basel) 6:3281–3295

Gerhards N, Matuschek M, Wallwey C, Li SM (2015) Genome mining of ascomycetous fungi reveals their genetic potential for ergot alkaloid production. Arch Microbiol 197:701–713

Goetz KE, Coyle CM, Cheng JZ, O'Connor SE, Panaccione DG (2011) Ergot cluster-encoded catalase is required for synthesis of chanoclavine-I in *Aspergillus fumigatus*. Curr Genet 57:201–211

Gröger D, Floss HG (1998) Biochemistry of ergot alkaloids–achievements and challenges. Alkaloids 50:171–218

Haarmann T, Machado C, Lübbe Y, Correia T, Schardl CL, Panaccione DG, Tudzynski P (2005) The ergot alkaloid gene cluster in *Claviceps purpurea*: extension of the cluster sequence and intra species evolution. Phytochemistry 66:1312–1320

Haarmann T, Ortel I, Tudzynski P, Keller U (2006) Identification of the cytochrome P450 monooxygenase that bridges the clavineand ergoline alkaloid pathways. Chembiochem 7:645–652
Haarmann T, Rolke Y, Giesbert S, Tudzynski P (2009) Ergot: from witchcraft to biotechnology. Mol Plant Pathol 10:563–577
Han JO, Naeger NL, Hopkins BK, Sumerlin D, Stamets PE, Carris LM, Sheppard WS (2021) Directed evolution of *Metarhizium* fungus improves its biocontrol efficacy against *Varroa* mites in honey bee colonies. Sci Rep 11:10582
Havemann J, Vogel D, Loll B, Keller U (2014) Cyclolization of D-lysergic acid alkaloid peptides. Chem Biol 21:146–155
Hill NS, Thompson FN, Dawe DL, Stuedemann JA (1994) Antibody binding of circulating ergot alkaloids in cattle grazing tall fescue. Am J Vet Res 55:419–424
Hofmann A (1980) LSD–my problem child. McGraw-Hill, New York, NY
Hofmann A, Tscherter H (1960) Isolierung von Lysergsäure-Alkaloiden aus der mexikanischen Zauberdroge Ololiuqui (*Rivea corymbosa* (L.) Hall. f.). Experientia 16:414
Hu D, Li M (2017) Three new ergot alkaloids from fruiting bodies of *Xylaria nigripes* (KL.) Sacc. Chem Biodivers 14:e1600173
Hu S, Bidochka MJ (2021) Root colonization by endophytic insect-pathogenic fungi. J Appl Microbiol 130:570–581
Hudson D, Mace W, Popay A, Jensen J, McKenzie C, Cameron C, Johnson R (2021) Genetic manipulation of the ergot alkaloid pathway in *Epichloë festucae* var. *lolii* and its effect on black beetle feeding deterrence. Toxins 13:76
Hussein N, Abdel-Hafez SI, Abdel-Sater MA, Ismail MA, Eshraq AL (2017) *Aspergillus homomorphus*, a first global record from millet grains. Curr Res Environ Appl Mycol 7:82–89
Jakubczyk D, Caputi L, Hatsch A, Nielsen CA, Diefenbacher M, Klein J, Molt A, Schröder H, Cheng JZ, Naesby M, O'Connor SE (2015) Discovery and reconstitution of the cycloclavine biosynthetic pathway–enzymatic formation of a cyclopropyl group. Angew Chem Int Ed Engl 54:5117–5121
Jakubczyk D, Caputi L, Stevenson CE, Lawson DM, O'Connor SE (2016) Structural characterization of EasH (*Aspergillus japonicus*)—an oxidase involved in cycloclavine biosynthesis. Chem Commun 52:14306–14309
Jones AM, Panaccione DG (2023) Ergot alkaloids contribute to the pathogenic potential of the fungus *Aspergillus leporis*. Appl Environ Microbiol 89:e00415–e00423
Jones AM, Steen CR, Panaccione DP (2021) Independent evolution of a lysergic acid amide in *Aspergillus* species. Appl Environ Microbiol 87:e01801–e01821
Kaur N, Cooper WR, Duringer JM, Badillo-Vargas IE, Esparza-Dıaz G, Rashed A et al (2018) Survival and development of potato psyllid (Hemiptera: Triozidae) on Convolvulaceae: effects of a plant-fungus symbiosis (Periglandula). PLoS One 13:e0201506
Kleinerova E, Kybal J (1973) Ergot alkaloids IV. Contribution to the biosynthesis of lysergic acid amides. Folia Microbiol 18:390–392
Körber K, Song D, Rheinheimer J, Kaiser F, Dickhaut J, Narine A, Culbertson DL, Thompson S, Rieder J (2014) Cycloclavine and derivatives thereof for controlling invertebrate pests. WIPO WO2014096238A1
Kozlovsky AG (1999) Producers of ergot alkaloids out of *Claviceps* genus. In: Kren V, Cvak L (eds) Ergot: the genus *Claviceps*. Harwood Academic Publishers, The Netherlands, pp 479–499
Krell V, Unger S, Jakobs-Schoenwandt D, Patel AV (2018) Endophytic *Metarhizium brunneum* mitigates nutrient deficits in potato and improves plant productivity and vitality. Fungal Ecol 34:43–49
Kucht S, Groβ J, Hussein Y, Grothe T, Keller U, Basar S, Konig WA, Steiner U, Leistner E (2004) Elimination of ergoline alkaloids following treatment of *Ipomoea asarifolia* (Convolvulaceae) with fungicides. Planta 219:619–625
Kuhnert E, Navarro-Muñoz JC, Becker K, Stadler M, Collemare J, Cox RJ (2021) Secondary metabolite biosynthetic diversity in the fungal family Hypoxylaceae and *Xylaria hypoxylon*. Stud Mycol 99:100118
Latgé J-P (1999) *Aspergillus fumigatus* and aspergillosis. Clin Microbiol Rev 12:310–350
Leadmon CE, Sampson JK, Maust MD, Macias AM, Rehner SA, Kasson MT, Panaccione DG (2020) Several *Metarhizium* species produce ergot alkaloids in a condition-specific manner. Appl Environ Microbiol 86:e00373–e00320
Lee S, Na SH, Chung YA, Jeong H, Song IU (2019) Efficacy of nicergoline treatment in Parkinson's disease associated with dementia. J Clin Neurol 70:136–139
Lev-Yadun S, Halpern M (2007) Ergot (*Claviceps purpurea*)– an aposematic fungus. Symbiosis 43:105–108
Lewis WH, Elvin-Lewis M (1990) Obstetrical use of the parasitic fungus *Balansia cyperi* by Amazonian Jívaro women. Econ Bot 44:131–133
Li Y-X, Himaya SWA, Dewapriya P, Zhang C, Kim S-K (2013) Fumigaclavine C from a marine-derived fungus *Aspergillus fumigatus* induces apoptosis in MCF-7 breast cancer cells. Mar Drugs 11:5063–5086
Liao X, O'Brien TR, Fang W, St Leger RJ (2014) The plant beneficial effects of *Metarhizium* species correlate with their association with roots. Appl Microbiol Biotechnol 98:7089–7096
Lin HC, Chiou G, Chooi YH, McMahon TC, Xu W, Garg NK, Tang Y (2015) Elucidation of the concise biosynthetic pathway of the communesin indole alkaloids. Angew Chem 54:300–3007
Liu M, Panaccione DG, Schardl CL (2009a) Phylogenetic analyses reveal monophyletic origin of the ergot alkaloid gene *dmaW* in fungi. Evol Bioinforma 5:15–30

Liu X, Wang L, Steffan N, Yin W-B, Li S-M (2009b) Ergot alkaloid biosynthesis in *Aspergillus fumigatus*: FgaAT catalyses the acetylation of fumigaclavine B. Chembiochem 10:2325–2328

Lorenz N, Wilson EV, Machado C, Schardl CL, Tudzynski P (2007) Comparison of ergot alkaloid biosynthesis gene clusters in *Claviceps* species indicates loss of late pathway steps in evolution of *C. fusiformis*. Appl Environ Microbiol 73:7185–7191

Lorenz N, Olšovská J, Šulc M, Tudzynski P (2010) Alkaloid cluster gene *ccsA* of the ergot fungus *Claviceps purpurea* encodes chanoclavine I synthase, a flavin adenine dinucleotide-containing oxidoreductase mediating the transformation of N-methyl-dimethylallyltryptophan to chanoclavine I. Appl Environ Microbiol 76:1822–1830

Lund F, Filtenborg O, Frisvad JC (1995) Associated mycoflora of cheese. Food Microbiol 12:173–180

Lund F, Neilsen AB, Skouboe P (2003) Distribution of *Penicillium commune* isolates in cheese dairies mapped using secondary metabolite profiles, morphotypes, RAPD and AFLP fingerprinting. Food Microbiol 20:725–734

Luttrell ES (1981) Tissue replacement diseases caused by fungi. Annu Rev Phytopathol 19:373–389

Makarieva TN, Ilyin SG, Stonik VA, Lyssenko KA, Denisenko VA (1999) Pibocin, the first ergoline marine alkaloid from the Far-Eastern ascidian *Eudistoma* sp. Tetrahedron Lett 40:1591–1594

Makarieva TN, Dmitrenok AS, Dmitrenok PS, Grebnev BB, Stonik VA (2001) Pibocin B, the first N-O-methylindole marine alkaloid, a metabolite from the Far-Eastern ascidian *Eudistoma* species. J Nat Prod 64:1559–1561

Mantle PG, Waight ES (1968) Dihydroergosine: a new naturally occurring alkaloid from the sclerotia of *Sphacelia sorghi* (McRae). Nature 218:581–582

Martín JF, Álvarez-Álvarez R, Liras P (2017) Clavine alkaloids gene clusters of *Penicillium* and related fungi: evolutionary combination of prenyltransferases, monooxygenases and dioxygenases. Genes (Basel) 8:E342

Matossian MK (1989) Poisons of the past: molds, epidemics, and history. Yale University Press, New Haven, CT

Matuschek M, Wallwey C, Xie X, Li SM (2011) New insights into ergot alkaloid biosynthesis in *Claviceps purpurea*: an agroclavine synthase EasG catalyses, via a non-enzymatic adduct with reduced glutathione, the conversion of chanoclavine-I aldehyde to agroclavine. Org Biomolec Chem 9:4328–4335

Melo dos Santos V, Dorner JW, Carreira F (2002) Isolation and toxigenicity of *Aspergillus fumigatus* from moldy silage. Mycopathologia 156:133–138

Montastruc F, Moulis F, Araujo M, Chebane L, Rascol O, Montastruc J-L (2017) Ergot and non-ergot dopamine agonists and heart failure in patients with Parkinson's disease. Eur J Clin Pharmacol 73:99–103

Nagaoka A, Kikuchi K, Nagawa Y (1972) I. Pharmacological studies of new indole alkaloids, rugulovasine A and B hydrochloride. II. Hypotensive mechanism of both alkaloids in the anesthetized cats. Arzneimittelforschung 22:137–146

Nielsen CA, Folly C, Hatsch A, Molt A, Schröder H, O'Connor SE, Naesby M (2014) The important ergot alkaloid intermediate chanoclavine-I produced in the yeast *Saccharomyces cerevisiae* by the combined action of EasC and EasE from *Aspergillus japonicus*. Microb Cell Factories 13:95

Nishi O, Sato H (2017) Species diversity of the entomopathogenic fungi *Metarhizium anisopliae* and *M. flavoviride* species complexes isolated from insects in Japan. Mycoscience 58:472–479

O'Hanlon KA, Gallagher L, Schrettl M, Jöchl C, Kavanagh K, Larsen TO, Doyle S (2012) Nonribosomal peptide synthetase genes *pesL* and *pes1* are essential for fumigaclavine C production in *Aspergillus fumigatus*. Appl Environ Microbiol 78:3166–3176

Olaranont Y, Stewart AB, Songnuan W, Traiperm P (2022) How and where *Periglandula* fungus interacts with different parts of *Ipomoea asarifolia*. J Fungi 8:823

Ortel I, Keller U (2009) Combinatorial assembly of simple and complex D-lysergic acid alkaloid peptide classes in the ergot fungus *Claviceps purpurea*. J Biol Chem 284:6650–6660

Panaccione DG (2005) Origins and significance of ergot alkaloid diversity in fungi. FEMS Microbiol Lett 251:9–17

Panaccione DG, Arnold SA (2017) Ergot alkaloids contribute to virulence in an insect model of invasive aspergillosis. Sci Rep 7:8930

Panaccione DG (2023) Derivation of the multiply-branched ergot alkaloid pathway of fungi. Microb Biotechnol 16:742–756

Panaccione DG, Coyle CM (2005) Abundant respirable ergot alkaloids from the common airborne fungus *Aspergillus fumigatus*. Appl Environ Microbiol 71:3106–3111

Panaccione DG, Johnson RD, Wang J, Young CA, Damrongkool P, Scott B, Schardl CL (2001) Elimination of ergovaline from a grass-*Neotyphodium* endophyte symbiosis by genetic modification of the endophyte. Proc Natl Acad Sci USA 98:12820–12825

Panaccione DG, Tapper BA, Lane GA, Davies E, Fraser K (2003) Biochemical outcome of blocking the ergot alkaloid pathway of a grass endophyte. J Agric Food Chem 51:6429–6437

Panaccione DG, Cipoletti JR, Sedlock AB, Blemings KP, Schardl CL, Machado C, Seidel GE (2006) Effects of ergot alkaloids on food preference and satiety in rabbits, as assessed with gene knockout endophytes in perennial ryegrass (*Lolium perenne*). J Agric Food Chem 54:4582–4587

Parish JA, McCann MA, Watson RH, Paiva NN, Hoveland CS, Parks AH, Upchurch BL, Hill NS, Bouton JH (2003a) Use of nonergot alkaloid-producing endophytes for alleviating tall fescue toxicosis in stocker cattle. J Am Sci 81:2856–2868

Parish JA, McCann MA, Watson RH, Hoveland CS, Hawkins LL, Hill NS, Bouton JH (2003b) Use of nonergot alkaloid-producing endophytes for alle-

viating tall fescue toxicosis in sheep. J Anim Sci 81:1316–1322
Pertz H (1996) Naturally occurring clavines: antagonism/partial agonism at 5-HT_{2A} receptors and antagonism at α1-adrenoceptors in blood vessels. Planta Med 62:387–392
Pertz H, Eich E (1999) Ergot alkaloids and their derivatives as ligands for serotoninergic, dopaminergic, and adrenergic receptors. In: Kren V, Cvak L (eds) Ergot: the genus *Claviceps*. Harwood Academic Publishers, The Netherlands, pp 411–440
Pitt JI, Lange L, Lacey AE, Vuong D, Midgley DJ, Greenfield P, Bradbury MI, Lacey E, Busk PK, Pilgaard B, Chooi YH, Piggott AM (2017) *Aspergillus hancockii* sp. nov., a biosynthetically talented fungus endemic to southeastern Australian soils. PLoS One 12:e0170254
Porter JK, Thompson FN Jr (1992) Effects of fescue toxicosis on reproduction in livestock. J Anim Sci 70:1594–1603
Potter DA, Stokes JT, Redmond CT, Schardl CL, Panaccione DG (2008) Contribution of ergot alkaloids to suppression of a grass-feeding caterpillar assessed with gene-knockout endophytes in perennial ryegrass. Entomol Exp Appl 126:138–147
Powell RG, Plattner RD, Yates SG, Clay K, Leuchtmann A (1990) Ergobalansine, a new ergot-type peptide alkaloid isolated from *Cenchrus echinatus* (sandbur grass) infected with *Balansia obtecta*, and produced in liquid cultures of *B. obtecta* and *Balansia cyperi*. J Nat Prod 53:1272–1279
Rebek J, Shue YK, Tai DF (1984) The rugulovasines: synthesis, structure and interconversions. J Org Chem 49:3540–3545
Reddy GVP, Zhao Z, Humber RA (2014) Laboratory and field efficacy of entomopathogenic fungi for the management of the sweetpotato weevil, *Cylas formicarius* (Coleoptera: Brentidae). J Invertebr Pathol 122:10–15
Reeves EP, Reiber K, Neville C, Scheibner O, Kavanagh K, Doyle S (2006) A nonribosomal peptide synthetase (Pes1) confers protection against oxidative stress in *Aspergillus fumigatus*. FEBS J 273:3038–3053
Riederer B, Han M, Keller U (1996) D-Lysergyl peptide synthetase from the ergot fungus *Claviceps purpurea*. J Biol Chem 271:27524–27530
Rigbers O, Li S-M (2008) Ergot alkaloid biosynthesis in *Aspergillus fumigatus*: overproduction and biochemical characterization of a 4-dimethylallyltryptophan N-methyltransferase. J Biol Chem 283:26859–26868
Roberto G, Raschi E, Piccini C, Conti V, Vignatelli L, D'Alessandro R, De Ponti F, Poluzzi E (2015) Adverse cardiovascular events associated with triptans and ergotamines for treatment of migraine: systematic review of observational studies. Cephalalgia 35:118–131
Robinson SL, Panaccione DG (2012) Chemotypic and genotypic diversity in the ergot alkaloid pathway of *Aspergillus fumigatus*. Mycologia 104:804–812
Robinson SL, Panaccione DG (2014) Heterologous expression of lysergic acid and novel ergot alkaloids in *Aspergillus fumigatus*. Appl Environ Microbiol 80:6465–6472
Robinson SL, Panaccione DG (2015) Diversification of ergot alkaloids in natural and modified fungi. Toxins (Basel) 7:201–218
Russell CW, Ugine TA, Hajek AE (2010) Interactions between imidacloprid and *Metarhizium brunneum* on adult Asian longhorned beetles (*Anoplophora glabripennis* (Motschulsky) (Coleoptera: Cerambycidae)). J Invert Pathol 105:305–311
Ryan KL, Moore CT, Panaccione DG (2013) Partial reconstruction of the ergot alkaloid pathway by heterologous gene expression in *Aspergillus nidulans*. Toxins 5:445–455
Ryan KL, Akhmedov NG, Panaccione DG (2015) Identification and structural elucidation of ergotryptamine, a new ergot alkaloid produced by genetically modified *Aspergillus nidulans* and natural isolates of *Epichloë*species. J Agric Food Chem 63:61–67
Saikkonen K, Young CA, Helander M, Schardl CL (2016) Endophytic *Epichloë* species and their grass hosts: from evolution to applications. Plant Mol Biol 90:665–675
Sanchez-Monedero MA, Stentiford EI (2003) Generation and dispersion of airborne microorganisms from composting facilities. PSEP 81:166–170
Schade R, Andersohn F, Suissa S, Haverkamp W, Garbe E (2007) Dopamine agonists and the risk of cardiac-valve regurgitation. N Engl J Med 356:29–38
Schardl CL, Phillips TD (1997) Protective grass endophytes: where are they from and where are they going? Plant Dis 81:430–437
Schardl CL, Leuchtmann A, Tsai H-F, Collett MA, Watt DM, Scott DB (1994) Origin of a fungal symbiont of perennial ryegrass by interspecific hybridization of a mutualist with the ryegrass choke pathogen *Epichloë typhina*. Genetics 136:1307–1317
Schardl CL, Leuchtmann A, Spiering MJ (2004) Symbioses of grasses with seedborne fungal endophytes. Annu Rev Plant Biol 55:315–340
Schardl CL, Panaccione DG, Tudzynski P (2006) Ergot alkaloids– biology and molecular biology. Alkaloids 62:45–86
Schardl CL, Florea S, Pan J, Nagabhyru P, Bec S, Calie PJ (2013) The epichloae: alkaloid diversity and roles in symbiosis with grasses. Curr Opin Plant Biol 16:480–488
Schultes RE, Hofmann A (1973) The botany and chemistry of hallucinogens. Charles C Thomas, Springfield, pp 240–257
Scott PM (1981) Toxins of *Penicillium* species used in cheese manufacture. J Food Prot 44:702–710
Sheng H, McNamara PJ, St. Leger RJ (2022) *Metarhizium*: an opportunistic middleman for multitrophic lifestyles. Curr Opin Microbiol 69:102176
Shimshoni JA, Cuneah O, Sulyok M, Krska R, Sionov E, Barel S, Meller Harel Y (2017) Newly discovered

ergot alkaloids in sorghum ergot *Claviceps africana* occurring for the first time in Israel. Food Chem 219:459–467

Shymanovich T, Faeth SH (2018) Anti-insect defenses of *Achnatherum robustum* (sleepygrass) provided by two *Epichloë* endophytespecies. Entomol Exp Appl 166:474–482

Shymanovich T, Saari S, Lovin ME, Jarmusch AK, Jarmusch SA, Musso AM, Charlton ND, Young CA, Cech NB, Faeth SH (2015) Alkaloid variation among Epichloid endophytes of sleepygrass (*Achnatherum robustum*) and consequences for resistance to insect herbivores. J Chem Ecol 41:93–104

Shymanovich T, Crowley G, Ingram S, Steen C, Panaccione DG, Young CA, Watson W, Poore M (2020) Endophytes matter: variation of dung beetle performance across different endophyte-infected tall fescue cultivars. Appl Soil Ecol 152:103561

Smakosz A, Kurzyna W, Rudko M, Dasal M (2021) The usage of ergot (*Claviceps purpurea* (fr.) Tul.) in obstetrics and gynecology: a historical perspective. Toxins 13:492

St. Leger RJ, Wang JB (2020) *Metarhizium*: jack of all trades, master of many. Open Biol 10:200307

States JS, Christensen M (1966) *Aspergillus leporis*, a new species related to *A. flavus*. Mycologia 58:738–742

States JS, Christensen M (2001) Fungi associated with biological crusts in desert grasslands of Utah and Wyoming. Mycologia 93:432–439

Stauffacher D, Niklaus P, Tscherter H, Weber HP, Hofmann A (1969) Cycloclavin, ein neues alkaloid aus *Ipomoea hildebrandtii* Vatke–71: Mutterkornalkaloide. Tetrahedron 25:5879–5887

Steen CR, Sampson JK, Panaccione DG (2021) A Baeyer-Villiger monooxygenase gene involved in the synthesis of lysergic acid amides affects the interaction of the fungus *Metarhizium brunneum* with insects. Appl Environ Microbiol 87:e00748–e00721

Steiman R, Guiraud P, Sage L, Seigle-Murandi F (1995) New strains from Israel in the *Aspergillus niger* group. System Appl Microbiol 17:620–624

Steiner U, Ahimsa-Müller MA, Markert A, Kucht S, Groβ J, Kauf N, Kuzma M, Zych M, Lamshöft M, Furmanowa M, Knoop V, Drewke C, Leistner E (2006) Molecular characterization of a seed transmitted clavicipitaceous fungus occurring on dicotyledonous plants (*Convolvulaceae*). Planta 224:533–544

Steiner U, Leibner S, Schardl CL, Leuchtmann A, Leistner E (2011) *Periglandula*, a new fungal genus within the Clavicipitaceae and its association with Convolvulaceae. Mycologia 103:1133–1145

Steiner U, Hellwig nee Kucht S, Ahimsa-Müller MA, Grundmann N, Li S-M, Drewke C, Leistner E (2015) The key role of peltate glandular trichomes in symbiota comprising Clavicipitaceous fungi of the genus *Periglandula* and their host plants. Toxins 7:1355–1373

Stiles CE, Lloyd G, Bhattacharyya S, Steeds RP, Boomla K, Bestwick JP, Drake WM (2021) Incidence of cabergoline-associated valvulopathy in primary care patients with prolactinoma using hard cardiac endpoints. J Clin Endocrinol Metab 106:e711–e720

Tanaka A, Takemoto D, Chujo T, Scott B (2012) Fungal endophytes of grasses. Curr Opin Plant Biol 15:462–468.

Tasker NR, Wipf P (2021) Biosynthesis, total synthesis, and biological profiles of ergot alkaloids. Alkaloids 85:1–112.

Tfelt-Hansen PC, Koehler PJ (2008) History of the use of ergotamine and dihydroergotamine in migraine from 1906 and onward. Cephalalgia 28:877–886

Tor-Agbidye J, Blythe LL, Craig AM (2001) Correlation of endophyte toxins (ergovaline and lolitrem B) with clinical disease: fescue foot and perennial ryegrass staggers. Vet Human Toxicol 43:140–146

Tsai H-F, Wang H, Gebler JC, Poulter CD, Schardl CL (1995) The *Claviceps purpurea* gene encoding dimethylallyltryptophan synthase, the committed step for ergot alkaloid biosynthesis. Biochem Biophys Res Comm 216:119–125

Tudzynski P, Scheffer J (2004) *Claviceps purpurea*: molecular aspects of a unique pathogenic lifestyle. Mol Plant Pathol 5:377–388

Tudzynski P, Holter K, Correia T, Arntz C, Grammel N, Keller U (1999) Evidence for an ergot alkaloid gene cluster in *Claviceps purpurea*. Mol Gen Genet 261:133–141

Uhlig S, Petersen D (2008) Lactam ergotalkaloids (ergopeptams) as predominant alkaloids in sclerotia of *Claviceps purpurea* from Norwegian wild grasses. Toxicon 52:175–185

Unsöld IA, Li S-M (2005) Overproduction, purification and characterization of FgaPT2, a dimethylallyltryptophan synthase from *Aspergillus fumigatus*. Microbiology 151:1499–1505

Unsöld IA, Li S-M (2006) Reverse prenyltransferase in the biosynthesis of fumigaclavine C in *Aspergillus fumigatus*: gene expression, purification, and characterization of fumigaclavine C synthase FgaPT1. Chembiochem 7:158–164

Varga J, Frisvad JC, Kocsubé S, Brankovics B, Tóth B, Szigeti G, Samson RA (2011) New and revisited species in *Aspergillus* section *Nigri*. Stud Mycol 69:1–17

Vázquez MJ, Roa AM, Reyes F, Vega A, Rivera-Sagredo A, Thomas DR, Díez E, Hueso-Rodríguez JA (2003) A novel ergot alkaloid as a 5-HT1A inhibitor produced by *Dicyma* sp. J Med Chem 46:5117–5120

Wali PP, Wali PR, Saikkonen K, Tuomi J (2013) Is the pathogenic ergot fungus a conditional defensive mutualist for its host grass? PLoS One 8:e69249

Wallwey C, Li S-M (2011) Ergot alkaloids: structure diversity, biosynthetic gene clusters and functional proof of biosynthetic genes. Nat Prod Rep 28:496–510

Wallwey C, Matuschek M, Li S-M (2010a) Ergot alkaloid biosynthesis in *Aspergillus fumigatus*: conversion of chanoclavine-I to chanoclavine-I aldehyde catalyzed by a short-chain alcohol dehydrogenase FgaDH. Arch Microbiol 92:127–134

Wallwey C, Matuschek M, Xie X-L, Li S-M (2010b) Ergot alkaloid biosynthesis in *Aspergillus fumigatus*: conversion of chanoclavine-I aldehyde to festuclavine by the festuclavine synthase FgaFS in the presence of the old yellow enzyme FgaOx3. Org Biomol Chem 8:3500–3508

Walton JD, Panaccione DG, Hallen H (2004) Peptide synthesis without ribosomes. In: Tkacz J, Lange L (eds) Advances in fungal biotechnology for industry, agriculture, and medicine. Kluwer Academic/Plenum Publishers, New York, pp 127–162

Walzel B, Riederer B, Keller U (1997) Mechanism of alkaloid cyclopeptide synthesis in the ergot fungus *Claviceps purpurea*. Chem Biol 4:223–230

Wang S, Fang W, Wang C, St. Leger RJ (2011) Insertion of an esterase gene into a specific locust pathogen (*Metarhiziumacridum*) enables it to infect caterpillars. PLoS Pathog 7:e1002097

Wangeline AL, Valdez JR, Lindblom SD, Bowling KL, Reeves FB, Pilon-Smits EA (2011) Characterization of rhizosphere fungi from selenium hyperaccumulator and nonhyperaccumulator plants along the eastern Rocky Mountain front range. Am J Bot 98:1139–1147

Wu X-F, Fei M-J, Shu R-G, Tan R-X, Xu Q (2005) Fumigaclavine C, a fungal metabolite, improves experimental colitis in mice via downregulating Th1 cytokine production and matrix metalloproteinase activity. Int Immunopharmacol 5:1543–1553

Xu YJ, Luo F, Li B, Shang Y, Wang C (2016) Metabolic conservation and diversification of *Metarhizium* species correlate with fungal host-specificity. Front Microbiol 7:2020

Yao Y, An C, Evans D, Liu W, Wang W, Wei G, Ding N, Houk KN, Gao S-S (2019) Catalase involved in oxidative cyclization of the tetracyclic ergoline of fungal ergot alkaloids. J Am Chem Soc 141:17517–17521

Young CA, Hume DE, McCulley RL (2013) Forages and pastures symposium: fungal endophytes of tall fescue and perennial ryegrass: pasture friend or foe? J Anim Sci 91:2379–2394

Young CA, Schardl CL, Panaccione DG, Florea S, Takach JE, Charlton ND, Moore N, Webb JS, Jaromczyk J (2015) Genetics, genomics and evolution of ergot alkaloid diversity. Toxins (Basel) 7:1273–1302

Zajdel P, Bednarski M, Sapa J, Nowak G (2015) Ergotamine and nicergoline—facts and myths. Pharmacol Rep 67:360–363

Zanettini R, Antonini A, Gatto G, Gentile R, Tesei S, Pezzoli G (2007) Valvular heart disease and the use of dopamine agonists for Parkinson's disease. N Engl J Med 356:3–46

Zhao Y, Liu J, Wang J, Wang L, Yin H, Tan R, Xu Q (2010) Fumigaclavine C improves concanavalin A-induced liver injury in mice mainly via inhibiting TNF-α production and lymphocyte adhesion to extracellular matrices. J Pharm Pharmacol 56:775–782

5 Ectomycorrhizae for the Rescue: Ectomycorrhizosphere Signaling in Agroforestry and Afforestation

Erika Kothe, Olga Bogdanova, Oluwatosin Abdulrahman Abdulsalam, Marycolette Ndidi Ezediokpu, and Katrin Krause

Abstract

This chapter focuses on the different roles of ectomycorrhizal fungi, and thereby mostly mushroom-forming basidiomycetes and their importance for agroforestry or afforestation in times of climate change. To reduce carbon dioxide emissions, production of lignocellulose would be an option. However, the question of fuel over food poses a serious problem that needs to be considered carefully. Therefore, the use of degraded land, e.g., with high metal loads resulting from mining activities, could provide a solution, given growth of trees could be supported without further increasing agrochemical use. Here, ectomycorrhizal fungi might provide a solution. The same holds true for afforestation programs that consider future more serious droughts, as mycorrhiza can redistribute water in the entire forest ecosystem. To understand these ecosystem functions of mycorrhizal fungi, the impact on the soil properties, on the bacterial community composition and on the host are revisited. Ectomycorrhizae tend to be host specific, which involves early signaling through volatile organic compounds, specifically terpenes, as well as production of phytohormones and their influence on the host tree. *In planta*, the mutually symbiotic relationship involves the formation of highly branched mycelia between plant root cells, the Hartig' net, which is discussed with respect of cell wall covering fungal hydrophobins. Taken together, the molecular biology of the ectomycorrhizal fungi can explain for which applications in agroforestry these association may be helpful—including the use of degraded land.

This chapter is an updated version of "Ectomycorrhiza-Specific Gene Expression" published in Vol. 11: Agricultural Applications, 2013, ISBN 978-3-642-36820-2, https://link.springer.com/book/10.1007/978-3-642-36821-9.

E. Kothe (✉) · K. Krause
Institute of Microbiology, Friedrich Schiller University Jena, Jena, Germany
e-mail: erika.kothe@uni-jena.de; http://www.mikrobiologie.uni-jena.de

O. Bogdanova
Institute of Microbiology, Friedrich Schiller University Jena, Jena, Germany

BioNTech SE, Jena, Germany

O. A. Abdulsalam
Institute of Microbiology, Friedrich Schiller University Jena, Jena, Germany

Bain and Company, Düsseldorf, Germany

M. N. Ezediokpu
Institute of Microbiology, Friedrich Schiller University Jena, Jena, Germany

McMaster University, Hamilton, ON, Canada

S. Grüttner et al. (eds.), *Agricultural and Industrial Applications*, The Mycota 16,
https://doi.org/10.1007/978-3-031-81904-9_5

Keywords

Mutual symbiosis · Ectomycorrhizophere · Climate change · Microbiome · Host tree specificity · Terpene signals · Phytohormone response · Hydrophobins · Bioremediation · Metal contamination

5.1 Introduction

Sustainable agriculture needs to cope with stress factors such as low soil fertility, metal toxicity, low soil moisture, or low nutrient availability, to name but a few. To overcome high dosages of synthetic fertilizer, pesticides, herbicides, and so on, biotic amendments have been introduced. Among those plant health and growth promoting agents, mycorrhizal fungi are ranking highly, with arbuscular mycorrhiza receiving increasing attention (Chen et al. 2018; da Silva Campos et al. 2017; Hu et al. 2019). With respect to ectomycorrhizal fungi, less attention has been received.

More than 5000 ectomycorrhizal fungal species are known from basidiomycetes, ascomycetes, and a few zygomycetes. Especially, basidiomycete agaricomycetes form mutually beneficial symbiosis with approx. 3% of wooded plants, including trees and some shrubs of the families Pinaceae, Fagaceae, Myrtaceae, Dipterocarpaceae, and Caesalpinoidaceae (Brundrett et al. 1996; Smith and Read 2008). These fungi, in addition to being able to form ectomycorrhizae, are good saprophytes that can spread in soil, but usually require signals from their host to form mushrooms—and hence sexual spores for dispersal (Jones et al. 2009; Bonfante and Genre 2010). Due to the high prevalence of boreal and temperate forests, they considerably shape ecosystems and are of significant economic importance for timber production in forestry. For one example, up to 2000 fungal species have been proposed to associate with Douglas fir, interconnecting single trees in a forest, which led to introduction of the concept of a "wood wide web" (Helgason et al. 1998; Smith and Read 2008). In addition, they will have to communicate with co-occurring soil fungi and bacteria (Simard et al. 1997).

To form symbiotic structures, the soil mycelium is attracted to the plant root leading to formation of a hyphal mantle enveloping short roots of the host that already acts as a barrier against phytopathogen attack and can reduce stress exerted by soil pollutants (Smith and Read 2008; Chot and Reddy 2022). Especially, strigolactones and indole acetic acid induce branching in mantle hyphae allowing for the invasion between root cells using hydrolytic enzymes as well as physical force. Thus, a highly interconnected mycelial plane around the host cells, the Hartig' net, is formed. The invasion is causing initial expression of plant defense genes. The symbiotic fungus reacts with production of signaling compounds and effector proteins that repress the plant defense (Plett et al. 2011). The intimate association in the Hartig' net then form the basis of nutrient exchange, where the fungus supplies the plant with, i.e., nitrogen, phosphate, sulfur, and micronutrients, while the plant returns up to 50% of its photosynthates (Smith and Read 2008; McDowell et al. 2001; Nehls 2008). These symbiotic associations at short roots, the morphotypes, are specific for a plant–fungus interaction and is used for identification (Agerer 2001; Iotti and Zambonelli 2006).

From ectomycorrhizal root tips, the fungus explores the surrounding soil by the formation of extensive extramatrical mycelium. Since these soil hyphae can reach smaller soil pores than roots or even root hairs, the fungus can supply more water to the tree (Lehto and Zwiazek 2011). The interconnection between different trees through those hyphal networks allows for better water distribution in the entire forest ecosystem, becoming more important in case of extended droughts (Shah et al. 2016). In addition to soluble nutrients, the fungi can mobilize nitrogen as well as phosphate enzymatically, and the transport over long distances can be enhanced with rhizomorphs, bundles of soil hyphae (Högberg et al. 2020; Bruns et al. 2002; Smith and Read 2008).

Ectomycorrhizal fungi exhibit different host ranges, with examples for broad host range, but also with high host specificity. A nice example

Fig. 5.1 Three-part image showing: A) Mushrooms of the ectomycorrhizal *Tricholoma vaccinum*; B) micrograph of morphotypes of the symbionts with the whitish hyphal mantle around the shortened lateral roots; C) the host tree Norway spruce (P. abies; Henke 2015) at the site where fruiting bodies found

for the latter group is *Tricholoma vaccinum* (Fig. 5.1), which is highly specific to spruce (*Picea abies*), and with only about 5% of associations found in nature with pine (*Pinus sylvestris*). The fungal species is sequenced, can be cultivated in the laboratory axenically and in mycorrhizae, as well as being reasonably well amenable to molecular techniques (Krause et al. 2015; Mankel et al. 2002; Sammer et al. 2016; Abdulsalam et al. 2022).

5.2 Climate Change and Land Use: Why Ectomycorrhizal Fungi?

The changes in climate are assumed to lead to longer and more severe drought events (including the risk of forest fires), intermitted by heavy rain and storm events. In this scenario, it might be advisable to change forestry practice. Both reforestation with mixed forests including more drought-tolerant species is already widely discussed (e.g., Rudel et al. 2020).

In contrast to other mycorrhizae, ectomycorrhizae are showing a higher host specificity. The concept dividing mycorrhizal fungi into groups based on their ability to cooperate with a broad range of host plants assigns two groups: r-strategists assuring high reproduction across a diverse environment, or k-strategists with a very narrow host range up to being limited to one specific host. Examples for an r-strategist mycorrhizal fungus would be *Pisolithus tinctorius*, an ectomycorrhiza fungus that forms ectomycorrhiza with roots of 46 different tree species belonging to 20 different genera (Cairney and Chambers 1997), while *Tricholoma vaccinum* is an example of a k-strategist, found in nature compatible with spruce (*Picea abies*) and low-compatibility with pine, *Pinus sylvestris* (Krieglsteiner 2000). *Cortinarius* sp. and *Tricholoma* sp. form late-stage mycorrhizae with extensive hydrophobic rhizomorphs (Lilleskov et al. 2011). A significant correlation with soil nitrogen contents has been reported (Staudenrausch et al. 2005; Hrynkiewicz et al. 2008; Krpata et al. 2008; Bierza et al. 2020).

Another parameter that is to be considered when proposing cures that withstand future changing climatic condition is the soil hyphae associated with the different morphotypes

(Agerer 2001). These can be distinguished by their exploration strategies (Hobbie and Agerer 2010; Table 5.1). For eutrophic conditions, contact and short-distance exploration are prevalent (Lilleskov et al. 2011; Moeller et al. 2014). In more spatially dispersed nutrient supply conditions or with complex organic matter, long-distance, hydrophobic rhizomorphs are prevalent, supporting long-distance transport. With respect to the use of (metal) contaminated sites, observations or short-, medium-, and long-distance exploration types being prevalent have been reported, likely associated with other soil factors such as soil structure, distribution of contaminants, or additional stressors being present at individual sites (Hrynkiewicz et al. 2008; Rudawska et al. 2011; Bierza et al. 2020; Rosinger et al. 2018).

Another classification refers to early- versus late-stage fungi (Mason et al. 1983). These are more linked to changes in the fungal community during ecosystem development: from the dominance of *Inocybe*, *Laccaria*, and *Thelephora* at early stages of succession, *Lactarius*, *Cortinarius*, *Russula*, or *Tricholoma* replace these taxa at later developmental stages. Here, different site-specific environmental conditions, however, become more dominating, especially at advanced successional stages with changes of nutrient flux between the symbiotic partners (Newton 1992; Iordache et al. 2009; Stuart and Plett 2020).

With respect to the inevitable plant defense initiated as soon as hyphae grow into the root, fast growing fungi may overcome easier the initially mounted alkaloid and flavonoid production of the plant (Mauch-Mani et al. 2017; Pieterse et al. 2014; Luo et al. 2009). The microbially induced priming effect contributes to increased pathogen resistance of, e.g., *P. abies* against the pathogenic fungi *Botrytis cinerea* or *Heterobasidion annosum* (Wagner et al. 2019). There is, however, a sustained phytoalexin production observed (Walling 2000; Jeandet 2015; Hasegawa et al. 2014; Sanzani et al. 2014). The fast fungal growth, on the long run, would end in a relationship that resembles more pathogenic interactions, since the fungus multiplies much faster than the plant. The plant defense, therefore, must remain at a higher level in these interactions to control fungal propagation *in planta*, which

Table 5.1 Characteristics of exploration types with putative function in host protection from soil contaminants like toxic metals (see Agerer 2001)

Exploration type	Mantle and soil hyphae	Substrate contact	Protective value	Typical fungal genera
Contact	Smooth mantle, rare emanating hyphae, hydrophilic, no rhizomorphs	Close, via mantle	Protective sheath	*Lactarius*
Short distance	Different mantle types, voluminous emanating hyphae, hydrophilic, no rhizomorphs	Close, via mantle and emanating hyphae	Protection by dense hyphal network	*Cenococcum*, *Hebeloma*
Medium distance fringe subtype	Voluminous mantle, fans of ramified, hydrophilic, interconnected rhizomorphs	Extended through rhizomorphs	Protection by biofilter function, selection of ions supplied at Hartig' net	*Cortinarius*, *Tricholoma*
Medium distance mat subtype	Mats of densely packed, hydrophobic undifferentiated rhizomorphs			*Boletopsis*
Medium distance smooth subtype	Smooth mantle, hydrophobic, rare hyphae, rhizomorphs slightly differentiated with a thick central hypha			*Amanita*, *Laccaria*, *Thelephora*, *Tricholoma*
Long distance	Smooth mantle, hydrophobic, emanating hyphae present, rare and highly differentiated rhizomorphs	Long range distribution in soil	Biofilter function	*Boletus*, *Paxillus*, *Pisolithus*, *Tricholoma*

over time becomes detrimental to the fast-growing fungus, albeit its efforts to suppress plant defense (Plett et al. 2014a, b; Camehl et al. 2010; Langin et al. 2020; Plett et al. 2011; Voß et al. 2018). Hence, for a longer relationship in the mutually beneficial ectomycorrhiza, a slow-growing fungus would be easier to sustain over fast-growers due to a lower plant defense cost. The high host specificity usually found with slow growers would be an argument for additionally reduced costs towards the host. While the fast-growing, early-stage ectomycorrhizal fungi have been studied in more detail, slow-growing, late-stage, and likely very host specific fungi have been less researched and, therefore, need to be stressed to find solutions for long-standing relationships that can sustain tree growth in afforestation or agroforestry.

Ectomycorrhizal fungi have been used in the fortification of forests (Read and Perez-Moreno 2003; Simard et al. 1997; Wagner et al. 2019), and the fungi are used as a kick-start in transplantation of forest tree seedlings (Repáč 2007). Aside from the direct benefits, ectomycorrhizal fungi have been shown to be heavily involved in soil organic matter cycling of forest habitats (Phillips et al. 2014). The new paradigm in agriculture, sustainable intensification, is focusing on beneficial soil microorganisms for their role in reducing the need for chemical fertilizers and plant protection agents and improving plant nutrition and health (Avio et al. 2018). Thus, a closer look at ectomycorrhizal fungi is required to establish agroforestry and sustainable forestry practices. Artificial mycorrhization has been shown to stimulate plant growth, intensify photosynthetic rate and root respiration, increase accumulation of nitrogen and phosphorus, and to raise the survival of trees planted during afforestation (Cairney and Chambers 1997; Baxter and Dighton 2001; Menkis et al. 2007; Makita et al. 2012; Oliveira et al. 2012; Sanchez-Zabala et al. 2013). Resilience of the fungal partner against abiotic stress could specifically aid application (Dodd and Thomson 1994; Cullings and Makhija 2001).

5.3 No One Lives Alone: Rhizosphere and Ectomycorrhizae

The root exudates render the soil directly surrounding the root surface nutrient-rich (Bais et al. 2006). The attracted and supported microbiome thus defines a complex ecosystem (Venturi and Keel 2016). Community-level analyses have shown that plants actively select different rhizosphere microbial communities depending on plant species (Hacquard et al. 2017; Mhlongo et al. 2018; Stegen et al. 2012; Yang et al. 2017). Ectomycorrhizae redistribute the nutrients and add exudates of their own, and hence will lead to an enlarged zone influenced by both tree and fungus (Fig. 5.2). Since root exudates further stimulate fungal growth and development, and the fungus takes up some of the nutrients, this zone becomes even more distinct (Rambelli 1973). This leads to a shift in diversity and activity of the microbiome (Garbaye 1991; Barea et al. 2002; Nazir et al. 2009). The ectomycorrhizosphere shows an altered soil structure and water-holding capacity with significantly higher bacterial count

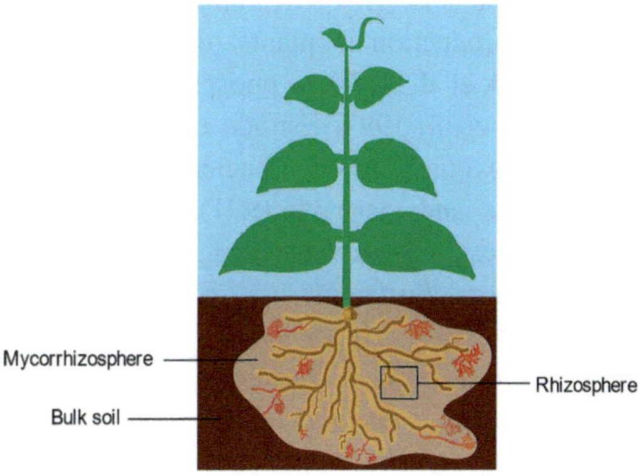

Fig. 5.2 Illustration of a plant with labeled sections of the soil environment. The plant has a green stem and leaves above ground. Below ground, the root system is shown with three labeled areas: "Mycorrhizosphere", indicating the area influenced by mycorrhizal fungi; "Rhizosphere", the region around the roots; and "Bulk soil", the surrounding soil not directly influenced by the roots. The diagram highlights the interaction between plant roots and soil components.

and phosphate solubilization (Andrade et al. 1998). In addition, bacteria may use hyphal surfaces for nutrition (mycophagy, necrophagy, or extracellular as well as endocellular biotrophy), as well as the formation of microcolonies (Nurmiaho-Lassila et al. 1997; Mogge et al. 2000; Vik et al. 2013; Marupakula et al. 2017; Emmett et al. 2021; Leveau and Preston 2008). Several studies demonstrated that the chemical composition of fungal exudates explained fungal selectivity for bacterial taxa in the mycorrhizosphere (Rangel-Castro et al. 2002; Sun et al. 1999; Frey et al. 1997).

Mycorrhizosphere bacterial communities prominently include *Pseudomonas*, *Burkholderia*, and *Bacillus* (Linderman 1987; Uroz et al. 2012; Navarro-Ródenas et al. 2016). Selected mycorrhizosphere bacterial genera can promote plant growth through symbiotic or diazotrophic nitrogen fixation, such as *Azospirillum*, *Burkholderia*, *Azotobacter*, *Klebsiella*, *Nostoc*, *Anabaena*, or some methanogenic archaea (Bae et al. 2018; Smercina et al. 2019). Other plant growth-promoting traits include the production of auxins (Krause et al. 2015), siderophores (da Costa et al. 2014; Dimkpa et al. 2009), 1-aminocyclopropane 1-carboxylic acid (ACC) deaminase lowering of ethylene production in plants (de Souza et al. 2015; Glick et al. 2007), or phosphate solubilization (Kim et al. 1997; Estrada et al. 2013). In addition, mycorrhiza helper bacteria are enriched (Duponnois and Garbaye 1991), among those Proteobacteria (*Agrobacterium*, *Azospirillum*, *Azotobacter*, *Burkholderia*, *Bradyrhizobium*, *Enterobacter*, *Pseudomonas*, *Klebsiella*, *Rhizobium*, etc.), Firmicutes (*Bacillus*, *Brevibacillus*, *Paenibacillus*, etc.), and Actinobacteria (*Rhodococcus*, *Streptomyces*, *Arthrobacter*; Frey-Klett et al. 2007; Aspray et al. 2006; Duponnois et al. 1993; Poole et al. 2001). Moreover, endocellular bacteria may be present (Bertaux et al. 2005; Nurmiaho-Lassila et al. 1997).

5.4 Who Does the Job: Soil, Tree, or Fungi?—The Ectomycorrhizosphere Effect on Bacterial Population

As discussed above, both tree and mycorrhizal fungus contribute to the selection of a microbial community in the ectomycorrhizosphere. Moreover, associations with fungi change amount and composition of root exudates, which intensifies microbial communication in the mycorrhizosphere (Meier et al. 2013; Jousset et al. 2010). Therefore, the physiology of both tree and fungus, in turn, are depending on the respective soil types and chemical composition. Hence, the influences of abiotic and biotic factors selecting the microbial community will collaboratively decide on microbiome composition and function.

In a recent field study, Bogdanova et al. (2023) determined factors selecting microbial communities at a metal contaminated former mining site. Here, microbiomes of birch, oak, and pine differed, both with respect to ectomycorrhizal fungi and associated bacterial taxa. Of course, the site-specific differences in soil texture as well as bioavailable heavy metal loads had an overarching influence on trees, fungi, and bacterial taxa present. Still the metal-rich site led to a reduced microbiome (and mycobiome) that allowed to assign additional factors with more discriminating power as compared to pristine environments, where the picture becomes more obscure because of higher bacterial loads.

Thus, beyond the general differences it became apparent that the ectomycorrhizosphere soil showed different microbiomes compared to bulk soil not influenced by the roots. As a result, a quite clear picture emerged from the reanalyzed data. Here, fungi formed the central hubs to the respective microbiomes (Fig. 5.3). Some fungal taxa were found associated with only one tree species for pine and birch. Exclusively in birch

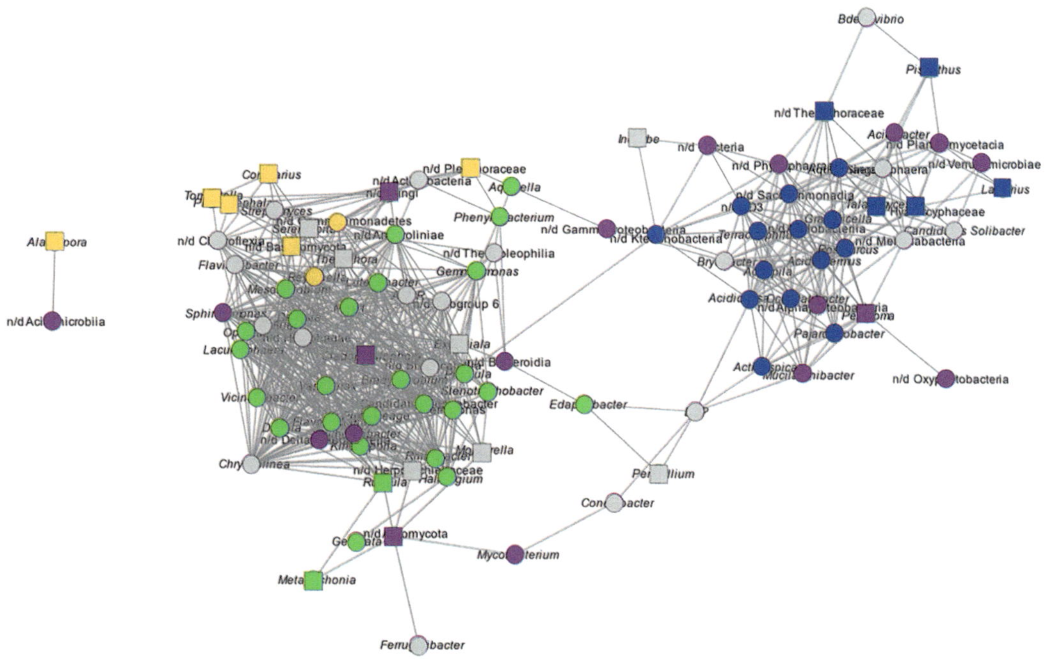

Fig. 5.3 Network graph illustrating microbial interactions with nodes and edges for co-occurring fungi (square symbols) and bacteria (round symbols) in the ectomycorrhizospheres of field plants at a former uranium mining site (reanalyzed from Bogdanova et al. 2023). Taxa present with all tree species (purple), or exclusively with birch (blue), oak (green), or pine (yellow) are highlighted

ectomycorrhizosphere, *Lactarius*, *Pisolithus*, and a Thelophoracean ectomycorrhizal species as well as two Ascomycetes, including the endophytic *Hyaloscyphus* described from ectomycorrhizal roots, were identified. Collectively, they were closest associated with members of the Verrucomicrobia, Planctomyces, and Acidibacter groups of bacterial taxa. For pine, some overlap with ectomycorrhizosphere mycobiomes was found with another *Thelephora*, while *Tomentella* and *Cortinarius* were specifically determined with pine. Here, *Streptomyces* was the most closely correlated bacterial taxon, followed by *Gemmatiomonades*, *Chloroflexa*, and *Anaerolinea*. The slightly richer community of oak seedlings again showed a typical ectomycorrhizal specialist, Russula, which was closely correlated with Flavobacteria and Deltaproteobacteria. Here, in addition, several Ascomycete and Mortierella soil fungi were present that were not exclusive to oak but also had been found with the other two tree species. These were more closely associated with several bacterial phyla typically found in soil.

Thus, this investigation indicated that the soil determined the trees, the trees select for the fungi with mycobiomes rather specific for each tree species. Here, specialists were identified within the ectomycorrhizal communities. This is in line with most fungal taxa found in the mycorrhizospheres being ectomycorrhizal species. The fungi then selected for associated bacterial groups, forming separate clusters in the network analysis.

5.5 Host Specificity of Ectomycorrhizae: Terpenes in Signaling with Host Trees

Host association requires communication between the symbionts. Hence, signals are constantly exchanged, decoded, translated into cellular reactions, and likely answered with new communication molecules. These signaling compounds could be diffusible with the water phase, or volatile. With respect to ectomycorrhizae, an early sensing seems indicated to allow for the mutual symbiosis early on, which is why a specific look into volatile infochemicals seems warranted (Ezediokpu et al. 2022; Abdulsalam et al. 2021; Klaschka 2008). Terpenoids are the most abundant natural products in fungi. When compared to ascomycetes, sesquiterpenes, diterpenes, and triterpenes are most abundant in basidiomycetes (Wang et al. 2021). Ectomycorrhizal fungi produce sesquiterpenes, which can alter root architecture and enable higher mycorrhization rates (Ditengou et al. 2015; Abdulsalam et al. 2020).

The role of terpenoids in trans-kingdom communication as allelochemicals might involve ecological function as allomones beneficial exclusively to the sender, kairomones benefiting the receiver only, or synomones beneficial to both partners since they can be recognized at a distance before any direct contact or even before diffusibles may be recognizable (Abd El-Ghany 2019).

Tricholoma vaccinum is compatible with its main host spruce, while low compatibility is seen with pine. Hence, this ectomycorrhizal fungus allows for a closer look into host specificity (Mankel et al. 2002; Wagner et al. 2015; Wagner et al. 2019; Sammer et al. 2016). The genome sequence allowed to attribute genetic information to pathways connected to host recognition and signaling with the host, including terpene synthase and terpene cyclase genes. Among volatiles produced by *T. vaccinum*, illudene, and a potential precursor ketone as well as geosmin have been identified to be produced during mycorrhization in addition to β-barbatene and limonene. Geosmin associated with the earthy, odor of, e.g., *Cortinarius herculeus* and with germination (Abdulsalam et al. 2021; Breheret et al. 1999; Ogura et al. 2000) is produced via terpene synthase Ges1. The gene *pie1* has been identified to be involved in illudene production (Ezediokpu et al. 2022). Terpene synthases may be recognized bioinformatically by their consensus sequences DDXXD/E and (N/H)DXX(S/T)XXE (Ringel et al. 2022). Most protoilludene producing basidiomycetes feature three to five protoilludane synthases in their genome (Quin et al. 2014). The cyclization of terpenes is catalyzed by terpene cyclases that allow prediction of the cyclization by phylogeny (Gressler et al. 2021). The sesquiterpene Δ6-protoilludene is, in addition, a precursor to fungal lactaranes, fommanosanes, and illudanes with antimicrobial function. Like *Tricholoma*, fungi involved in symbioses with spruce are producing protoilludane (Fig. 5.4). *Ophiostoma picea* enters the tree via bark beetle galleys. Wood-degrading *Amanita* or *Heterobasidion* associated with spruce are also known to be active protoilludane producers.

5.6 Ménage à Troi—Fungus-Fungus Interaction in Phytohormone Production

In addition to basidiomycetes, *Mucorales* fungi form terpenoids, which are of the trisporic acid family derived from β-carotenoids and used in intraspecific signaling for sexual development (Lee and Heitman 2014). An intermediate of trisporic acid production, D'orenone, has been shown to induce overproduction and excretion of the phytohormone indole-3-acetic acid in *T. vaccinum* (Wagner et al. 2016; Krause et al. 2015). This phytohormone alters plant root architecture allowing for better short root colonization by the ectomycorrhizal basidiomycete. At the same time, indole acetic acid induces branching in the mycorrhizal fungus, thus leading to the formation of the highly branched pseudoparenchymatic palmetti structures in the Hartig' net (Henke et al. 2016). Thus, interaction with molecules derived from soil hyphae environmental interaction could be shown to aid mycorrhization by *T. vaccinum* through modulation of plant–fungal mutual symbiosis (Wagner et al. 2015).

Fungi are known to produce a broad range of phytohormones, such as ethylene, abscisic acid, indole-3-acetic acid and some of its derivatives, and derivatives of jasmonates and salicylate

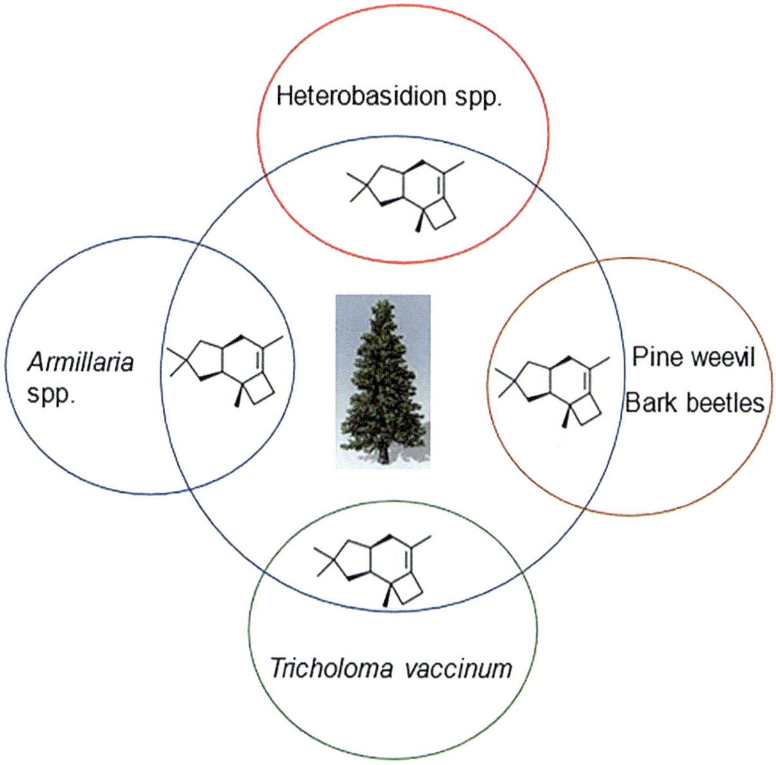

Fig. 5.4 A Venn diagram illustrating distribution of Δ6-protoilludene producing fungi associated with *Picea abies* shown in the middle with four groups of fungal life-styles represented: fungi associated with the bark beetles are *Ophistosoma canum*, *O. pluriannulatum,* and *O. picea; T. vaccinum* is an ectomycorrhizal associate; *Armillaria* and *Heterobasidion* are two wood-degrading spruce pathogens

(Wagner et al. 2016). In the mycorrhizal symbiosis the profound morphological changes and the intimate cellular relationship between the fungus and the plant require a constant, bi-directional exchange of signals, and indole acetic acid was described as important signaling molecule in mycorrhiza (Raudaskoski and Kothe 2015). It might facilitate and stimulate the mycorrhiza establishment by alterations in the root morphology as well as affect the mycobiont's growth. Obviously, two perspectives must be differentiated even if they cannot be disconnected: the mycobiont and the plant are both producers of the phytohormone, as well as signal receivers (Harrison 2005; Pieterse et al. 2012). Indole acetic acid production was shown for other ectomycorrhizal fungi (Kumla et al. 2020), as well as for the endophyte *Serendipita indica*. For *T. vaccinum,* indole acetic acid is produced via the indole-3-pyruvate pathway, and the aldehyde dehydrogenase gene, *ald1*, catalyzes the final step (Henke et al. 2016). The excretion of the phytohormone is achieved via a multidrug-and-toxic-extrusion transporter of the MATE family, *mte1*, that shows no similarity to the plant indole acetic acid PIN transporters (Zazímalová et al. 2007; Schlunk et al. 2015). Putting this picture together leads to a deeper insight into ectomycorrhizosphere signaling connected to ectomycorrhizal symbiosis (Fig. 5.5).

5.7 Building a Relationship: Fungal Hydrophobins

After the initial contact between the future symbiont via volatiles, the closer contact with mantle and Hartig' net formation needs to be initiated. However, the direct interface in the apoplastic space of the short root will require the hyphal cell wall being admissible to form the highly branched palmetti structures, which allow close enough contact for nutrient exchange.

The basidiomycete cell wall consists of chitin, glucans, and in contact with a hydrophilic–hydrophobic interface such as air—or lignin inside the

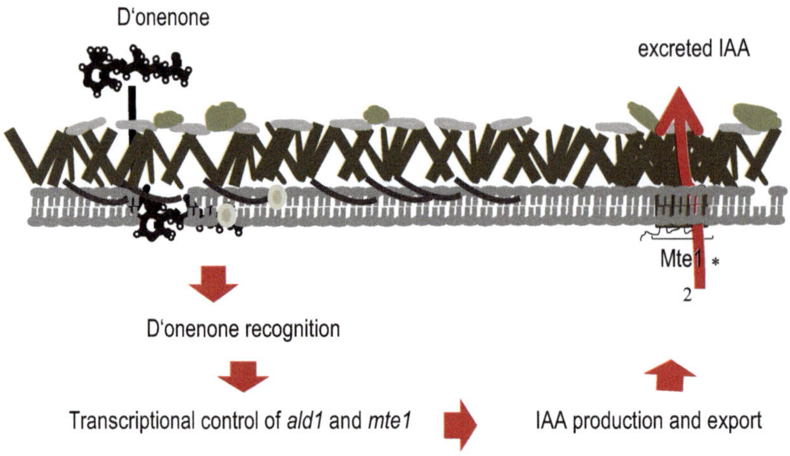

Fig. 5.5 Diagram illustrating the control of indole acetic acid production and excretion through recognition of D'orenone. The presence of Mucoralean fungi and their morphogen in the ectomycorrhizosphere leads to upregulation of the biosynthesis of the phytohormone by the ectomycorrhizal *T. vaccinum* and release of the phytohormone through upregulation of the gene *mte1* coding for the transporter able to export the phytohormone into the mycorrhizal spruce root

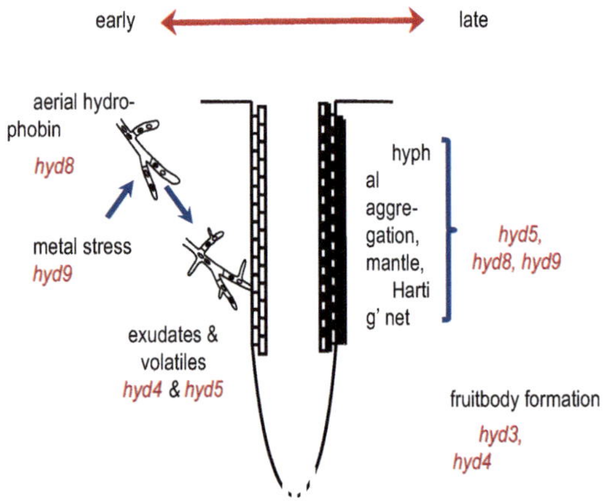

Fig. 5.6 Diagram illustrating the differential expression of hydrophobin genes *hyd1* through *hyd9* in *T. vaccinum* with respect to mycorrhiza development

apoplast—is covered by an outer layer of small amphipathic proteins, the hydrophobins. Hydrophobins of class I are present in basidiomycetes and ascomycetes, and in aerial hyphae allow growth out into the air by changing surface tension in the water phase (Wessels et al. 1991; de Vocht et al. 1998; Wösten et al. 1999). *T. vaccinum* possess nine hydrophobin genes, which are expressed at different stages of its development (Fig. 5.6; Sammer et al. 2016; Wagner et al. 2015). Hyd8 is the predominantly produced hydrophobin covering aerial hyphae in axenic conditions. However, when root exudates of *P. abies* (or *P. sylvestris*) are present, another gene, *hyd5*, becomes highly upregulated (Sammer et al. 2016). This led to the idea that hydrophobins may be used to strengthen the attachment between plant primary cell wall and fungal cell wall in the apoplastic space.

This view has been contested. The mere naming of hydrophobin seems to indicate an inner hydrophilic and outer hydrophobic side of a cover along the entire hyphae. However, this picture has not turned out to be true. Rather, rodlets of hydrophobin are visible by scanning electron microscopy on the aerial hyphal surface (Scholtmeijer et al. 2009; Fan et al. 2006). These rodlets have a diameter of approx. 7 nm, which coincides with the predicted secondary structure of the protein in assembly. Between the single rods, sufficient space remains to allow for diffusion. That this is permissible for proteins as well as secondary metabolites becomes visible with guttation droplets formed by *T. vaccinum* grown on metal-containing media.

Guttation is known mostly from plants, where especially salt tolerance is associated with the extrusion of salt-rich fluid that becomes visible after active production on the leaves as small droplets (Formann 2016; Ezediokpu et al. 2022). The formation of guttation droplets on hyphal mats grown on solid media corresponds to this mechanism. Since the droplets contain sugars like inositol and trehalose not contained in the medium, proteins, peptides, as well as metal ions or complexes and secondary metabolites show that the excretion is an active process. The same has been observed for, e.g., *Aspergillus* (Harpke et al. 2022).

With the potential of forming a hydrophobic interface inside the apoplastic space, the connection between both partners cell walls can be strengthened and still, allow for nutrient exchange.

5.8 Potential Use in Bioremediation of Heavy Metal Contaminated Land

The differential expression of metal-specific hydrophobin genes is a hint at ectomycorrhizal fungi, protecting their host trees from abiotic stress. The spatial redistribution of nutrient elements and water, as well as the protection from root pathogens, make this symbiosis prone to benefit the host, especially in adverse conditions such as heavy metal contamination. Metals are natural part of the fungal habitat, originating from weathering of rocks, or by anthropogenic activities including, e.g., industrial production, traffic, power plants, current and former mining areas, wastewater, sewage sludge, metallo-pesticides, or certain fertilizers (Chot and Reddy 2022). Elevated metal concentrations compromise ecosystem integrity and functioning, threaten human health, agricultural production, and drinking water resources; several heavy metals were among the ten most threatening chemicals to human health.

Ectomycorrhiza can lower metal loads in the host and change root–shoot transfer, showing a filtering effect (Turnau et al. 2001). The fungal sheath functions as a physical barrier and can lead to altered nutrient availability, enhanced plant antioxidant production, as well as the fungal selective uptake and delivery of elements (Chot and Reddy 2022).

However, fungal protection and benefits for the plant could not always been demonstrated with clarity and direct evidence. In some cases, the symbiosis promoted the accumulation of heavy metals in the plants with simultaneous increase in plant tolerance to oxidative stress caused by heavy metals (Chot and Reddy 2022). The response is dependent on the fungal and the plant species as well as the severity of stress, and the type of individual responses of each symbiont (Veselkin 2004).

Another option would be to include agroforestry with short-rotation coppice. This planting of tree species that allows for cutting down every 5–7 years would present a better manageable system that can avoid both threats of tree loss during drought that effects higher trees because of water transport into the higher parts of the crown, and still prevent wind and water erosion of (agricultural) soil. Nevertheless, the increasing world population must be sustained as well, and, therefore, the use of agricultural land is not seen favorable. On the other hand, the use of barren land that is, for example, contaminated with heavy metals because of mining activities, or highly compacted land, could be useful (Spasić et al. 2021; Gebhardt et al. 2007). Even if the yields are not reaching those of short rotation coppice on good farmland, the potential recreational value, the long-term soil fertility increase, and the use of the harvest for lignocellulose production might be good arguments in favor of such practice.

Especially, poplar (*Populus* sp.), willow (*Salix* sp.), locust (*Robinia* sp.), birch (*Betula* sp.), alder (*Alnus* sp.), oak (*Quercus* sp.), or pine (*Pinus* sp.) have been used since they are well adapted to grow in post-mining soils. They produce high biomass, significantly contribute to nutrient cycling, and promote ecosystem stability (Pietrzykowski 2019; Li et al. 2014; Pulford and Watson 2003; Johnson et al. 1997; Pulford and Watson 2003; Capuana 2011; Šnajdr et al. 2013). Ectomycorrhizal interactions, both with the host as well as with co-occurring microbes, need to be understood in some detail.

Other species often associated with degraded land reforestation include *Meliniomyces bicolor*, belonging to the *Hymenoscyphus ericae* aggregate, that can form both ericoid and dark-colored ectomycorrhizal symbiosis on *Picea* sp., *Pinus* sp., *Betula* sp., *Populus* sp., *Quercus* sp., and *Salix* sp. (Grelet et al. 2010; Martino et al. 2018; Vrålstad et al. 2000). It has been found prominently especially under unfavorable local conditions (Fogarty and Tobin 1996). The dark-colored, ubiquitous ascomycete *Cenococcum geophillum* is known to tolerate low soil water availability (Mexal and Reid, 1973; Carcaillet et al. 2020). The melanin that is responsible for the dark color can not only increase water holding capacity but also bind copper or iron (Senesi et al. 1987). *Thelephoraceae* are widely distributed in coniferous forests and often found in heavy metal contaminated areas (Kõljalg et al. 2000; Hrynkiewicz et al. 2008; Krpata et al. 2008). The brown color of thelephoroid morphotypes again might indicate melanization under adverse conditions (Bogdanova et al. 2023). Melanized representatives of *Helotiales* isolated from *Piceirhiza bicolorata* roots of trees growing at Cu-mine spoils add to the putative connection of melanization with metal-rich environmental occurrence (Vrålstad et al. 2002). *Lactarius* sp. as well as *Pisolithus tinctorius* could tolerate high levels of aluminum not passing those on to the host tree (Egerton-Warburton 2015; Moyer-Henry et al. 2005; Gu et al. 2021). *Inocybe* including the subgenus *Mallocybe* were observed on *Salix*, *Betula*, and adult *P. sylvestris*, and have been found in metal-affected soils as well (Krpata et al. 2008; Moser et al. 2008). The high ability of *Rhizopogon* spores to maintain viability during a long period of time was discussed previously (Horton et al. 1998). Moreover, it was mentioned that *Rhizopogon* could successfully colonize small, disturbed habitats because of its ability to germinate from spores very fast (Bruns 1995).

5.9 Conclusions: The Water–Soil-Microbes-Plant Continuum

Metabolites play important roles in the physiological and morphological shifts that lead towards the establishment of ectomycorrhizal symbiosis. The role of phytohormone indole acetic acid (IAA) has been reported for host plants and fungal partners. Ectomycorrhizal (ECM) fungi are known to produce IAA and their derivatives, for example, *Tricholoma vaccinum* produces IAA that facilitates hyphal ramification, elongation, and increased mantle development (Krause et al. 2015). Similar observation has been reported for auxin over producer mutant *Hebeloma cylindrosporum* (Gay et al. 1994).

Host plants produce and release metabolites into the rhizosphere. Released metabolites contribute to the microbial consortia that are present in the rhizosphere. Flavonoids form part of the major component of most plant root exudate and are reportedly known to function as signaling molecules (Lagrange et al. 2001), as well as the expression of effector proteins in fungi.

With respect to agricultural systems, the current public concerns are about side effects of agrochemicals, and evidence for soil degradation resulting in the loss of arable land is a global challenge. This draws attention to research areas concerning biological balance in soil, microbial diversity, or microbial dynamics in soil. Sustainable agricultural approaches, namely low-input systems and organic cultivation, are seeking for managing the benefit of microorganisms, including mycorrhizal fungi, as biofertilizers, biopesticides, and/or biostimuli (Bonfante and Genre 2010). The practical use of mycorrhiza is already addressed in various European projects (e.g., Kothe and Büchel 2014; Loreto et al. 2014). In forestry, mycorrhization of seedlings and cuttings before planting is a promising tool for successful reforestation by increased survival, better plant development, and enhanced productivity (Jeffries et al. 2003). Efforts should focus on user-friendly formulations and a broader product range. In agroforestry, where crop plants are cultivated beneath tree canopies often under tropic and poor soil conditions, soil fertilizing effects, and allopathic associations by mycorrhization are of major concern. New strategies of microbially assisted bioremediation (phytoextraction, phytostabilization, and phytodegradation) are proposed to replace the formally preceded ex situ decontamination for large areas of (metal) contamination mainly due to anthropogenic activity (Jeffries et al. 2003). In vitro ectomycorrhization is further interesting for culinary use of fruiting body formation (e.g., truffles), and mycorrhizal fungi might also be successful for metabolic engineering approaches in biotechnology (Loreto et al. 2014). To cope with limitations due to species' specificities and multi-organismic processes in the soil, further clarification is urgently needed (Jeffries et al. 2003). Plant breeding activities should focus on mycorrhizal association capacities.

References

Abd El-Ghany NM (2019) Semiochemicals for controlling insect pests. J Plant Protect Res 59:1–12

Abdulsalam O, Wagner K, Wirth S, Kunert M, David A, Kallenbach M, Boland W, Kothe E, Krause K (2020) Phytohormones and volatile organic compounds, like geosmin, in the ectomycorrhiza of Tricholoma vaccinum and Norway spruce (Picea abies). Mycorrhiza 31:173–188

Abdulsalam O, Wagner K, Wirth S, Kunert M, David A, Kallenbach M, Boland W, Kothe E, Krause K (2021) Phytohormones and volatile organic compounds, like geosmin, in the ectomycorrhiza of *Tricholoma vaccinum* and Norway spruce (*Picea abies*). Mycorrhiza 31:173–188

Abdulsalam O, Ueberschaar N, Krause K, Kothe E (2022) Geosmin synthase *ges1* knock-down by siRNA in the dikaryotic fungus *Tricholoma vaccinum*. J Basic Microbiol 62:109–115

Agerer R (2001) Exploration types of ectomycorrhizae. Mycorrhiza 11:107–114

Andrade GK, Mihara L, Linderman RG, Bethlenfalvay GJ (1998) Soil aggregation status and rhizobacteria in the mycorrhizosphere. Plant Soil 202:89–96

Aspray TJ, Jones EE, Whipps JM, Bending GD (2006) Importance of mycorrhization helper bacteria cell density and metabolite localization for the *Pinus sylvestris-Lactarius rufus* symbiosis. FEMS Microbiol Ecol 56:25–33

Avio L, Turrini A, Giovannetti M, Sbrana C (2018) Designing the ideotype mycorrhizal symbionts for the production of healthy food. Front Plant Sci 9:1089

Bae H-S, Morrison E, Chanton JS, Ogram A (2018) Methanogens are major contributors to nitrogen fixation in soils of the Florida everglades. App Env Microbiol 84:e02222–e02217

Bais HP, Weir TL, Perry LG, Gilroy S, Vivanco JM (2006) The role of root exudates in rhizosphere interactions with plants and other organisms. Annu Rev Plant Biol 57:233–266

Barea J-M, Azcón R, Azcón-Aguilar V (2002) Mycorrhizosphere interactions to improve plant fitness and soil quality. Ant Leeuw 81:343–351

Baxter J, Dighton J (2001) Ectomycorrhizal diversity alters growth and nutrient acquisition of grey birch (*Betula populifolia*) seedlings in host-symbiont culture conditions. New Phytol 152:139–149

Bertaux J, Schmid M, Hutzler P, Hartmann A, Garbaye J, Frey-Klett P (2005) Occurrence and distribution of endobacteria in the plant-associated mycelium of the ectomycorrhizal fungus *Laccaria bicolor* S238N. Env Microbiol 7:1786–1795

Bierza W, Bierza K, Trzebny A, Greń I, Dabert M, Ciepał R, Trocha LK (2020) The communities of ectomycorrhizal fungal species associated with *Betula pendula* Roth and *Pinus sylvestris* L. growing in heavy-metal contaminated soils. Plant Soil 457:321–338

Bogdanova O, Kothe E, Krause K (2023) Ectomycorrhizal community shifts at a former uranium mining site. J Fungi 9:483

Bonfante P, Genre A (2010) Mechanisms underlying beneficial plant–fungus interactions in mycorrhizal symbiosis. Nat Commun 1:48. https://doi.org/10.1038/ncomms1046

Breheret S, Talou T, Rapior S, Bessière J-M (1999) Geosmin, a sesquiterpenoid compound responsible for the musty-earthy odor of *Cortinarius herculeus, Cystoderma amianthinum*, and *Cy. carcharias*. Mycologia 91:117–120

Brundrett MC, Bougher NL, Dell B, Grove TS, Malajczuk N (1996) Working with mycorrhizas in forestry and agriculture. ACIAR Monograph 32:347–373

Bruns TD (1995) Thoughts on the processes that maintain local species diversity of ectomycorrhizal fungi. Plant Soil 170:63–73

Bruns TD, Bidartondo MJ, Taylor DL (2002) Host specificity in ectomycorrhizal communities: what do the exceptions tell us? Integr Compar Biol 42:352–359

Cairney J, Chambers S (1997) Interactions between *Pisolithus tinctorius* and its hosts: a review of current knowledge. Mycorrhiza 7:117–131

Camehl I, Sherameti I, Venus Y, Bethke G, Varma A, Lee J, Oelmüller R (2010) Ethylene signalling and ethylene-targeted transcription factors are required to balance beneficial and nonbeneficial traits in the symbiosis between the endophytic fungus *Piriformospora indica* and *Arabidopsis thaliana*. New Phytol 185:1062–1073

Capuana M (2011) Heavy metals and woody plants - biotechnologies for phytoremediation. iForest—Biogeosci Forestry 4:7–15

Carcaillet C, Desponts M, Robin V, Bergeron Y (2020) Long-term steady-state dry boreal forest in the face of disturbance. Ecosystems 23:1075–1092

Chen M, Arato M, Borghi L, Nouri E, Reinhardt D (2018) Beneficial services of arbuscular mycorrhizal fungi—from ecology to application. Front Plant Sci 9:1270

Chot E, Reddy MS (2022) Role of ectomycorrhizal symbiosis behind the host plants ameliorated tolerance against heavy metal stress. Front Microbiol 13:855473

Cullings K, Makhija S (2001) Ectomycorrhizal fungal associates of *Pinus contorta* in soils associated with a hot spring in Norris Geyser Basin, Yellowstone

National Park, Wyoming. Appl Env Microbiol 67:5538–5543

da Costa PB, Granada CE, Ambrosini A, Moreira F, de Souza R, dos Passos JFM, Arruda L, Passaglia LMP (2014) A model to explain plant growth promotion traits: a multivariate analysis of 2,211 bacterial isolates. PLoS One 9:e116020

da Silva Campos MA, da Silva FSB, Yano-Melo AM, de Melo NF, Maia LC (2017) Application of arbuscular mycorrhizal fungi during the acclimatization of *Alpinia purpurata* to induce tolerance to *Meloidogyne arenaria*. Plant Pathol J 33:329–336

de Souza R, Ambrosini A, Passaglia LMP (2015) Plant growth-promoting bacteria as inoculants in agricultural soils. Genet Mol Biol 38:401–419

de Vocht ML, Scholtmeijer K, van der Vegte EW, de Vries OM, Sonveaux N, Wösten HA, Ruysschaert JM, Hadziloannou G, Wessels JG, Robillard GT (1998) Structural characterization of the hydrophobin SC3, as a monomer and after self-assembly at hydrophobic/hydrophilic interfaces. Biophys J 74:2059–2068

Dimkpa CO, Merten D, Svatoš A, Büchel G, Kothe E (2009) Metal-induced oxidative stress impacting plant growth in contaminated soil is alleviated by microbial siderophores. Soil Biol Biochem 41:154–162

Ditengou FA, Müller A, Rosenkranz M, Felten J, Lasok H, Miloradovic van Doorn M, Legué V, Palme K, Schnitzler JP, Polle A (2015) Volatile signalling by sesquiterpenes from ectomycorrhizal fungi reprogrammes root architecture. Nat Commun 6:6279

Dodd JC, Thomson BD (1994) The screening and selection of inoculant arbuscular- mycorrhizal and ectomycorrhizal fungi. Plant Soil 159:149–158

Duponnois R, Garbaye J (1991) Mycorrhization helper bacteria associated with the Douglas fir-*Laccaria laccata* symbiosis: effects in aseptic and in glasshouse conditions. Ann Forest Sci 48:239–251

Duponnois R, Garbaye J, Bouchard D, Churin JL (1993) The fungus-specificity of mycorrhization helper bacteria (MHBs) used as an alternative to soil fumigation for ectomycorrhizal inoculation of bare-root Douglas-fir planting stocks with *Laccaria laccata*. Plant Soil 157:257–262

Egerton-Warburton LM (2015) Aluminum-tolerant *Pisolithus* ectomycorrhizas confer increased growth, mineral nutrition, and metal tolerance to *eucalyptus* in acidic mine spoil. Appl Env Soil Sci 2015:803821

Emmett BD, Lévesque-Tremblay V, Harrison MJ (2021) Conserved and reproducible bacterial communities associate with extraradical hyphae of arbuscular mycorrhizal fungi. ISME J 15:2276–2288

Estrada GA, Baldani VLD, de Oliveira DM, Urquiaga S, Baldani JI (2013) Selection of phosphate-solubilizing diazotrophic *Herbaspirillum* and *Burkholderia* strains and their effect on rice crop yield and nutrient uptake. Plant Soil 369:115–129

Ezediokpu MN, Krause K, Kunert M, Hoffmeister D, Boland W, Kothe E (2022) Ectomycorrhizal influence on the dynamics of sesquiterpene release by *Tricholoma vaccinum*. J Fungi 8:555

Fan H, Wang X, Zhu J, Robillard GT, Mark AE (2006) Molecular dynamics simulations of the hydrophobin SC3 at a hydrophobic/hydrophilic interface. Proteins 64:863–873

Fogarty RV, Tobin JM (1996) Fungal melanins and their interactions with metals. Enzym Microb Technol 19:311–317

Formann S (2016) Sequestration of heavy metals and radionuclides in ectomycorrhiza. Dissertation, Friedrich Schiller University Jena

Frey P, Frey-Klett P, Garbaye J, Berge O, Heulin T (1997) Metabolic and genotypic fingerprinting of fluorescent Pseudomonads associated with the Douglas Fir-Laccaria bicolor Mycorrhizosphere. Appl Environ Microbiol 63(5):1852–1860. https://doi.org/10.1128/aem.63.5.1852-1860.1997

Frey-Klett P, Garbaye J, Tarkka M (2007) The mycorrhiza helper bacteria revisited. New Phytol 176:22–36

Garbaye J (1991) Biological interactions in the mycorrhizosphere. Experientia 47:370–375

Gay G, Normand L, Marmeisse R, Sotta B, Debaud JC (1994) Auxin overproducer mutants of *Hebeloma cylindrosporum Romagnesi* have increased mycorrhizal activity. New Phytol 128:645–657

Gebhardt S, Neubert K, Wöllecke J, Münzenberger B, Hüttl RF (2007) Ectomycorrhiza communities of red oak (*Quercus rubra* L.) of different age in the Lusatian lignite mining district, East Germany. Mycorrhiza 17:279–290

Glick B, Cheng RZ, Czarny J, Duan J (2007) Promotion of plant growth by ACC deaminase-producing soil bacteria. Eur J Plant Pathol 119:329–339

Grelet G, Johnson AD, Vrålstad T, Alexander IJ, Anderson IC (2010) New insights into the mycorrhizal *Rhizoscyphus ericae* aggregate: spatial structure and co-colonization of ectomycorrhizal and ericoid roots. New Phytol 188:210–222

Gressler M, Löhr NA, Schäfer T, Lawrinowitz S, Seibold PS, Hoffmeister D (2021) Mind the mushroom: natural product biosynthetic genes and enzymes of Basidiomycota. Nat Prod Rep 38:702–722

Gu X, Jiang A, Wang X, Jia H, Li J, Cui Y, Hu J, Mao Q, He X (2021) Differences in aluminum tolerance and immobilization between two indigenous ectomycorrhizal fungi *Lactarius deliciosus* and *Pisolithus tinctorius* from Southwest China's forest stands. Ecotox Env Safety 213:112042

Hacquard S, Spaepen S, Garrido-Oter R, Schulze-Lefert P (2017) Interplay between innate immunity and the plant microbiota. Annu Rev Phytopathol 55:565–589

Harpke M, Pietschmann S, Ueberschaar N, Krüger T, Kniemeyer O, Brakhage AA, Nietzsche S, Kothe E (2022) Salt and metal tolerance involves formation of guttation droplets in species of the *Aspergillus versicolor* complex. Genes 13:1631

Harrison MJ (2005) Signaling in the arbuscular mycorrhizal symbiosis. Ann Rev Microbiol 59:19–42

Hasegawa M, Mitsuhara I, Seo S, Okada K, Yamane H, Iwai T, Ohashi Y (2014) Analysis on blast fungus-responsive characters of a flavonoid phytoalexin sakuranetin; accumulation in infected rice leaves, antifungal activity and detoxification by fungus. Molecules 19:11404–11418

Helgason T, Daniell TJ, Husband R, Fitter AH, Young JP (1998) Ploughing up the wood-wide web? Nature 394:431

Henke C (2015) Ectomycorrhizal signaling: The role of indole-3-acetic acid and dehydrogenases in the basidiomycete Tricholoma vaccinum. Dissertation, Friedrich Schiller University Jena

Henke C, Jung EM, Voit A, Kothe E, Krause K (2016) Dehydrogenase genes in the ectomycorrhizal fungus *Tricholoma vaccinum*: a role for Ald1 in mycorrhizal symbiosis. J Basic Microbiol 56:162–174

Hobbie EA, Agerer R (2010) Nitrogen isotopes in ectomycorrhizal sporocarps correspond to belowground exploration types. Plant Soil 327:71–83

Högberg MN, Skyllberg U, Högberg P, Knicker H (2020) Does ectomycorrhiza have a universal key role in the formation of soil organic matter in boreal forests? Soil Biol Biochem 140:107635

Horton T, Cázares RE, Bruns TD (1998) Ectomycorrhizal, vesicular-arbuscular and dark septate fungal colonization of bishop pine (*Pinus muricata*) seedlings in the first 5 months of growth after wildfire. Mycorrhiza 8:11–18

Hrynkiewicz K, Haug I, Baum C (2008) Ectomycorrhizal community structure under willows at former ore mining sites. Eur J Soil Biol 44:37–44

Hu Z-H, Zhuo F, Jing S-H, Li X, Yan T-X, Lei L-L, Lu R-R, Zhang X-F, Jing Y-X (2019) Combined application of arbuscular mycorrhizal fungi and steel slag improves plant growth and reduces Cd, Pb accumulation in *Zea mays*. Int J Phytoremediation 21:857–865

Iordache V, Gherghel F, Kothe E (2009) Assessing the effect of disturbances on ectomycorrhiza diversity. Int J Env Res Public Health 6:414–432

Iotti M, Zambonelli A (2006) A quick and precise technique for identifying ectomycorrhizas by PCR. Mycol Res 110:60–65

Jeandet P (2015) Phytoalexins: current progress and future prospects. Molecules 20:2770–2774

Jeffries P, Gianinazzi S, Perotto S et al (2003) The contribution of arbuscular mycorrhizal fungi in sustainable maintenance of plant health and soil fertility. Biol Fertil Soils 37:1–16. https://doi.org/10.1007/s00374-002-0546-5

Johnson NC, Graham JH, Smith FA (1997) Functioning of mycorrhizal associations along the mutualism–parasitism continuum. New Phytol 135:575–585

Jones DL, Nguyen C, Finlay RL (2009) Carbon flow in the rhizosphere: carbon trading at the soil–root interface. Plant Soil 321:5–33

Jousset A, Rochat L, Lanoue A, Bonkowski M, Keel C, Scheu S (2010) Plants respond to pathogen infection by enhancing the antifungal gene expression of root-associated bacteria. Mol Plant-Microbe Int 24:352–358

Kim KY, Jordan D, Krishnan HB (1997) *Rahnella aquatilis*, a bacterium isolated from soybean rhizosphere, can solubilize hydroxyapatite. FEMS Microbiol Lett 153:273–277

Klaschka U (2008) The infochemical effect—a new chapter in ecotoxicology. Environ Sci Pollut Res 15:452–462

Kõljalg U, Dahlberg A, Taylor FS, Larsson E, Hallenberg N, Stenlid J, Larsson KH, Fransson PM, Kårén O, Jonsson L (2000) Diversity and abundance of resupinate thelephoroid fungi as ectomycorrhizal symbionts in Swedish boreal forests. Mol Ecol 9:1985–1996

Kothe E, Büchel G (2014) UMBRELLA: using MicroBes for the REgulation of heavy metaL mobiLity at ecosystem and landscape scAle. Environ Sci Pollut Res 21:6761–6764

Krause K, Henke C, Asiimwe T, Ulbricht A, Klemmer S, Schachtschabel S, Boland W, Kothe E (2015) Biosynthesis and secretion of indole-3-acetic acid and its morphological effects on *Tricholoma vaccinum*-spruce ectomycorrhiza. Appl Env Microbiol 81:7003–7011

Krieglsteiner GJ (2000) Die Großpilze Baden-Württembergs. Ulmer, Stuttgart

Krpata D, Peintner U, Langer I, Fitz WJ, Schweiger P (2008) Ectomycorrhizal communities associated with *Populus tremula* growing on a heavy metal contaminated site. Mycol Res 112:1069–1079

Kumla J, Suwannarach N, Matsui K, Lumyong S (2020) Biosynthetic pathway of indole-3-acetic acid in ectomycorrhizal fungi collected from northern Thailand. PLoS One 15:e0227478

Lagrange H, Jay-Allgmand C, Lapeyrie F (2001) Rutin, the phenolglycoside from eucalyptus root exudates, stimulates *Pisolithus* hyphal growth at picomolar concentrations. New Phytol 149:349–355

Langin G, Gouguet P, Üstün S (2020) Microbial effector proteins; a journey through the proteolytic landscape. Trends Microbiol 28:523–535

Lee SC, Heitman J (2014) Sex in the Mucoralean fungi. Mycoses 57S3:18–24

Lehto T, Zwiazek JJ (2011) Ectomycorrhizas and water relations of trees: a review. Mycorrhiza 21:71–90

Leveau JHJ, Preston GM (2008) Bacterial mycophagy: definition and diagnosis of a unique bacterial-fungal interaction. New Phytol 177:859–876

Li Y, Wen H, Chen L, Yin T (2014) Succession of bacterial community structure and diversity in soil along a chronosequence of reclamation and re-vegetation on coal mine spoils in China. PLoS One 9:e115024

Lilleskov EA, Hobbie EA, Horton TR (2011) Conservation of ectomycorrhizal fungi: exploring the linkages between functional and taxonomic responses to anthropogenic N deposition. Fungal Ecol 4:174–183

Linderman RG (1987) Mycorrhizal interactions with the rhizosphere microflora: the mycorrhizosphere effect. Phytopathology 78:366–371

Loreto F, Dicke M, Schnitzler JP, Turlings TC (2014) Plant volatiles and the environment. Plant Cell Environ 37:1905–1908

Luo Z-B, Janz D, Jiang X, Göbel C, Wildhagen H, Tan Y, Rennenberg H, Feussner I, Polle A (2009) Upgrading root physiology for stress tolerance by ectomycorrhizas: insights from metabolite and transcriptional profiling into reprogramming for stress anticipation. Plant Physiol 151:1902

Makita N, Hirano Y, Yamanaka T, Yoshimura K, Kosugi Y (2012) Ectomycorrhizal-fungal colonization induces physio-morphological changes in *Quercus serrata* leaves and roots. J Plant Nutr Soil Sci 175:900–906

Mankel A, Krause K, Kothe E (2002) Identification of a hydrophobin gene that is developmentally regulated in the ectomycorrhizal fungus *Tricholoma terreum*. Appl Environ Microbiol 68:1408–1413

Martino E, Morin E, Grelet G-A, Kuo A, Kohler A, Daghino S, Barry KW, Cichocki N, Clum A, Dockter RB, Hainaut M, Kuo RC, LaButti K, Lindahl BD, Lindquist EA, Lipzen A, Khouja H-R, Magnuson J, Murat C, Ohm RA, Singer SW, Spatafora JW, Wang M, Veneault-Fourrey C, Henrissat B, Grigoriev IV, Martin FM, Perotto S (2018) Comparative genomics and transcriptomics depict ericoid mycorrhizal fungi as versatile saprotrophs and plant mutualists. New Phytol 217:1213–1229

Marupakula S, Mahmood S, Jernberg J, Nallanchakravarthula S, Fahad Z, Finlay RD (2017) Bacterial microbiomes of individual ectomycorrhizal *Pinus sylvestris* roots are shaped by soil horizon and differentially sensitive to nitrogen addition. Env Microbiol 19:4736–4753

Mason PA, Wilson J, Last FT, Walker C (1983) The concept of succession in relation to the spread of sheathing mycorrhizal fungi on inoculated tree seedlings growing in unsterile soils. Plant Soil 71:247–256

Mauch-Mani B, Baccelli I, Luna E, Flors V (2017) Defense priming: an adaptive part of induced resistance. Annu Rev Plant Biol 68:485–512

McDowell NG, Balster NJ, Marshall JD (2001) Belowground carbon allocation of Rocky Mountain Douglas-fir. Can J For Res 31:1425–1436

Meier IC, Avis PG, Phillips RP (2013) Fungal communities influence root exudation rates in pine seedlings. FEMS Microbiol Ecol 83:585–595

Menkis A, Vasiliauskas R, Taylor AFS, Stenlid J, Finlay R (2007) Afforestation of abandoned farmland with conifer seedlings inoculated with three ectomycorrhizal fungi - impact on plant performance and ectomycorrhizal community. Mycorrhiza 17:337–348

Mexal J, Reid CPP (1973) The growth of selected mycorrhizal fungi in response to induced water stress. Can J Bot 51:1579–1588

Mhlongo MI, Piater LA, Madala NE, Labuschagne N, Dubery IA (2018) The chemistry of plant–microbe interactions in the rhizosphere and the potential for metabolomics to reveal signaling related to defense priming and induced systemic resistance. Front Plant Sci 9:112

Moeller HV, Peay KG, Fukami T (2014) Ectomycorrhizal fungal traits reflect environmental conditions along a coastal California edaphic gradient. FEMS Microbiol Ecol 87:797–806

Mogge B, Loferer C, Agerer R, Hutzler P, Hartmann A (2000) Bacterial community structure and colonization patterns of *Fagus sylvatica* L. ectomycorrhizospheres as determined by fluorescence *in situ* hybridization and confocal laser scanning microscopy. Mycorrhiza 9:271–278

Moser AM, Frank JL, D'Allura JA, Southworth D (2008) Ectomycorrhizal communities of *Quercus garryana* are similar on serpentine and nonserpentine soils. Plant Soil 315:185

Moyer-Henry K, Silva I, Macfall J, Johannes E, Allen N, Goldfarb B, Rufty T (2005) Accumulation and localization of aluminium in root tips of loblolly pine seedlings and the associated ectomycorrhiza *Pisolithus tinctorius*. Plant Cell Environ 28:111–120

Navarro-Ródenas AL, Berná M, Lozano-Carrillo C, Andrino A, Morte A (2016) Beneficial native bacteria improve survival and mycorrhization of desert truffle mycorrhizal plants in nursery conditions. Mycorrhiza 26:769–779

Nazir R, Warmink JA, Boersma H, van Elsas JD (2009) Mechanisms that promote bacterial fitness in fungal-affected soil microhabitats. FEMS Microbiol Ecol 71:169–185

Nehls U (2008) Mastering ectomycorrhizal symbiosis: the impact of carbohydrates. J Exp Bot 59:1097–1108

Newton AC (1992) Towards a functional classification of ectomycorrhizal fungi. Mycorrhiza 2:75–79

Nurmiaho-Lassila E, Timonen LS, Haahtela K, Sen R (1997) Bacterial colonization patterns of intact *Pinus sylvestris* mycorrhizospheres in dry pine forest soil: an electron microscopy study. Can J Microbiol 43:1017–1035

Ogura T, Sunairi M, Nakajima M (2000) 2-methylisoborneol and geosmin, the main sources of soil odor, inhibit the germination of Brassicaceae seeds. Soil Sci Plant Nutrit 46:217–227

Oliveira RS, Franco AR, Castro PML (2012) Combined use of *Pinus pinaster* plus and inoculation with selected ectomycorrhizal fungi as an ecotechnology to improve plant performance. Ecol Eng 43:95–103

Phillips LA, Ward V, Jones MD (2014) Ectomycorrhizal fungi contribute to soil organic matter cycling in sub-boreal forests. ISME J 8:699–713

Pieterse CM, Van der Does D, Zamioudis C, Leon-Reyes A, Van Wees SC (2012) Hormonal modulation of plant immunity. Annu Rev Cell Dev Biol 28:489–521

Pieterse CMJ, Zamioudis C, Berendsen RL, Weller DM, Van Wees SCM, Bakker PAHM (2014) Induced systemic resistance by beneficial microbes. Annu Rev Phytopathol 52:347–375

Pietrzykowski M (2019) Tree species selection and reaction to mine soil reconstructed at reforested post-mine sites: central and eastern European experiences. Ecol Eng 3:100012

Plett JM, Kemppainen M, Kale Shiv D, Kohler A, Legué V, Brun A, Tyler Brett M, Pardo Alejandro G, Martin F (2011) A secreted effector protein of Laccaria bicolor is required for symbiosis development. Curr Biol 21:1197–1203

Plett JM, Daguerre Y, Wittulsky S, Vayssières A, Deveau A, Melton SJ, Kohler A, Morrell-Falvey JL, Brun A, Veneault-Fourrey C, Martin F (2014a) Effector

MiSSP7 of the mutualistic fungus *Laccaria bicolor* stabilizes the *Populus* JAZ6 protein and represses jasmonic acid (JA) responsive genes. Proc Natl Acad Sci USA 111:8299

Plett JM, Khachane A, Ouassou M, Sundberg B, Kohler A, Martin F (2014b) Ethylene and jasmonic acid act as negative modulators during mutualistic symbiosis between *Laccaria bicolor* and *Populus* roots. New Phytol 202:270–286

Poole EJ, Bending GD, Whipps JM, Read DJ (2001) Bacteria associated with *Pinus sylvestris–Lactarius rufus* ectomycorrhizas and their effects on mycorrhiza formation *in vitro*. New Phytol 151:743–751

Pulford ID, Watson C (2003) Phytoremediation of heavy metal-contaminated land by trees—a review. Env Intern 29:529–540

Quin MB, Flynn CM, Schmidt-Dannert C (2014) Traversing the fungal terpenome. Nat Prod Rep 31:1449–1473

Rambelli A (1973) The rhizosphere of mycorrhizae. In: Marks GC, Kozlowski TT (eds) Ectomycorrhizae. Academic Press, New York

Rangel-Castro J, Danell IE, Pfeffer PE (2002) A ^{13}C-NMR study of exudation and storage of carbohydrates and amino acids in the ectomycorrhizal edible mushroom *Cantharellus cibarius*. Mycologia 94:190–199

Raudaskoski M, Kothe E (2015) Novel findings on the role of signal exchange in arbuscular and ectomycorrhizal symbioses. Mycorrhiza 25:243–252

Read DJ, Perez-Moreno J (2003) Mycorrhizas and nutrient cycling in ecosystems—a journey towards relevance? New Phytol 157:475–492

Repáč I (2007) Ectomycorrhiza formation and growth of *Picea abies* seedlings inoculated with alginate-bead fungal inoculum in peat and bark compost substrates. Forestry 80:517–530

Ringel M, Dimos N, Himpich S, Haack M, Huber C, Eisenreich W, Schenk G, Loll B, Brück T (2022) Biotechnological potential and initial characterization of two novel sesquiterpene synthases from Basidiomycota *Coniophora puteana* for heterologous production of δ-cadinol. Microbial Cell Fact 21:64

Rosinger C, Sandén H, Matthews B, Mayer M, Godbold DL (2018) Patterns in ectomycorrhizal diversity, community composition, and exploration types in European beech, pine, and spruce forests. Forests 9:445

Rudawska M, Leski T, Stasińska M (2011) Species and functional diversity of ectomycorrhizal fungal communities on Scots pine (*Pinus sylvestris* L.) trees on three different sites. Ann Forest Sci 68:5–15

Rudel TK, Meyfroidt P, Chazdon R, Bongers F, Sloan S, Grau HR, Van Holt T, Schneider L (2020) Whither the forest transition? Climate change, policy responses, and redistributed forests in the twenty-first century. Ambio 49:74–84

Sammer D, Krause K, Gube M, Wagner K, Kothe E (2016) Hydrophobins in the life cycle of the ectomycorrhizal basidiomycete *Tricholoma vaccinum*. PLoS One 1:e0167773

Sanchez-Zabala J, Majada J, Martín-Rodrigues N, Gonzalez-Murua C, Ortega U, Alonso-Graña M, Arana O, Duñabeitia MK (2013) Physiological aspects underlying the improved outplanting performance of *Pinus pinaster* Ait. seedlings associated with ectomycorrhizal inoculation. Mycorrhiza 23:627–640

Sanzani SM, Schena L, Ippolito A (2014) Effectiveness of phenolic compounds against citrus green mould. Molecules 19:12500–12508

Schlunk I, Krause K, Wirth S, Kothe E (2015) A transporter for abiotic stress and plant metabolite resistance in the ectomycorrhizal fungus *Tricholoma Vaccinum*. Environ Sci Pollut Res 22:19384–19393

Scholtmeijer K, de Vocht ML, Rink R, Robillard GT, Wösten HA (2009) Assembly of the fungal SC3 hydrophobin into functional amyloid fibrils depends on its concentration and is promoted by cell wall polysaccharides. J Biol Chem 284:26309–26314

Senesi N, Sposito G, Martin JP (1987) Copper (II) and iron (III) complexation by humic acid-like polymers (melanin) from soil fungi. Sci Total Env 62:241–252

Shah F, Nicolás C, Bentzer J, Ellström M, Smits M, Rineau F, Canbäck B, Floudas D, Carleer R, Lackner G, Braesel J, Hoffmeister D, Henrissat B, Ahrén D, Johansson T, Hibbett DS, Martin F, Persson P, Tunlid A (2016) Ectomycorrhizal fungi decompose soil organic matter using oxidative mechanisms adapted from saprotrophic ancestors. New Phytol 209(4):1705–1719. https://doi.org/10.1111/nph.13722

Simard SW, Jones MD, Durall DM, Perry DA, Myrold DD, Molina R (1997) Reciprocal transfer of carbon isotopes between ectomycorrhizal *Betula papyrifera* and *Pseudotsuga menziesii*. New Phytol 137:529–542

Smercina DN, Evans SE, Friesen ML, Tiemann LK, Cann I (2019) To fix or not to fix: controls on free-living nitrogen fixation in the rhizosphere. Appl Env Microbiol 85:e02546–e02518

Smith SE, Read D (2008) Mycorrhizal symbiosis. Academic Press, London

Šnajdr J, Dobiášová P, Urbanová M, Petránková M, Cajthaml T, Frouz J, Baldrian P (2013) Dominant trees affect microbial community composition and activity in post-mining afforested soils. Soil Biol Biochem 56:105–115

Spasić M, Boruvka L, Vacek O, Drabek O, Tejnecký V (2021) Pedogenesis problems on reclaimed coal mining sites. Soil Water Res 16:137–150

Staudenrausch S, Kaldorf M, Renker C, Luis P, Buscot F (2005) Diversity of the ectomycorrhiza community at a uranium mining heap. Biol Ferti Soils 41:439–446

Stegen JC, Lin X-K, Konopka AE, Fredrickson JK (2012) Stochastic and deterministic assembly processes in subsurface microbial communities. ISME J 6:1653–1664

Stuart EK, Plett KL (2020) Digging deeper: in search of the mechanisms of carbon and nitrogen exchange in ectomycorrhizal symbioses. Front Plant Sci 10:1658–1658

Sun Y-P, Unestam T, Lucas SD, Johanson K, Kenne L, Finlay R (1999) Exudation-reabsorption in a mycorrhizal fungus, the dynamic interface for interaction with soil and soil microorganisms. Mycorrhiza 9:137–144

Turnau K, Berger A, Loewe A, Einig W, Hampp R, Chalot M, Dizengremel P, Kottke I (2001) Carbon dioxide concentration and nitrogen input affect the C and N storage pools in *Amanita muscaria-Picea abies* mycorrhizae. Tree Physiol 21:93–99

Uroz S, Oger P, Morin E, Frey-Klett P (2012) Distinct ectomycorrhizospheres share similar bacterial communities as revealed by pyrosequencing-based analysis of 16S rRNA genes. Appl Env Microbiol 78:3020–3024

Venturi V, Keel C (2016) Signaling in the rhizosphere. Trends Plant Sci 21:187–198

Veselkin DV (2004) Anatomical structure of Ectomycorrhiza in Abies sibirica Ledeb. and Picea obovata Ledeb. under conditions of forest ecosystems polluted with emissions from copper-smelting works. Russ J Ecol 35:71–78. https://doi.org/10.1023/B:RUSE.0000018930.17569.09

Vik U, Logares R, Blaalid R, Halvorsen R, Carlsen T, Bakke I, Kolstø A-B, Økstad O-A, Kauserud H (2013) Different bacterial communities in ectomycorrhizae and surrounding soil. Sci Rep 3:3471

Voß S, Betz R, Heidt S, Corradi N, Requena N (2018) RiCRN1, a crinkler effector from the arbuscular mycorrhizal fungus *Rhizophagus irregularis*, functions in arbuscule development. Front Microbiol 9:2068

Vrålstad T, Fossheim T, Schumacher T (2000) *Piceirhiza bicolorata*—the ectomycorrhizal expression of the *Hymenoscyphus ericae* aggregate? New Phytol 145:549–563

Vrålstad T, Myhre E, Schumacher T (2002) Molecular diversity and phylogenetic affinities of symbiotic root-associated ascomycetes of the Helotiales in burnt and metal polluted habitats. New Phytol 155:131–148

Wagner K, Linde J, Krause K, Gube M, Koestler T, Sammer D, Kniemeyer O, Kothe E (2015) *Tricholoma vaccinum* host communication during ectomycorrhiza formation. FEMS Microbiol Ecol 91:fiv120

Wagner K, Krause K, David A, Kai M, Jung E-M, Sammer D, Kniemeyer O, Boland W, Kothe E (2016) Influence of zygomycete-derived D'orenone on IAA signalling in *Tricholoma*-spruce ectomycorrhiza. Env Microbiol 18:2470–2480

Wagner K, Krause K, Gallegos-Monterrosa R, Sammer D, Kovács AT, Kothe E (2019) The ectomycorrhizospheric habitat of Norway Spruce and *Tricholoma vaccinum*: promotion of plant growth and fitness by a rich microorganismic community. Front Microbiol 10:307

Walling LL (2000) The myriad plant responses to herbivores. J Plant Growth Regul 19:195–216

Wang Q, Cao R, Zhang Y, Qi P, Wang L, Fang S (2021) Biosynthesis and regulation of terpenoids from basidiomycetes: exploration of new research. AMB Express 11:150

Wessels J, De Vries O, Asgeirsdottir SA, Schuren F (1991) Hydrophobin genes involved in formation of aerial hyphae and fruit bodies in *Schizophyllum*. Plant Cell 3:793–799

Wösten HA, Richter M, Willey JM (1999) Structural proteins involved in emergence of microbial aerial hyphae. Fungal Genet Biol 27:153–160

Yang Y, Wang N, Guo X, Zhang Y, Ye B (2017) Comparative analysis of bacterial community structure in the rhizosphere of maize by high-throughput pyrosequencing. PLoS One 12:e0178425

Zazímalová E, Krecek P, Skůpa P, Hoyerová K, Petrásek J (2007) Polar transport of the plant hormone auxin—the role of PIN-FORMED (PIN) proteins. Cell Mol Life Sci 64:1621–1637

6 Your Death, My Life: Understanding the Success Story of Necrotrophic Plant Pathogenic Ascomycota

Michele C. Malvestiti and Jan A. L. van Kan

Abstract

Necrotrophic plant pathogenic Ascomycota require dead host plant tissue to obtain nutrients for growth and reproduction. In interactions between plants and necrotrophic fungi, successful infection relies on the programmed cell death-inducing activity of compounds secreted by the fungus, which are referred to as necrotrophic effectors. These compounds can either be proteins, small modified peptides, or secondary metabolites. Their cell death-inducing capability allows the fungus to hijack the plant immune system for its own benefit and contribute to fungal virulence in a quantitative manner. In this chapter we first describe the effector repertoire in several genera of necrotrophic Ascomycota. Next, we discuss case studies that enabled to identify the cellular targets of specific effectors in different plant species. Finally, we present an overview of tools and methodologies to identify novel necrotrophic effectors and to assess their importance for fungal virulence. This knowledge will contribute to the selection of plant genotypes displaying increased resistance to diseases caused by necrotrophic Ascomycota.

M. C. Malvestiti · J. A. L. van Kan (✉)
Laboratory of Phytopathology, Wageningen University, Wageningen, The Netherlands
e-mail: jan.vankan@wur.nl

Keywords

Effector · Necrotrophic pathogen · Phytotoxin · Plant cell death · Susceptibility gene · Virulence factor

6.1 Lifestyle of Phytopathogenic Necrotrophic Ascomycota

No matter whether they settle in natural habitats or in agricultural systems, plants live in environments that teem with infectious fungi, among which many belong to the phylum Ascomycota. These fungi have evolved different lifestyles to acquire nutrients from their host plants. Some of them require living plant tissue to obtain nutrients for growth and they are classified as biotrophic fungi (biotrophs). These Ascomycota are able to colonize the plant by suppressing the host immune responses when growing in the invaded tissue. Biotrophs have evolved specialized feeding structures to acquire carbon and nitrogen and therefore it is essential for them to keep their host cells alive. Differently from biotrophs, other phytopathogenic Ascomycota can acquire nutrients for growth and reproduce exclusively from dead plant tissue. Successful infection is achieved upon active killing of host cells and, therefore, Ascomycota with such lifestyle are classified as necrotrophic fungi (necrotrophs). In the final decades of the twentieth century, necrotrophs

S. Grüttner et al. (eds.), *Agricultural and Industrial Applications*, The Mycota 16,
https://doi.org/10.1007/978-3-031-81904-9_6

were thought to colonize their hosts in a rather brutal manner. It was considered that they first kill plant cells through the activity of phytotoxins and cell wall degrading enzymes (CWDEs), and subsequently grow saprophytically on dead host tissue. In the past two decades, however, evidence has accumulated that the interaction between necrotrophs and their host plants is more subtle and sophisticated than previously assumed. Necrotrophs have evolved strategies to exploit the plant immune system for their own benefit. They can co-opt dominant host susceptibility genes (HSGs) to actively trigger programmed cell death (PCD) in the host plant, thereby achieving disease development and plant susceptibility in an inverse gene-for-gene manner (Hammond-Kosack and Rudd 2008; Oliver and Solomon 2010; Kabbage et al. 2013; Friesen and Faris 2012, 2021). More specifically, necrotrophs are able to drive the plant cell toward apoptosis-like PCD (Dickman et al. 2017; Kabbage et al. 2017), which differs from autophagy-like PCD that results in a resistance response during plant–biotroph interactions and is referred to as hypersensitive response (HR) (Minina et al. 2014). Autophagy- and apoptosis-like PCD are regulated by different genes and these two types of cell death can be experimentally distinguished by cellular and biochemical hallmarks. Nevertheless, detailed descriptions of the hallmarks that distinguish autophagy from apoptosis-like PCD are beyond the scope of this chapter. For readers interested in learning more comprehensively about the different types of PCD, several reviews are recommended (Minina et al. 2014; Dickman et al. 2017; Kabbage et al. 2017; Balint-Kurti 2019). In plant–necrotroph interactions, apoptosis-like PCD is induced by the activity of fungal-secreted compounds, which are defined here as necrotrophic effectors (NEs). Effector recognition may occur in the apoplast upon effector binding to the extracellular domain of a receptor protein of the receptor-like protein (RLP)/receptor-like kinase (RLK) type. Alternatively, the effector may first be internalized via endocytosis into the plant cell and subsequently trigger PCD induction upon direct or indirect recognition by a cytoplasmic nucleotide binding-leucin-rich repeat (NB-LRR)-type receptor, which is responsible for PCD induction (Fig. 6.1).

NEs comprise proteinaceous compounds and phytotoxic secondary metabolites. In older literature, they have often been referred to as host selective toxins (HSTs) (Yoder 1980; Markham and Hille 2001; Wolpert et al. 2002; Friesen et al. 2008a; Oliver and Solomon 2010) since they were detected only in certain fungal species or even isolates, and confer pathogenicity only to a single host plant species or a specific genotype within a species. Moreover, many of these PCD-inducing compounds were of non-proteinaceous nature. Nevertheless, to avoid confusion in this chapter, all compounds involved in PCD induction and required for pathogenicity in plant–necrotroph interactions fall under the definition of NEs, no matter whether they are ribosomal synthetized proteins or secondary metabolites, and regardless whether they might be fungal isolate-specific or operate in a host-specific context. PCD induction by means of NEs is a key step in disease development and successful plant colonization (Kabbage et al. 2017; Veloso and van Kan 2018; Kariyawasam et al. 2023) (Fig. 6.1). The cell death response occurring upon NE recognition often relies on perception and signaling mechanisms analogous to the recognition of biotrophic effectors, but in this case, these lead to host susceptibility instead of resistance (Lorang et al. 2007; Friesen and Faris 2012, 2021). Depending on the number of NEs and the corresponding targets present in the plant, the level of host susceptibility as well as the virulence of the fungal isolate can vary in a quantitative manner. The more functional NEs are produced by a given fungal isolate, and the more corresponding functional NE targets are present in a given plant genotype, the more aggressive is the fungus and the more susceptible is the host (Fig. 6.2). Importantly, even though the expression of a specific HSG is required for PCD induction in the plant genotype, thereby conferring plant sensitivity to the NE, the effector itself does not necessarily need to directly interact with the HSG protein, in accordance to the "guard hypothesis" (van der Hoorn and Kamoun 2008).

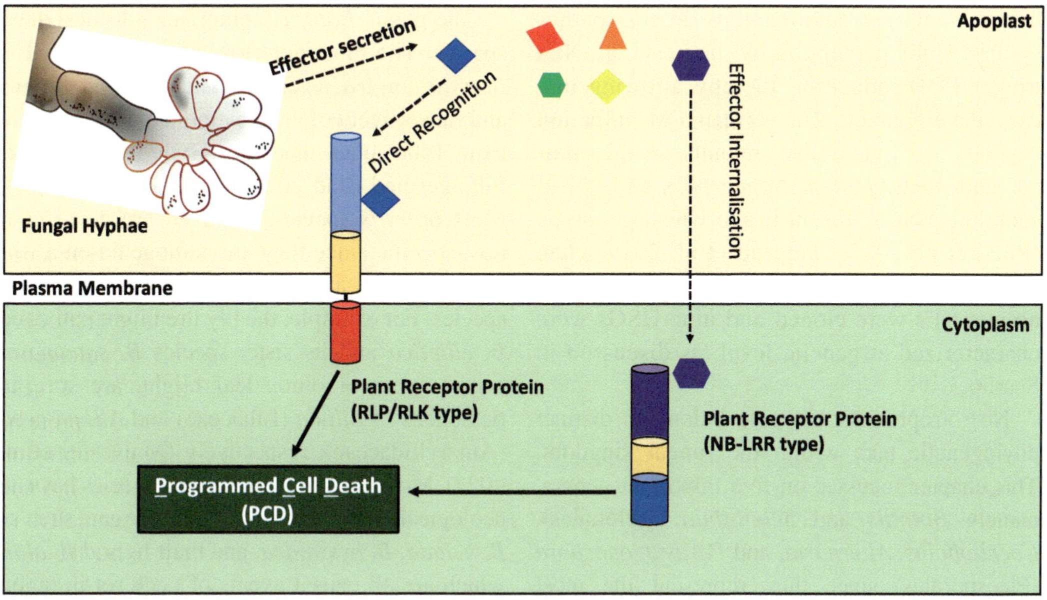

Fig. 6.1 Schematic illustration of secretion of necrotrophic effectors (indicated by colored shapes) into the plant tissue, some of which remain apoplastic, while others are translocated into the plant cytoplasm. Effectors can be recognized by extracellular RLP/RLK-type receptors or by intracellular NB-LRR type receptors, leading to Programmed Cell Death in the host and disease susceptibility

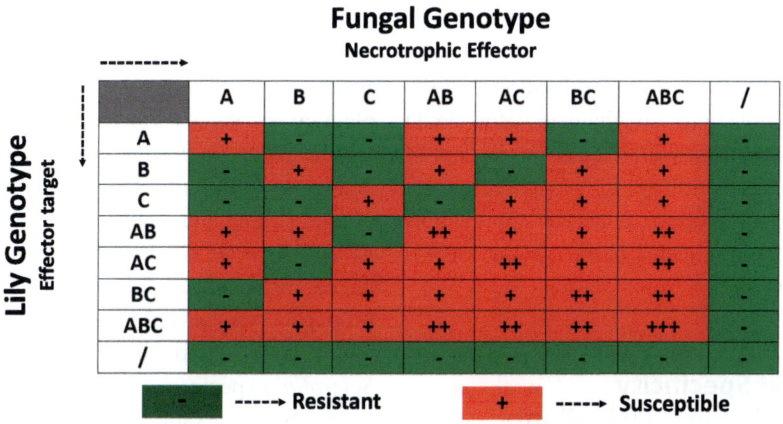

	A	B	C	AB	AC	BC	ABC	/
A	+	-	-	+	+	-	+	-
B	-	+	-	+	-	+	+	-
C	-	-	+	-	+	+	+	-
AB	+	+	-	++	+	+	++	-
AC	+	-	+	+	++	+	++	-
BC	-	+	+	+	+	++	++	-
ABC	+	+	+	++	++	++	+++	-
/	-	-	-	-	-	-	-	-

Fig. 6.2 Schematic representation of the inverse gene-for-gene interactions between fungal NEs and plant NE targets, which govern the disease susceptibility in plant–necrotroph interactions. The columns represent fungal genotypes containing effector genes A, B or C, or combinations thereof. The column with a forward slash represents a fungal genotype lacking effectors. The rows represent lily genotypes carrying the corresponding receptor proteins A, B or C, or combinations thereof. The row with a forward slash represents a lily genotype lacking receptor proteins. Red cells with a plus sign indicate susceptibility, while green cells with a minus sign indicate resistance. Note that susceptibility (of lily genotypes) and virulence (of B. elliptica isolates) are quantitative traits, reflected by the number of plus signs

The best example to illustrate such a scenario results from studies on the necrotrophic fungus *Parastagonospora nodorum* (Phaeosphaeriaceae), causal agent of Septoria Nodorum Blotch (SNB) in wheat (*Triticum aestivum*, Poaceae). Secreted NEs are key determinants for pathogenicity in this pathosystem. Upon perception by the host cell, NEs trigger PCD induction, thereby allowing disease development. The severity of infection depends in a quantitative manner on the number and identity of pathogen NEs and corresponding HSGs present in the wheat genotype (Ruud et al. 2017; Haugrud et al. 2019; Phan et al. 2021). In this pathosystem, nine proteinaceous NEs were cloned and nine HSGs were characterized at genetic level, as discussed in Sect. 6.3.

Necrotrophic Ascomycota belong to distinct phylogenetic taxa within the fungal kingdom. This chapter focusses on five important genera, namely *Botrytis* and *Sclerotinia* (Helotiales), *Cochliobolus*, *Alternaria*, and *Parastagonospora* (Pleosporales) since they represent the most widespread and best-studied necrotrophic fungi and have significant economic impact on cultivated plants. We will first present the host range and the phylogeny of these taxa and describe the repertoire of NEs for each fungal genus. Subsequently, we will elaborate on a number of case studies that have unraveled the plant cellular targets of specific NEs, their localization, and the underlying molecular mechanisms for PCD induction. Finally, we will discuss the concept of quantitative susceptibility in plant–necrotroph interactions and propose concrete applications for the agricultural sector based on the knowledge about NEs.

6.2 Host Specificity and Phylogeny in Plant–Necrotroph Interactions

Host specificity is defined as the range and diversity of host species that a parasitic organism is capable to infect (Wells and Clark 2019). Four of the five genera of necrotrophic fungi mentioned above contain species with a broad host range (i.e., occurring on phylogenetically distant host plants), as well as host-specific species that occur only on a limited number or even on a single host plant species. These are fungi from the genus *Botrytis*, *Sclerotinia*, *Cochliobolus*, and *Alternaria*.

The genus *Botrytis* comprises ~40 described species. The most notorious and best studied is *Botrytis cinerea*, a generalist necrotrophic fungus and causal agent of the gray mold disease in more than 1400 plant species (Hyde et al. 2014; Fillinger and Elad 2016; Garfinkel et al. 2017). Most *Botrytis* species, however, are considered host-specific since they are pathogenic on a single host or on a few taxonomically related host species. For example, the lily fire blight pathogen *B. elliptica* and its sister species *B. squamosa*, causal agent of onion leaf blight, are specific pathogens of *Lilium* (Liliaceae) and *Allium cepa* (Amaryllidaceae), respectively (Malvestiti et al. 2022). Moreover, several *Botrytis* species have an ecological niche in a specific plant organ, such as *B. aclada*, *B. byssoidea*, and their hybrid *B. allii*, which are all causal agents of neck rot in onion bulbs (Nielsen and Yohalem 2001; Chilvers and du Toit 2006). *Botrytis* species are phylogenetically divided in two clades. Clade I contains species that infect exclusively dicots and includes the generalists *B. cinerea* and *B. pseudocinerea*. By contrast, Clade II harbors only host-specific *Botrytis* spp., whereby the majority of them is pathogenic on a specific bulbiferous monocot plant taxon (Valero-Jiménez et al. 2020).

The genus *Sclerotinia* is the sister taxon of the genus *Botrytis* and comprises seven described species (Liang and Rollins 2018). Their peculiarity is that none of the *Sclerotinia* species produces asexual conidiospores. Instead, they propagate via sexual ascospores dispersed by wind and rain and via mycelial fragments. All *Sclerotinia* species can cause white rot on multiple hosts. For example, *S. trifoliorum* is pathogenic on Fabaceae such as *Medicago sativa*, *Trifolium pratense*, and *T. repens* (Lithourgidis et al. 2003), while *S. borealis* can infect plants of 17 different genera from six families (Mardanov et al. 2014). In this chapter, we will focus on the best-studied species of this genus, *S. sclerotiorum*, which can infect more than 600 host plants worldwide (Liang and Rollins 2018).

The genus *Cochliobolus* (anamorph *Bipolaris*) includes more than 55 species, each distinguished by a host-specific interaction with a member of the Poaceae family (Manamgoda et al. 2011).

The most relevant and best-studied members include the necrotrophic *Zea mays* pathogens *C. heterostrophus* and *C. carbonum*, the *Oryza sativa* pathogen *C. miyabeanus* as well as the *Avena sativa* pathogen, *C. victoriae* (van Bockhaven et al. 2015; Condon et al. 2013).

The genus *Alternaria* is characterized by complex phylogenetic relationships. Morphological studies supported by phylogenetic analysis on ITS regions, secondary metabolite biosynthetic gene clusters, and CWDEs identified 26 *Alternaria* sections (Lawrence et al. 2012, 2013, 2014, 2016; Woudenberg et al. 2013, 2015). Most notably, *Alternaria* sect. *Alternaria* contains the majority of important plant pathogens including the broad host range pathogen *A. alternata*, the Brassicaceae specialist *A. brassicicola*, and *A. solani*, which can infect several plants in the Solanaceae family (Woudenberg et al. 2015).

Finally, the necrotrophic Ascomycete *Parastagonospora nodorum*, causal agent of the devastating SNB, is known to infect only the plant genus *Triticum*. This fungus is prevalent in warm areas of South Asia, North Africa, and North and South America and is known to coexist with the necrotrophic Ascomycota *Pyrenophora tritici-repenstis* causal agent of the tan spot disease in certain wheat-growing regions (Katoch et al. 2022).

6.3 Repertoire of Necrotrophic Effectors in Fungal Genera

6.3.1 *Botrytis* and *Sclerotinia*

Fungi of the genus *Botrytis* initiate their disease cycle when wind and rain-dispersed conidia land on the host surface. Under moist conditions the conidia germinate, form a germ tube that develops into a unicellular appressorium or into a multicellular structure described as infection cushion to facilitate host tissue penetration (Choquer et al. 2021). The initial infection stage of *Botrytis* is characterized by a short biotrophic phase, which is crucial for the outcome of the interaction between *Botrytis* and its host. For successful disease progression, the fungus must suppress the induction of autophagy-like PCD and drive the host cell toward an apoptosis-like PCD (Kabbage et al. 2017; Veloso and van Kan 2018; Bi et al. 2022). Autophagy suppression gives the fungus time to feed within the host tissue and accumulate biomass for the transition into the necrotrophic phase. At this point, compounds secreted by the fungus induce apoptosis-like PCD and mark the formation of necrotic lesions (Veloso and van Kan 2018). Vigorous outgrowth of the necrotic lesions causes rapid maceration of plant tissue, on which the fungus eventually sporulates to produce inoculum for the next infection. A common feature of (probably all) *Botrytis* species is their capacity to secrete a wide array of proteinaceous NEs during host colonization (van Baarlen et al. 2004; Malvestiti et al. 2022; Bi et al. 2022; Leisen et al. 2022; Steentjes et al. 2022). Traditionally, CWDEs are not considered as NEs, given the assumption that CWDEs are exclusively involved in hydrolysis of plant cell wall polysaccharides. However, several CWDEs were shown to be capable of PCD induction upon effector recognition in the plant apoplast, without involving their hydrolytic activity. For example, *B. cinerea* endopolygalacturonases (PGs) induce cell death upon protein infiltration in certain *Arabidopsis* lines (Zhang et al. 2014). Other described proteinaceous CWDEs with NE activity in *B. cinerea* are the xylanases BcXyn11a, BcXyl1, and the xyloglucanase BcXyg1, which induce PCD in tomato (*Solanum lycopersicum*, Solanaceae), tobacco (*Nicotiana tabacum*, Solanaceae), and bean (*Phaseolus* spp., Fabaceae) leaf tissue (Noda et al. 2010; Zhu et al. 2017; Yang et al. 2018a). Infiltration of the glycoprotein BcGs1 caused a necrotic response and ROS accumulation in tomato and tobacco leaves (Zhang et al. 2015). The transglycolase BcCrh1 has been identified as NE since its translocation into the plant cytoplasm induces PCD and defense responses in *Arabidopsis* (Bi et al. 2021). In addition to CWDEs with cell death-inducing activity, *B. cinerea* also secretes NEs that lack an annotated enzymatic domain. For example, the necrosis and ethylene-inducing proteins BcNep1 and BcNep2, the cerato-platanin protein BcSpl1, the hypersensitive response inducing protein BcHip1, and the glycoprotein BcIEB1 could induce PCD upon pure protein infiltration or transient expression in plant leaf tissue (Cuesta Arenas et al. 2010; Frías et al. 2011, 2016; Jeblick et al. 2022). Besides proteinaceous NEs, *Botrytis* species also secrete secondary metabolites that contribute to virulence. The two best-studied secondary metabolites are botrydial and botcinic acid, which can cause chlorosis and cell collapse when applied onto leaf, flower, and fruit tissues of different hosts (Colmenares et al. 2002). These compounds are synthetized by biosynthetic gene clusters that are present in the genomes of several, but not all, *Botrytis* species (Valero-Jiménez et al. 2020).

As mentioned before, *Sclerotinia* species lack the production of asexual conidia. Their dispersal and host plant inoculation is achieved mostly by sexual ascospores but in some cases also via mycelial tissue. In nature, inoculation by mycelium occurs mostly when flowers have first been infected by ascospores and subsequently drop onto lower plant parts. The mycelium in the infected tissue can then spread and directly infect other part of the plant through the formation of infection cushions (Derbyshire and Denton-Giles 2016). Analogous to *Botrytis*, the infection process in *S. sclerotiorum* can be divided in a short biotrophic stage, where the fungus penetrates the host surface, forms multicellular appressoria and downregulates host defense mechanisms, followed by a necrotrophic phase where the pathogen kills host cells and extensive tissue maceration takes place (Liang and Rollins 2018). *S. sclerotiorum* disease development is inhibited by the occurrence of autophagy-like PCD and requires the induction of apoptosis-like PCD. The importance of apoptosis-like PCD became apparent by the observation that transgenic plants expressing heterologous anti-apoptotic genes from animals, including the nematode *CED-9* gene, the human *Bcl-xl* gene, and the Baculovirus *OP-IAP* gene, all displayed resistance to *S. sclerotiorum* and *B. cinerea* (Dickman et al. 2001). To achieve apoptosis-like PCD, *S. sclerotiorum* utilizes an array of secreted proteinaceous and non-proteinaceous NEs. As for *Botrytis*, SsNEP1 and SsNEP2 and endopolygalacturonases cause a cell death response when infiltrated as pure proteins or transiently expressed in various plants such as soybean (*Glycine max*, Fabaceae), *Arabidopsis*, and tobacco (Zuppini et al. 2005; Dallal Bashi et al. 2010; Bashi et al. 2013). More recently, two small secreted proteins, one translocated into host cell cytoplasm (SsVP1) and one localized in the apoplastic space (SsCP1), were shown to cause cell death when transiently expressed in *Nicotiana benthamiana* (Lyu et al. 2016; Yang et al. 2018b).

6.3.2 *Alternaria* and *Cochliobolus*

Fungi of the genera *Alternaria* and *Cochliobolus* use similar strategies to infect their respective host plants and to spread disease (Saharan et al. 2016; Manamgoda et al. 2011). As many other leaf pathogens, these fungi are seed borne and the primary infection can arise from infected seeds or crop residues (Reis 1991; Lanoiselet et al. 2005; Saharan et al. 2016). At the end of the growing season, *Alternaria* and *Cochliobolus* form chlamydospores and melanized conidia to overwinter on plant debris (Chand et al. 2002; Saharan et al. 2016). Histological and transcriptomic studies revealed that *Alternaria* and *Cochliobolus* both go through a short biotrophic phase in which appressorium formation and host tissue penetration occur (Gupta et al. 2018; Mandal et al. 2018; Brouwer et al. 2021). Once within the host tissue, the epidermal cells are fully invaded, the mycelium ramifies through and between the mesophyll and palisade cells, and the fungus colonizes the entire leaf (Aggarwal et al. 2008; Sahu et al. 2016). PCD is induced in the post-penetration phase, the invaded epidermal cells become necrotic and the parenchyma tissue collapses behind the advancing invasive hyphae (Kumar et al. 2002; Ibeagha et al. 2005; Lenz et al. 2011; Brouwer et al. 2021). PCD induction is achieved by means of secreted NEs. In *Alternaria* and *Cochliobolus*, NEs have often been found to be isolate-specific, allowing disease development on one single host genotype.

In *Alternaria* species, the majority of known NEs is non-proteinaceous and they are involved in PCD induction in a specific host. For example, the polyketide ACR(L-) toxin I blocks mitochondrial respiration (Ohtani et al. 2002), is able to cause formation of necrotic leaf spots in rough lemon (*Citrus jambhiri*, Rosaceae), and is produced by the lemon-specific *A. alternata* f. sp. *citri jambhiri* (Gardner et al. 1985; Akimitsu et al. 1989; Izumi et al. 2012). The AAL-toxins from *A. alternata* f. sp. *lycopersici* (Bottini and Gilchrist 1981) are amylopentol esters and act by disturbing the sphingolipid metabolisms of plant cells. Given their structural analogy to sphinganines, AAL-toxins inhibit ceramide synthase activity in the endoplasmic reticulum (ER), thereby interfering with the breakdown of ceramide-containing lipids, which induces apoptotic-like PCD (Caldas et al. 1994; Abbas et al. 1994; Gilchrist 1997, 1998; Spassieva et al. 2002; Liang et al. 2003; Michaelson et al. 2016). ACT-, AF-, and AK-toxins are epoxy-decatrieonic esters, respectively, produced from *A. alternata* pathovars affecting tangerine (*Citrus tangerina*, Rosaceae), strawberry (*Fragaria* spp., Rosaceae), and Japanese pear (*Pyrus pyrifolia*, Rosaceae) and that induce PCD by causing plasma membrane permeabilization in the respective hosts (Nakashima et al. 1985; Nakatsuka et al. 1986; Kohmoto et al. 1993). In addition to phytotoxic secondary metabolites, several cyclic depsipeptides have been reported to act as NEs in *Alternaria* fungi. For example, the cyclic depsipeptides AM-toxins, produced by the apple-specific pathogen (*Malus domestica*, Rosaceae) *A. alternata* f. sp. *Mali*, induce ultra-structural changes in the grana lamellae of chloroplasts, which leads to reduction of photosynthetic activity and PCD induction (Ueno et al. 1975; Park et al. 1981; Kohmoto et al. 1982; Miyashita et al. 1998, 2003). Proteinaceous NEs have also been identified in *Alternaria* fungi. The tetrapeptides AS-I toxin and Maculosin were isolated from culture filtrate of *A. alternata* pathogenic on sunflower (*Helianthus annuus*, Asteraceae; Liakopoulou-Kyriakides et al. 1997) and *Centaurea maculosa* (Asteraceae; Stierle et al. 1988) and cause chlorosis or necrosis on leaf tissue of the respective host plants.

Necrotrophic fungi of the genus *Cochliobolus* produce isolate-specific NEs, which are required for pathogenicity on their respective hosts. The first identified PCD-inducing compound was victorin, which confers pathogenicity to *C. victoriae* on oat (*Avena sativa*) genotypes carrying the *Vb* locus, which acts as a susceptibility determinant. The *Vb* locus was genetically inseparable from the *Pc-2* locus that is required for resistance to the biotrophic rust fungus, *Puccinia coronata* (Scheffer et al. 1967; Mayama et al. 1995; Lorang et al. 2004). The first evidence that necrotrophic fungi can hijack the plant immune system for their benefit was obtained in 2007 when the *LOV*1 gene was cloned from a victorin-sensitive Arabidopsis genotype. The *LOV1* gene encodes a NB-LRR receptor protein that structurally resembles *Arabidopsis* receptor proteins that are involved in resistance against biotrophic pathogens (Lorang et al. 2007, 2012; Wolpert and Lorang 2016). Victorin consists of a mixture of heterodetic cyclic hexapeptides comprising victorin B, C, D, E, victoricine, and HV-toxin M (Macko et al. 1985; Wolpert et al. 1986; Kono et al. 1989). Recently, victorin was shown to be a member of the ribosomally synthesized and posttranslationally modified peptides (RiPPs, Kessler et al. 2020). The core peptides that form the backbones of several RiPPs are encoded in multiple repeats within a single precursor peptide gene (Arnison et al. 2013). After cleavage from the precursor peptide by Kexin-type endopeptidases, the core peptides are cyclized and posttranslationally modified via enzymes encoded by the corresponding gene clusters (Umemura 2020; Outram et al. 2021a, b). Several phytotoxic secondary metabolites that act as NEs have been identified in other host-specific *Cochliobolus* spp. For example, *C. heterostrophus* race T possesses polyketide synthases to produce the secondary metabolite T-toxin, a host-specific NE that confers pathogenicity in *Zea mays* that possesses the plastid-mitochondrial gene *T-urf13* that confers cytoplasmic male sterility. The biosynthetic gene cluster required for T-toxin production is absent in *C. heterostrophus* race O (nonpathogenic in maize) and is located on a 1.2 Mb DNA segment (Kodama et al. 1999; Turgeon and Baker 2007). T-toxin binds to the mitochondrial membrane protein produced by the *T-urf13* gene, leading to changes in oxidative phosphorylation, leakage of nutrients, cessation of ATP synthesis, and ultimately cell death (Turgeon and Baker 2007). Intriguingly, the gene cluster required for T-toxin biosynthesis is not restricted to *C. heterostrophus* race T (Condon et al. 2018). In addition to phytotoxic secondary metabolites, *C. heterostrophus* secretes proteinaceous Nes, which were identified in an effector study among different maize-pathogenic and endophytic *Cochliobolus* isolates (Condon et al. 2013). One of these proteinaceous NEs could induce light-dependent PCD in maize mesophyll cells (Lu et al. 2015) and showed sequence similarity to the PCD-inducing *P. nodorum* ToxA protein (Ciuffetti et al. 1997, 2010). The *C. heterostrophus* ToxA-like protein is predicted to adopt a similar structure, even if lacking a sequence motif required for endocytotic uptake of *P. nodorum* ToxA into host mesophyll cells (Manning et al. 2008; Lu et al. 2015). In contrast to the dominant genes required for susceptibility to *C. heterostrophus* and *C. victoriae*, susceptibility to *C. carbonum* is conferred by a homozygous recessive maize gene (Johal and Briggs 1992; Multani et al. 1998). *C. carbonum* race 1 produces the cyclic-tetrapeptide HC-toxin, which confers pathogenicity in maize genotypes harboring the nonfunctional carbonyl reductases Hm1 and Hm2, which are incapable of inactivating the fungal effector (Johal and Briggs 1992). HC-toxin inhibits histone deacetylase causing hyperacetylated nucleosomal histones and altering expression of genes involved in plant defense responses (Ransom and Walton 1997; Walton 2006). Despite its importance for fungal virulence, there is no evidence that HC-toxin is directly involved in host PCD induction.

6.3.3 *Parastagonospora*

The fungal species *Parastagonospora nodorum* is the causal agent of SNB, a devastating disease in wheat. After spore germination, invasive hyphae penetrate the host leaf tissues and grow initially at subcuticular level. This first biotrophic phase is characterized by the swelling of hyphal tips and development of lateral hyphae, resembling appressoria formation (Zinkernagel et al. 1988). The formation of primary lesions marks the initiation of the necrotrophic phase where the fungus has killed the first mesophyll cells (Solomon et al. 2006). When the infected tissue turns yellow the fungus produces pycnidia in the collapsed tissue and continues to expand through the leaf (Solomon et al. 2006). Secreted NEs are key determinants for pathogenicity in the *P. nodorum*–wheat interaction. Upon perception by the host cell, they trigger PCD, thereby mediating disease development. In total, nine proteinaceous NEs have been characterized at molecular level in this pathosystem (Table 6.1).

Some of the nine described NEs appear to be unequally distributed among the *P. nodorum* isolates worldwide. A population study performed in 2013 (McDonald et al. 2013) showed that functional SnToxA, SnTox1, and SnTox3 alleles were present only in certain isolates and that they displayed substantial diversity at the amino acid sequence level. A total of 18, 11, and 14 protein isoforms were reported for SnTox1, SnTox3, and SnToxA, respectively (McDonald et al. 2013). The high rate of non-synonymous substitutions in the different alleles indicates that each of the three effectors is undergoing positive selection, in the assumption that a high rate of non-synonymous mutations occurring in pathogen effector genes reflects diversifying selection, which promotes the evolution of isoforms that are altered in their recognition by HSGs (Stukenbrock and McDonald 2007). This assumption finds evidence in a study about effector activity of SnToxA isoforms present in a *P. nodorum* isolate population (Tan et al. 2012). SnToxA isoforms H2 and H4 were the most active when infiltrated in *Tsn1* carrying wheat lines. The SnToxA isoforms H2 and H4 were conserved in South African *P. nodorum* isolates, which were among the most aggressive isolates in the population and which also displayed the highest sporulation rate, indicating a beneficial effect on fungal fitness and virulence (Tan et al. 2012; Stukenbrock and McDonald 2007).

Table 6.1 Effector–susceptibility gene interactions in the *P. nodorum*–wheat pathosystem

P. nodorum NE	Wheat HSG	NE isoforms	NE target	Orthologous NE genes in other fungi	Study
SnToxA	*Tsn-1*	14	TaNHL10 ToxABP1 Plastocyanin PR-1-5	*Phaeosphaeria avenaria* f. sp. *tritici* 1 *Pyrenophora tritici-repentis* *Cochliobolus heterostrophus*	Dagvadorj et al. (2022)
SnTox1	*Snn1*	18	Snn1 WAK receptor	*Phaeosphaeria avenaria* f. sp. *tritici* 1	Shi et al. (2016b), Liu et al. (2016)
SnTox2	*Snn2*	?	?	?	Zhang et al. (2009)
SnTox3	*Snn3-B1* and *Snn3-D1*	11	Snn3-D1 PK-MSD	*Phaeosphaeria avenaria* f. sp. *tritici 1*	Outram et al. (2021a)
SnTox4	*Snn4*	?	?	?	Abeysekara et al. (2012)
SnTox5	*Snn5*	20	?	?	Kariyawasam et al. (2022)
SnTox6	*Snn6*	?	?	?	Gao et al. (2015)
SnTox7	*Snn7*	?	?	?	Shi et al. (2015)
SnTox267	*Snn2*, *Snn6*, and *Snn7*	32	?	?	Richards et al. (2022)

WAK wall-associated kinase, *PK-MSD* protein kinase-major sperm protein, *PR* pathogenesis-related protein

6.4 Plant Cellular Targets of NEs

Plants possess a wide range of receptors that operate in the surveillance for the plant immune system, by recognizing the so-called immunogenic patterns (IPs), in both the apoplast and the cytoplasm (Gust et al. 2017; van der Burgh and Joosten 2019). Triggers of plant immunity include damage-associated molecular patterns (DAMPs) (Boller and Felix 2009), structural components of the pathogen (microbe-associated molecular patterns, MAMPs), and pathogen-secreted effectors (Kanyuka and Rudd 2019). Fungal NEs are secreted during plant infection and their target may reside in the apoplast or in the cytoplasm of plant cells. Apoplastic triggers of immunity are referred to as extracellular immunogenic patterns (ExIPs) (van der Burgh and Joosten 2019). These are perceived by transmembrane pattern recognition receptors (PRRs) located on the cell surface, which are grouped into receptor-like kinases (RLKs; containing a cytoplasmic kinase domain) and receptor-like proteins (RLPs; which lack such a domain) (Zipfel 2014; Gust et al. 2017). By contrast, cytoplasmic triggers of plant immunity are referred to as intracellular immunogenic patterns (InIPs) and are recognized by cytoplasmic receptors, usually from the family of NB-LRRs (van der Burgh and Joosten 2019). Extra- and intracellular receptors possess highly variable LRR domains that confer specific binding capacity to different IPs (van der Burgh and Joosten 2019). Upon NE recognition and interaction with the corresponding plant cellular targets, several immune responses may be activated. These responses comprise alkalization of the apoplast (Granado et al. 1995), cell wall reinforcement via callose and lignin deposition (Luna Diez et al. 2010; Voigt 2014), accumulation of oxygen and nitrogen radicals, an increase in intracellular calcium (Grant and Loake 2000), biosynthesis of pathogenesis-related (PR) proteins (Jain and Khurana 2018) and phytoalexins (Jeandet 2015), and the release of small RNAs to interfere with the transcriptional machinery of the invading pathogen (Cai et al. 2018). The downstream signaling pathways of these cellular responses may eventually lead to PCD induction. In this paragraph, some examples of NE targets in plants will be presented and the molecular mechanisms underlying PCD induction will be discussed.

6.4.1 Apoplastic Effector Targets

The plant apoplast comprises cell walls components such as lignin, (hemi-)cellulose, and pectin, and defines the interface between attacking fungal mycelium and defending plant cells (Doehlemann and Hemetsberger 2013). The apoplast is the location where NEs are initially delivered by invasive hyphae and appressoria to induce host PCD and enable disease development. Several *Botrytis cinerea* endopolygalacturonases (PGs) can induce defense responses in plants when applied as single proteins. For example, exogenous BcPG1 application to *Vitis vinifera* cell suspension caused the activation of two mitogen-activated protein kinases, the production of active oxygen species, and phytoalexins (Poinssot et al. 2003). In addition, BcPG2 protein infiltration in *Arabidopsis* Col-0 plants carrying the RLP-type receptor RBPG1 induced PCD independently from the catalytic activity (Zhang et al. 2014). Coimmunoprecipitation (Co-IP) assays showed that RBPG1 and BcPG2 interact in the apoplast and that the intracellular RLKs signaling of the *Arabidopsis* SOBIR1-BAK1 complex is required for PCD induction (Zhang et al. 2014; van der Burgh et al. 2019). Analogously, the xylanase BcXyl1 and xyloglucanase BcXYG1 were able to induce PCD in *N. benthamiana* and tomato independently from their enzymatic activity and the signal transmission, leading to PCD required the RLK complex BAK1-SOBIR1 (Zhu et al. 2017; Yang et al. 2018a). Likewise, the *Botrytis cinerea* xylanase BcXyn11A possesses PCD-inducing capability in plants and contributes to virulence. This property does not require the hydrolytic activity of the enzyme since *B. cinerea* Xyl11A mutant complemented with catalytically inactive Xyn11A regained the full virulence in the host plant (Brito et al. 2006; Noda et al. 2010). More recently, it was shown that a 25-residue peptide (named Xyn25) from BcXyn11A is sufficient to elicit the immune responses and PCD induction (Frías et al. 2019). It is likely that BcXyn11A recognition in the apoplast of tomato is mediated by two apoplastic RLP-type PRRs (named LeEIX1 and LeEIX2) that can interact with the xylanase EIX from fungus *Trichoderma viride* (Ron and Avni 2004), which is orthologous to *B. cinerea* BcXyn11A. EIX protein recognition activated immune responses and PCD induction in tomato and silencing of the *LeEIX* genes prevented the binding of EIX to cells and inhibited PCD induction (Ron and Avni 2004).

Apoplastic PRRs have been also identified for NEs without annotated enzymatic domains. For example, the *P. nodorum* effector protein SnTox1 interacts directly with the galacturonan-binding domain of the wheat Snn1 receptor protein, member of the wall-associated kinase (WAK) family of plant RLKs (Shi et al. 2016b). WAK proteins were initially described as receptors for oligogalacturonides (OGs), which act as DAMPs released by the plant cell upon pathogen attack (Brutus et al. 2010). However, the recognition of the SnTox1 effector by the Snn1 receptor results in PCD induction that leads to disease development instead of resistance. A recent study identified a novel plant apoplastic effector target in the wheat–*P. nodorum* pathosystem (Zhang et al. 2021). The transmembrane protein kinase major sperm protein (PK-MSP) encoded by the wheat HSG *Snn3-D1* gene is required for PCD induction upon recognition of the *P. nodorum* NE SnTox3. SnTox3 has features of a RiPP, since the pro-peptide undergoes posttranslational kexin-dependent cleavage in the fungal Golgi-apparatus before being delivered into the host apoplast as mature NE (Outram et al. 2021a). Importantly, the pro-domain is not required for the PCD-inducing capacity of SnTox3 when the effector is infiltrated as purified protein into Snn3 harboring wheat genotypes (Outram et al. 2021a). Structures of MSPs display similar folding as immunoglobulin-like domains, involved in protein–protein interactions (Tarr and Scott 2005). In plants, MSPs have been described as vesicle-associated membrane protein-associated proteins (VAPs), playing a role in vesicle transport. VAPs are integral membrane proteins made of an extracellular N-terminal MSP domain, a coiled-coil region, and a C-terminal transmembrane domain (Kamemura and Chihara 2019; Tarr and Scott 2005). *N. benthamiana* VAP27 interacts with the tomato RLP Cf-9, which confers resistance to the biotrophic Ascomycete *Cladosporium fulvum*-secreting effector Avr9 (Laurent et al. 2000). VAP27 participates in the protein complex between membrane and cytoplasmic components and presumably takes part in the signaling for PCD induction (Laurent et al. 2000).

NE activity has been also reported for proteins that perforate plasma membranes of plant cells, thereby disrupting cell integrity and inducing PCD. The best studied NEs of this type are the so-called necrosis- and ethylene-inducing peptide 1 (Nep1)-like proteins (NLPs) of *Alternaria*, *Botrytis*, and *Sclerotinia* (Duhan et al. 2021; Schouten et al. 2008; Steentjes et al. 2022; Dallal Bashi et al. 2010). NLPs belong to a protein superfamily widely distributed among phylogenetically unrelated microbes such as bacteria, fungi, and Oomycota (Seidl and van den Ackerveken 2019). NLPs are currently the only proteinaceous NEs that possess membranolytic activity in plant cells of both dicot and monocot species (Steentjes et al. 2022). NLPs can interact with plant plasma membranes because of their affinity to glycosylinositol phosphorylceramide (GIPCs) sphingolipids (Lenarčič et al. 2017; Steentjes et al. 2022). Recently, the membranolytic activity of NLPs has been elucidated as a multistep process that causes pore formation in the plasma membrane (Pirc et al. 2022). After secretion into the apoplast, NLPs target the GIPC-rich plant cell membrane. The low ionic strength in the apoplast enables an electrostatically driven initial interaction of NLPs with anionic GIPCs. NLPs then associate into clusters and cause the formation of membrane openings leading to leakage of the cell content. It has been proposed that clusters of NLP molecules interact with multiple GIPCs, which may cause disruption of GIPC homeostasis, leading to reassembling of membrane layers and pore formation (Pirc et al. 2022). In addition to their membranolytic activity, NLPs can be perceived in *Arabidopsis* by the RLP AtRLP23 through the binding of a 20 amino acid epitope, nlp20. After binding to nlp20, RLP23 forms a complex with the RLKs SOBIR1 and BAK1, causing the activation of immune responses but without triggering PCD induction, suggesting that its immunogenic activity is not linked to cell death or plasma membrane damage (Böhm et al. 2014; Albert et al. 2015).

6.4.2 Cytoplasmic Effector Targets

Necrotrophic Ascomycota do not possess structures for NE delivery into the host cell such as the bacterial Type III secretion system, stylets of nematodes or haustoria, and specialized feeding structures, which are developed by many (hemi-) biotrophic fungi and Oomycota. By contrast, NEs targeting plant cytoplasmic components are initially secreted in the apoplast and subsequently taken up by the plant cell. NE uptake can occur via endocytosis as in case of the *P. nodorum* effector SnToxA. The tripeptide RGD (Arg-Gly-Asp) motif is required for SnToxA internalization into host cells (Manning et al. 2008) and *Tsn1*-mediated PCD induction in wheat. The HSG *Tsn1* encodes a cytoplasmic NB-LRR receptor protein, carrying an N-terminal serine/threonine protein kinase domain, showing the typical architecture of classical receptor proteins involved in resistance against biotrophic pathogens (Faris et al. 2010). Nevertheless, there is no evidence of direct molecular SnToxA–Tsn1 interaction and the actual SnToxA receptor has been subject of debate until recently. Several studies tried to unravel the PCD-inducing mechanism of SnToxA through protein–protein interaction approaches (Lu et al. 2014; Manning et al. 2007). A study from 2007 showed that SnToxA associates with the chloroplast protein ToxABP1, a homolog to the Thylakoid formation 1 (Thf1) protein of *Arabidopsis* (Manning et al. 2007). SnToxA–ToxABP1 interaction was proposed to cause oxidative stress by interfering with the chloroplast metabolism, thereby inducing PCD (Ciuffetti et al. 2010). Another study identified SnToxA interaction with a chloroplast-localized plastocyanin (PC) via yeast two-hybrid (Y2H). Silencing the PC gene resulted in localized cell death similar to SnToxA-induced necrosis, indicating that the SnToxA–PC interaction may have a role in mediating PCD induction. In a more recent study, SnToxA was found to interact with the pathogenesis-related protein PR-1-5 (Lu et al. 2014). Targeted mutagenesis identified residues on both SnToxA and PR-1-5, which are essential for SnToxA–PR-1-5 interaction and PCD induction (Lu et al. 2014). Finally, a recent study demonstrated that SnToxA interacts with the apoplastic late embryogenesis abundant 2 (LEA2) domain of the TaNHL10 receptor protein. *In planta* Co-IP and confocal microscopy showed that SnToxA–TaNHL10 interaction is required for PCD induction in *Tsn1* harboring wheat (Dagvadorj et al. 2022). However, given the internalization of SnToxA, how could an extracellular domain of TaNHL10 be involved in Tsn1-mediated wheat sensitivity to SnToxA? The authors postulated that the initial recognition of the RGD motif in SnToxA occurs in the apoplast by TaNHL10, as well as mediates the internalization of SnToxA into host cells (Dagvadorj et al. 2022). In addition, PCD induction has never been observed upon transient co-expression of *SnToxA*, *TaNHL10*, and *Tsn1* in *N. benthamiana* or tobacco, indicating that these plants lack the downstream signaling component(s) required for Tsn1-mediated PCD induction (Dagvadorj et al. 2022). Analogously to the wheat HSG *Tsn1*, the cytoplasmic NB-LRR receptor protein LOV1 mediates PCD induction in victorin-sensitive *Arabidobsis* ecotypes and confers susceptibility to *Cochliobolus victoriae* (Lorang et al. 2004, 2007, 2012; Gilbert and Wolpert 2013, Wolpert and Lorang 2016). Victorin is a proteinaceous NE of the oat pathogen *C. victoriae* and is required for PCD induction and pathogenicity in oat genotypes carrying the dominant *Vb* HSG (Wolpert et al. 2011). In *Arabidopsis*, susceptibility to *C. victoriae* fits in the guard model of plant defense (van der Hoorn and Kamoun 2008). Instead of directly interacting with the cytoplasmic NB-LRR type receptor protein LOV1, victorin binds to thioredoxins, protein disulfide oxidoreductases that regulate cellular redox homeostasis (Sweat and Wolpert 2007). *LOV1* activation occurs when victorin binds the active site Cys^{39} of TRX-h5, which controls the redox status of NPR1, a master regulator of salicylic acid (SA)-mediated defense genes (Tada et al. 2008). When LOV1 is absent, victorin inhibits TRX-h5 and thereby compromises plant defense but it does not trigger PCD induction, which is required for *C. victoriae* pathogenicity and disease development in *Arabidopsis*. When LOV1 is present, victorin binding to TRX-h5 activates LOV1, thereby causing a PCD and conferring disease susceptibility.

Subcellular compartments such as mitochondria and plastids represent valuable targets for NEs as their functional perturbation may induce PCD in the host. For example, the *Alternaria alternata* f. sp. *citri jambhiri* phytotoxic secondary metabolite ACR-toxin 1 confers pathogenicity by targeting the mitochondrial membrane in rough lemon, thereby causing uncoupled oxidative phosphorylation, inhibition of mitochondrial respiration, Ca^{2+} and NAD^{+} leakage, and eventually PCD induction (Akimitsu et al. 1989; Ohtani et al. 2002). An analogous response was observed in maize due to the activity of the T-toxin produced by *C. heterostrophus* race T. T-toxin shares chemical similarities with ACR-toxin 1 of *Alternaria alternata* f. sp. *citri jambhiri* and causes pore formation in mitochondrial membranes of maize cells, thereby increasing ion permeability that disrupts mitochondrial respiration and triggers PCD induction (Levings and Siedow 1992). Similarly, the small secreted cysteine-rich NE of *Sclerotinia sclerotiorum* SSVP1 caused PCD when transiently expressed in *N. benthamiana* and was shown to be internalized both during fungal infection and in the absence of the pathogen (Lyu et al. 2016). SSVP1 can form homodimers and interacts with QCR8, a subunit of the mitochondrial cytochrome b–c1 complex (Lyu et al. 2016). Targeted mutagenesis at specific cysteine residues showed that disulfide bonds are required for homodimerization and for the molecular interaction with plant QRC8 (Lyu et al. 2016). QRC8 takes part in mitochondrial respiration but is synthetized in the cytoplasm. Co-localization assays showed that SsSSVP1 hijacks QCR8 in the cytoplasm and might prevent QCR8 translocation into mitochondria, thereby abrogating the assembly of a functional cytochrome b–c1 complex. Given the requirement of QCR8 for normal development of *N. benthamiana*, interference with the QRC8 biological function by SsSSVP1 hijacks plant cell metabolism and thereby causes PCD induction to facilitate *S. sclerotiorum* infection (Lyu et al. 2016).

Photochemical activity in chloroplasts represents the primary energy source for autotrophic organisms as plants. Chloroplast function must be tightly regulated to avoid ROS accumulation, which might damage the plant cell itself, thereby inducing PCD. Fungal pathogens may have evolved NEs to target chloroplast components to promote disease development. The majority of chloroplast-targeting effectors have been reported in biotrophic fungi or in relation to the downregulation of SA-dependent defense responses (reviewed in Littlejohn et al. 2021) but little is known about NEs inducing PCD by specifically disturbing chloroplast function. One example is reported on the activity of the AM-toxins, a group of cyclic depsipeptides produced by the apple fungus *Alternaria mali*. Electron microscope observations of apple leaf cells revealed that AM-toxin caused plasma membrane invagination and ER filaments extension from plasmodesmata, followed by fragmentation and vesiculation of the grana thylakoid membranes (Park et al. 1977). Exogenous application of AM toxin inhibited photochemical activity and CO_2 fixation, thereby leading to cell starvation and PCD induction (Shimomura et al. 1991).

6.5 Plant Susceptibility and Fungal Virulence as Quantitative Traits

As discussed above, notorious necrotrophic Ascomycota produce a wide array of NEs to target different plant cellular components. While the capacity of NEs to induce host apoptosis-like PCD is required for pathogenicity, none of the individual NEs in the repertoire of an individual fungus is essential for pathogenicity. Depending on the number of NEs and their targets in the plant, the level of host susceptibility as well as the virulence of the fungal isolate can quantitatively vary. The more functional NEs are produced by a given fungal isolate, and the more corresponding functional NE targets are present in a given plant genotype, the more susceptible is the host and the more virulent is the fungus (Fig. 6.2). The best example to illustrate such a scenario comes from studies on the wheat–*P. nodorum* interaction. SNB resistance was complex to be deciphered since it was found to be a polygenic trait. When searching for resistance loci in the wheat genome, QTLs were identified on almost every chromosome (Xu et al. 2004). The identification of the first NE in *P. nodorum* (Liu et al. 2004a, b) laid the foundation for the hypothesis that disease susceptibility was determined by NE–HSG interactions. Further studies revealed eight additional NE–HSG pairs (Table 6.1). This result was achieved by considering differences in aggressiveness among *P. nodorum* isolates and differences in SNB susceptibility among wheat genotypes. Genetic studies in *P. nodorum* populations showed that the *SnTox1* gene was present in 85% of the isolates (Liu et al. 2004b), while *SnToxA* was carried by only 36% of the isolates (Stukenbrock and McDonald 2007) and *SnTox3* by 60% of the isolates (Waters et al. 2011). Some of the inverse gene-for-gene interactions showed additive effects that reflect the quantitative character of wheat susceptibility to SNB. The necrotic response observed upon SnToxA and SnTox1 protein co-infiltration in *Tsn1–Snn1* harboring wheat lines was more pronounced and faster compared to the same wheat line infiltrated with either SnToxA or SnTox1 alone (Friesen et al. 2008b). Similarly, inoculation of *P. nodorum* isolates producing both SnToxA and SnTox1 caused more severe symptoms in wheat lines harboring both HSGs compared to wheat lines carrying either *Tsn1* or *Snn1* alone (Friesen et al. 2008b). By contrast, the SnToxA–Tsn1 interaction was epistatic to SnTox3–Snn3 since the symptoms were more severe when *Snn3–Tsn1* harboring wheat lines were inoculated with *P. nodorum* isolates lacking *SnToxA* (Friesen et al. 2008b). In addition, when considering the disease incidence in an area where a particular wheat cultivar is grown, a correlation was detected between the NE repertoire in the *P. nodorum* population in that geographical area and the presence of HSG alleles in the local wheat lines. For example, the homologous HSGs *Snn3-B1* and *Snn3-D1*, respectively, originate from *Aegilops speltoides* (Poaceae, B-genome diploid progenitor) and *A. tauschii* (the D-genome ancestor of wheat), but confer sensitivity to SnTox3 to different extent (Liu et al. 2009; Zhang et al. 2011). When *Snn3-D1* was introduced into cultivated wheat varieties, an increase in SNB disease incidence was observed in Western Australia where the majority of *P. nodorum* isolates carried SnTox3 (Oliver et al. 2009; Zhang et al. 2011). When *Snn3-D1* carrying wheat lines were introduced in North America, a lower disease incidence was observed since only a limited number of *P. nodorum* isolates carried SnTox3 (Oliver et al. 2009; Zhang et al. 2011). Interestingly, NE activity in *P. nodorum* appears to be also transcriptionally regulated. A recent study (Richards et al. 2022) identified a novel NE, SnTox267, that influenced the expression level of other NEs since *SnTox267* deletion caused transcriptional upregulation of *SnToxA*, *SnTox1*, and *SnTox3*. Analogously, *SnToxA* showed higher expression in *P. nodorum* mutant isolates lacking *SnTox1* (Peters et al. 2019), indicating that genetic compensation may be important in regulating NE gene expression in *P. nodorum*.

The quantitative contribution of NEs to fungal virulence and plant susceptibility may not only be restricted to the wheat–*P. nodorum* pathosystem. Given the broad arsenal of NEs in all the above necrotrophic fungi, it is reasonable to assume that other necrotrophic Ascomycota exploit the combined activity of NEs to induce PCD in a more effective manner, and that each effector is associated to a (specific?) target in the plant cell. For SNB, selection of wheat genotypes displaying insensitivity to NEs assisted in the successfully development of wheat lines displaying increased resistance to SNB (Ruud et al. 2017; Downie et al. 2018; Haugrud et al. 2019, 2022; Faris and Friesen 2020; Cowger et al. 2020). To achieve this goal, a so-called "effector-assisted breeding" strategy was implemented (Vleeshouwers and Oliver 2014; Faris and Friesen 2020). This strategy utilizes purified NE proteins to screen the plant germplasm and select genotypes displaying insensitivity to the NE protein. If the plant is insensitive to the PCD-inducing NE, the plant will display increased resistance to the fungus producing that NE since the fungus will not be able to exploit its molecular weapons to establish infection.

6.6 Effectoromics: From In Silico Prediction to Plant Germplasm Screening

Fungal NEs are essential for PCD induction and for disease establishment of necrotrophic fungi in their host plant. NEs can be used to screen the plant germplasm for the identification of NE-insensitive genotypes in a so-called effector-assisted breeding. The prediction is that NE-insensitive plants will display enhanced resistance to the disease. Therefore, the identification of NEs is the first step for successful implementation of effector-assisted breeding. In the field of plant pathology, the term "effectoromics" was proposed to describe a high-throughput, functional genomics approach that utilizes pathogen effectors for testing plant germplasm to identify host resistance and susceptibility genes and it currently represents a widespread methodology in modern breeding (Vleeshouwers and Oliver 2014; Todd et al. 2022). Effector candidates are usually selected based on the following criteria: presence of a signal peptide for secretion; amino acid sequence length below 300 amino acids; cysteine richness (>4 cysteine residues); absence of transmembrane and specific enzyme domains; limited taxonomic distribution and sequence homology to other organisms (Saunders et al. 2012; Raffaele and Kamoun 2012; Lo Presti et al. 2015; Carreón-Anguiano et al. 2020). The selected effector candidates can be prioritized according to an increased transcript level during plant infection. Due to the development of new-omics approaches and bioinformatic tools, the number of identified and functionally characterized effectors in plant pathogenic fungi has increased from 96 in 2018 to more than 300 in 2022 (Todd et al. 2022; Carreón-Anguiano et al. 2020; Jones et al. 2021). A turning-point for effector prediction coincided with the development of machine learning methods that predict effectors according to amino acid sequence characteristics shared among experimentally validated effectors. At the moment, EffectorP represents the most commonly used machine learning algorithm for the prediction of fungal effectors (Sperschneider and Dodds 2021). Additional bioinformatic tools for effector characterization include SignalP for detection of the secretion signal peptide (Teufel et al. 2022), LOCALIZER (Sperschneider et al. 2017) for predicting the subcellular localization of the effector, and DeepTMHMM (Krogh et al. 2001) for the prediction of transmembrane domains. Alternative computational search engines for effector prediction are represented by EffHunter (Carreón-Anguiano et al. 2020) and Predector (Jones et al. 2021), which integrate multiple software tools and methods to combine features that are relevant to fungal effectors, and enable ranking of effector candidates. Finally, AlphaFold allows to study the 3D structure of proteins and thus to group effector candidates based on structural similarities with other effectors even if they are unrelated in their protein sequence (Jumper et al. 2021; Rocafort et al. 2022; Varadi et al. 2022).

Depending on the lifestyle of the pathogen, the cell death response observed in effector-sensitive plant genotypes would correlate to plant resistance or susceptibility to the pathogen that produces that effector. In early stage of infection, hemi-biotrophic pathogens secrete effectors to avoid recognition by the plant immune system and effector perception events that trigger PCD will result in plant resistance. By contrast, necrotrophs utilize effector proteins to induce cell death and promote virulence, and effector recognition correlates therefore to plant susceptibility. Parallel to the progress achieved for in silico effector prediction, several experimental strategies have been developed for high-throughput functional validation of the candidate effectors that are predicted. Agrobacterium-mediated transient transformation (ATTA) and agroinfection are two transient expression systems for functional analysis of candidate effector genes *in planta* (Domazakis et al. 2017; Du and Vleeshouwers 2017; Du et al. 2014). Both methods are based on transient expression of pathogen genes in the plant. For ATTA, the bacterial pathogen *Agrobacterium tumefaciens* mediates the transfer of T-DNA from a plasmid, which carries the cloned candidate effector gene, into the nucleus of the plant cell. Upon infiltration of *Agrobacterium* suspensions into leaf tissue, plant cells will be transiently transformed and express the transgene(s) contained in the T-DNA in the infiltrated zone only (van der Hoorn et al. 2000). On the other hand, agroinfection relies on a systemic infection with a plant virus that serves as an expression cassette for the effector gene of interest. For agroinfection, RNA viruses represent ideal vectors for transient gene expression because of their proliferation rate and dispersal through the inoculated plant tissue. Among the plant RNA viruses, potato virus X (PVX) is the most widely used vector for effector validation (Kanneganti et al. 2007; Oh et al. 2009). Once *A. tumefaciens* has transferred the T-DNA into plant cells, the PVX genome contained in the plasmid is transcribed, translated into proteins, and begins to produce viral particles. Systemic spread of viral particles in the infected tissue causes homogeneous expression of the effector embedded in the viral genome.

Both in ATTA and agroinfection, recognition of the effector by the plant cell results in necrotic symptoms because of the PCD response that is induced. Both systems have been successfully implemented for the functional validation of pathogen effectors in Oomycota (Vleeshouwers et al. 2006, 2011; Oh et al. 2009; Bailey et al. 2011), Ascomycota (Hunziker et al. 2021), and Basidiomycota (Lorrain et al. 2019), as well as for the identification of genotypes harboring resistance genes in potato (*Solanum tuberosum*, Solanaceae) (Rietman et al. 2012; Lin et al. 2022), tomato (*S. lycopersicum*, Solanaceae) (Mesarich et al. 2018), lettuce (*Lactuca sativa*, Asteraceae) (Stassen et al. 2013), and banana (*Musa acuminata*, Musaceae) (Noar and Daub 2016; Nejat et al. 2017).

Besides by transient expression of effector candidates in a host plant, effectoromic studies can be conducted with purified effector proteins

(Vleeshouwers and Oliver 2014). In this approach, the candidate effector gene is cloned in an expression vector for heterologous protein production, thereby allowing not only functional validation but also to study the protein structure and its biological function. The yeast *Pichia pastoris* is often chosen as expression system for effector proteins because it possesses the eukaryotic posttranslational modification machinery of the endoplasmic reticulum (ER) and Golgi apparatus, which is required for proper protein folding and extracellular secretion (Kombrink 2012). On the other hand, excessive *N*- and *O*-linked glycosylation of the protein of interest can in some cases occur in *P. pastoris* fermentative culture. *N*-linked glycosylation begins in the ER with the transfer of an oligosaccharide core unit to the asparagine residue in the recognition sequence Asn-X-Ser/Thr. By contrast, *O*-linked glycosylation occurs when mannose residues are attached to the hydroxyl group of the amino acids Ser and Thr. Both *N*- and *O*-linked glycosylation may impact the protein activity, thereby causing a different response in the infiltrated plant tissue (Juturu and Wu 2018; Radoman et al. 2021). Alternative methods have been developed to express candidate effectors in *Escherichia coli* as in case of the cysteine-rich effectors AvrP, AvrP123, and AvrP4 from the biotrophic fungus *Melampsora lini* and the NE proteins SnToxA, SnTox1, and SnTox3 from *P. nodorum* (Zhang et al. 2021). Despite *E. coli* represents a widely used protein production system because of its relatively low cost, rapid application, high yield, and well-characterized genomics, large-scale production of disulfide bond-containing proteins in *E. coli* is challenging because the cytoplasmic redox potential of *E. coli* does not facilitate intracellular disulfide bond formation and thus, disulfide bond-containing proteins are often incorrectly folded after translation (Lobstein et al. 2012). To tackle this problem, several strategies can be implemented, such as forcing the secretion of proteins into the periplasm and the use of commercially available *E. coli* strains that have been engineered to enhance intracellular disulfide bond formation (Berkmen 2012). After protein expression and purification, the effector can be applied to the plant via infiltration. The *P. nodorum* effectors SnToxA, SnTox1, and SnTox3 were produced in *E. coli* and used to screen the wheat germplasm in Australia and to eliminate wheat genotypes that were highly susceptible to SNB, because these genotypes showed sensitivity to the fungal effectors due to the presence of the matching HSGs *Tsn1*, *Snn1*, and *Snn3*, respectively (Friesen et al. 2007; Liu et al. 2012; Shi et al. 2016a).

6.7 Future Perspectives

Studies on effectors of (hemi-)biotrophic fungal pathogens started in the 1990s. After the cloning of the *Avr9* effector gene of *C. fulvum* (van Kan et al. 1991), an ever-increasing number of effector genes was identified that serve as avirulence factor. The effector biology of (hemi-)biotrophic pathogens has evolved into a very broad and mature field to which hundreds of research groups contribute. By contrast, the field of effector biology in necrotrophic pathogens has long been dominated by pioneering studies in *P. nodorum*. The outstanding research in the past three decades by Wolpert and colleagues, and by Friesen and Faris in the United States, as well as by Oliver and Solomon in Australia has laid the conceptual foundation for the inverse gene-for-gene interaction, which is now widely accepted, and has set the example for similar strategies in other necrotrophic fungi. Their fundamental research has resulted in the development of effector-assisted breeding strategies that have already yielded tangible results for wheat breeders, in terms of the development of germplasm with increased resistance levels and reduced dependency on fungicide applications. This perspective also looms for other necrotrophs with serious economic repercussions, such as *Alternaria* and *Botrytis* species. We actively contribute to effector-assisted breeding programs by breeders of ornamental crops in the Netherlands to reduce the usage of chemical plant protective compounds by 2030 as required by European regulations.

References

Abbas HK, Tanaka T, Duke SO, Porter JK, Wray EM, Hodges L, Riley RT (1994) Fumonisin- and AAL-toxin induced disruption of sphingolipid metabolism with accumulation of free sphingoid bases. Plant Physiol 106:1085–1093. https://doi.org/10.1104/pp.106.3.1085

Abeysekara NS, Faris JD, Chao S, McClean PE, Friesen TL (2012) Whole genome QTL analysis of Stagonospora nodorum blotch resistance and validation of the SnTox4-Snn4 interaction in hexaploid wheat. Phytopathology 102:94–104

Aggarwal R, Das S, Jahani M, Singh DV (2008) Histopathology of spot blotch [*Bipolaris sorokiniana* (teleomorph: *Cochliobolus sativus*)] infection in wheat. Acta Phytopathologica 43:23–30

Akimitsu K, Kohmoto K, Otani H, Nishimura S (1989) Host-specific effect of toxin from the rough lemon pathotype of *Alternaria alternata* on mitochondria. Plant Physiol 89(3):925–931. https://doi.org/10.1104/pp.89.3.925

Albert I, Böhm H, Albert M, Feiler CE, Imkampe J, Wallmeroth N, Nürnberger T (2015) An RLP23–SOBIR1–BAK1 complex mediates NLP-triggered immunity. Nat Plants 1(10):1–9

Arnison PG, Bibb MJ, Bierbaum G, Bowers AA, Bugni TS, Bulaj G, Camarero JA, Campopiano DJ, Challis GL, Clardy J, Cotter PD, Craik DJ, Dawson M, Dittmann E, Donadio S, Dorrestein PC, Entian KD, Fischbach MA, Garavelli JS, Göransson U, Gruber CW, Haft DH, Hemscheidt TK, Hertweck C, Hill C, Horswill AR, Jaspars M, Kelly WL, Klinman JP, Kuipers OP, Link AJ, Liu W, Marahiel MA, Mitchell DA, Moll GN, Moore BS, Müller R, Nair SK, Nes IF, Norris GE, Olivera BM, Onaka H, Patchett ML, Piel J, Reaney MJ, Rebuffat S, Ross RP, Sahl HG, Schmidt EW, Selsted ME, Severinov K, Shen B, Sivonen K, Smith L, Stein T, Süssmuth RD, Tagg JR, Tang GL, Truman AW, Vederas JC, Walsh CT, Walton JD, Wenzel SC, Willey JM, van der Donk WA (2013) Ribosomally synthesized and post-translationally modified peptide natural products: overview and recommendations for a universal nomenclature. Nat Prod Rep 30(1):108–160. https://doi.org/10.1039/c2np20085f

Bailey K, Çevik V, Holton N, Byrne-Richardson J, Sohn M, Coates KH, Woods-Tör A, Aksoy HM, Hughes L, Baxter L, Jones JDG, Beynon J, Holub EB, Tör M (2011) Molecular cloning of ATR5Emoy2 from *Hyaloperonospora arabidopsidis*, an avirulence determinant that triggers RPP5-mediated defense in *Arabidopsis*. Mol Plant-Microbe Interact 24:827–838

Balint-Kurti P (2019) The plant hypersensitive response: concepts, control and consequences. Mol Plant Pathol 20(8):1163–1178

Bashi ZD, Rimmer SR, Khachatourians GG, Hegedus DD (2013) *Brassica napus* polygalacturonase inhibitor proteins inhibit *Sclerotinia sclerotiorum* polygalacturonase enzymatic and necrotizing activities and delay symptoms in transgenic plants. Can J Microbiol 59(2):79–86

Berkmen M (2012) Production of disulfide-bonded proteins in *Escherichia coli*. Protein Expr Purif 82:240–251

Bi K, Scalschi L, Jaiswal N, Mengiste T, Fried R, Sanz AB, Sharon A (2021) The *Botrytis cinerea* Crh1 transglycosylase is a cytoplasmic effector triggering plant cell death and defense response. Nat Commun 12(1):1–15

Bi K, Liang Y, Mengiste T, Sharon A (2022) Killing softly: a roadmap of *Botrytis cinerea* pathogenicity. Trends Plant Sci 28(2):211–222

Böhm H, Albert I, Oome S, Raaymakers TM, van den Ackerveken G, Nürnberger T (2014) A conserved peptide pattern from a widespread microbial virulence factor triggers pattern-induced immunity in *Arabidopsis*. PLoS Pathog 10(11):e1004491

Boller T, Felix G (2009) A renaissance of elicitors: perception of microbe-associated molecular patterns and danger signals by pattern-recognition receptors. Annu Rev Plant Biol 60:379–406

Bottini AT, Gilchrist DG (1981) Phytotoxins. I. A 1-aminodimethylheptadecapentol from *Alternaria alternata* f. sp. *lycopersici*. Tetrahedron Lett 22(29):2719–2722

Brito N, Espino JJ, González C (2006) The endo-β-1, 4-xylanase Xyn11A is required for virulence in *Botrytis cinerea*. Mol Plant-Microbe Interact 19(1):25–32

Brouwer SM, Brus-Szkalej M, Saripella GV, Liang D, Liljeroth E, Grenville-Briggs LJ (2021) Transcriptome analysis of potato infected with the necrotrophic pathogen *Alternaria solani*. Plan Theory 10(10):2212

Brutus A, Sicilia F, Macone A, Cervone F, De Lorenzo G (2010) A domain swap approach reveals a role of the plant wall-associated kinase 1 (WAK1) as a receptor of oligogalacturonides. Proc Natl Acad Sci USA. 107:9452–9457

Cai Q, Qiao L, Wang M, He B, Lin FM, Palmquist J, Jin H (2018) Plants send small RNAs in extracellular vesicles to fungal pathogen to silence virulence genes. Science 360:1126–1129

Caldas ED, Jones AD, Ward B, Winter CK, Gilchrist DG (1994) Structural characterization of three new AAL-toxins produced by *Alternaria alternata* f. sp. *Lycopersici*. J Agric Food Chem 42:327–333. https://doi.org/10.1021/jf00038a018

Carreón-Anguiano KG, Islas-Flores I, Vega-Arreguín J, Sáenz-Carbonell L, Canto-Canché B (2020) EffHunter: a tool for prediction of effector protein candidates in fungal proteomic databases. Biomol Ther 10:712

Chand R, Singh HV, Joshi AK, Duveiller E (2002) Physiological and morphological aspects of *Bipolaris sorokiniana* conidia surviving on wheat straw. Plant Pathol J 18(6):328–332

Chilvers MI, du Toit LJ (2006) Detection and identification of *Botrytis* species associated with neck rot, scape blight, and umbel blight of onion. Plant Health Progr 7:38

Choquer M, Rascle C, Gonçalves IR, de Vallée A, Ribot C, Loisel E, Poussereau N (2021) The infection cushion of *Botrytis cinerea*: a fungal 'weapon' of plant-biomass destruction. Environ Microbiol 23(4):2293–2314

Ciuffetti LM, Tuori RP, Gaventa JM (1997) A single gene encodes a selective toxin causal to the development of tan spot of wheat. Plant Cell 9:135–144

Ciuffetti LM, Manning VA, Pandelova I, Betts MF, Martinez JP (2010) Hostselective toxins, Ptr ToxA and Ptr ToxB, as NEs in the *Pyrenophora tritici-repentis*-wheat interaction. New Phytol 187:911–919

Colmenares A, Aleu J, Duran-Patron R, Collado I, Hernandez-Galan R (2002) The putative role of botrydial and related metabolites in the infection mechanism of *Botrytis cinerea*. J Chem Ecol 28(5):997–1005. https://doi.org/10.1023/A:1015209817830

Condon BJ, Leng Y, Wu D, Bushley KE, Ohm RA, Otillar R, Turgeon BG (2013) Comparative genome structure, secondary metabolite, and effector coding capacity across *Cochliobolus* pathogens. PLoS Genet 9(1):e1003233

Condon BJ, Elliott C, González JB, Yun SH, Akagi Y, Wiesner-Hanks T, Turgeon BG (2018) Clues to an evolutionary mystery: the genes for T-toxin, enabler of the devastating 1970 southern corn leaf blight epidemic, are present in ancestral species, suggesting an ancient origin. Mol Plant-Microbe Interact 31(11):1154–1165

Cowger C, Ward B, Brown-Guedira G, Brown J (2020) Role of effector sensitivity gene interactions and durability of quantitative resistance to Septoria nodorum blotch in Eastern US wheat. Front Plant Sci 11:155

Cuesta Arenas Y, Kalkman ER, Schouten A, Dieho M, Vredenbregt P, Uwumukiza B, van Kan JAL (2010) Functional analysis and mode of action of phytotoxic Nep1-like proteins of *Botrytis cinerea*. Physiol Mol Plant Pathol 74(5–6):376–386. https://doi.org/10.1016/j.pmpp.2010.06.003

Dagvadorj B, Outram MA, Williams SJ, Solomon PS (2022) The necrotrophic effector ToxA from *Parastagonospora nodorum* interacts with wheat NHL proteins to facilitate Tsn1-mediated necrosis. Plant J 110(2):407–418

Dallal Bashi ZAFER, Hegedus DD, Buchwaldt L, Rimmer SR, Borhan MH (2010) Expression and regulation of *Sclerotinia sclerotiorum* necrosis and ethylene-inducing peptides (NEPs). Mol Plant Pathol 11(1):43–53

Derbyshire MC, Denton-Giles M (2016) The control of *Sclerotinia* stem rot on oilseed rape (*Brassica napus*): current practices and future opportunities. Plant Pathol 65(6):859–877

Dickman MB, Park YK, Oltersdorf T, Li W, Clemente T, French R (2001) Abrogation of disease development in plants expressing animal antiapoptotic genes. Proc Natl Acad Sci USA 98:6957–6962

Dickman MB, Williams B, Li Y, de Figueiredo P, Wolpert T (2017) Reassessing apoptosis in plants. Nat Plants 3(10):773–779

Doehlemann G, Hemetsberger C (2013) Apoplastic immunity and its suppression by filamentous plant pathogens. New Phytol 198(4):1001–1016

Domazakis E, Lin X, Aguilera-Galvez C, Wouters D, Bijsterbosch G, Wolters PJ, Vleeshouwers VGAA (2017) Effectoromics-based identification of cell surface receptors in potato. In: Shan L, He P (eds) Plant pattern recognition receptors, Methods Mol Biol, vol 1578. Humana Press, New York, pp 337–353

Downie RC, Bouvet L, Furuki E, Gosman N, Gardner KA, Mackay IJ, Mantello CC, Mellers G, Phan H, Rose GA, Tan K-C, Oliver R, Cockram J (2018) Assessing European wheat sensitivities to *Parastagonospora nodorum* necrotrophic effectors and fine mapping of the Snn3-B1 locus conferring sensitivity to the effector SnTox3. Front Plant Sci 9:881

Du J, Rietman H, Vleeshouwers VGAA (2014) Agroinfiltration and PVX agroinfection in potato and *Nicotiana benthamiana*. J Vis Exp 83:e50971. https://doi.org/10.3791/50971

Du J, Vleeshouwers VGAA (2017) New strategies towards durable late blight resistance in potato. In: Kumar CS, Xie C, Kumar TJ (eds) The potato genome. Springer Cham, pp 161–169. ISBN 978-3-319-66135-3

Duhan D, Gajbhiye S, Jaswal R, Singh RP, Sharma TR, Rajarammohan S (2021) Functional characterization of the Nep1-Like protein effectors of the necrotrophic pathogen *Alternaria brassicae*. Front Microbiol 12:738617

Faris JD, Friesen TL (2020) Plant genes hijacked by necrotrophic fungal pathogens. Curr Opin Plant Biol 56:74–80

Faris JD, Zhang Z, Lu H, Lu S, Reddy L, Cloutier S, Fellers JP, Meinhardt SW, Rasmussen JB, Xu SS, Friesen TL (2010) A unique wheat disease resistance-like gene governs effector-triggered susceptibility to necrotrophic pathogens. Proc Natl Acad Sci USA 107:13544–13549

Fillinger S, Elad Y (2016) *Botrytis*—the fungus, the pathogen and its management in agricultural systems. Springer, pp 413–486

Frías M, González C, Brito N (2011) BcSpl1, a cerato-platanin family protein, contributes to *Botrytis cinerea* virulence and elicits the hypersensitive response in the host. New Phytol 192(2):483–495. https://doi.org/10.1111/j.1469-8137.2011.03802.x

Frías M, González M, González C, Brito N (2016) BcIEB1, a *Botrytis cinerea* secreted protein, elicits a defense response in plants. Plant Sci 250:115–124. https://doi.org/10.1016/j.plantsci.2016.06.009

Frías M, González M, González C, Brito N (2019) A 25-residue peptide from *Botrytis cinerea* xylanase BcXyn11A elicits plant defenses. Front Plant Sci 10:474

Friesen TL, Faris JD (2012) Characterization of plant-fungal interactions involving NE-producing plant pathogens. In: Bolton MD, Thomma BPHJ (eds) Plant fungal pathogens: methods and protocols. Humana Press, New York, pp 191–207

Friesen TL, Meinhardt SW, Faris JD (2007) The *Stagonospora nodorum*-wheat pathosystem involves multiple proteinaceous host-selective toxins and corresponding host sensitivity genes that interact in an inverse gene-for-gene manner. Plant J 51:681–692

Friesen TL, Faris JD, Solomon PS, Oliver RP (2008a) Host-specific toxins: effectors of necrotrophic pathogenicity. Cell Microbiol 10:1421–1428. https://doi.org/10.1111/J.1462-5822.2008.01153.X

Friesen TL, Zhang Z, Solomon PS, Oliver RP, Faris JD (2008b) Characterization of the interaction of a novel *Stagonospora nodorum* host-selective toxin with a wheat susceptibility gene. Plant Physiol 146:682–693

Friesen TL, Faris JD (2021) Characterization of effector-target interactions in necrotrophic pathosystems reveals trends and variation in host manipulation. Annu Rev Phytopathol 59:77–98. https://doi.org/10.1146/annurev-phyto-120320-012807

Gao Y, Faris JD, Liu Z, Kim YM, Syme RA, Oliver RP, Xu SS, Friesen TL (2015) Identification and characterization of the SnTox6-Snn6 interaction in the *Parastagonospora nodorum*-wheat pathosystem. Mol Plant Microbe Interact 28:615–625

Gardner JM, Kono Y, Tatum JH, Suzuki Y, Takeuchi S (1985) Structure of the major component of ACRL-toxins, host-specific pathotoxic compound produced by *Alternaria citri*. Agric Biol Chem 49:1235–1238. https://doi.org/10.1271/bbb1961.49.1235

Garfinkel AR, Lorenzini M, Zapparoli G, Chastagner GA (2017) *Botrytis euroamericana*, a new species from peony and grape in North America and Europe. Mycologia 109(3):495–507. https://doi.org/10.1080/00275514.2017.1354169

Gilbert BM, Wolpert TJ (2013) Characterization of the LOV1-mediated, victorin-induced, cell-death response with virus-induced gene silencing. Mol Plant-Microbe Interact 26(8):903–917

Gilchrist DG (1997) Mycotoxins reveal connections between plants and animals in apoptosis and ceramide signaling. Cell Death Differ 4(8):689–698. https://doi.org/10.1038/sj.cdd.4400312

Gilchrist DG (1998) Programmed cell death in plant disease: the purpose and promise of cellular suicide. Annu Rev Phytopathol 36:393–414. https://doi.org/10.1146/annurev.phyto.36.1.393

Granado J, Felix G, Boller T (1995) Perception of fungal sterols in plants (subnanomolar concentrations of ergosterol elicit extracellular alkalinization in tomato cells). Plant Physiol 107:485–490

Grant J, Loake G (2000) Role of reactive oxygen intermediates and cognate redox signaling in disease resistance. Plant Physiol 124:21–30

Gupta PK, Vasistha NK, Aggarwal R, Joshi AK (2018) Biology of *Bipolaris sorokiniana* (syn. *Cochliobolus sativus*) in genomics era. J Plant Biochem Biotechnol 27(2):123–138

Gust A, Pruitt R, Nürnberger T (2017) Sensing danger, key to activating plant immunity. Trends Plant Sci 22:779–791

Hammond-Kosack KE, Rudd JJ (2008) Plant resistance signalling hijacked by a necrotrophic fungal pathogen. Plant Signal Behav 3:993–995

Haugrud ARP, Zhang Z, Richards JK, Friesen TL, Faris JD (2019) Genetics of variable disease expression conferred by inverse gene-for-gene interactions in the wheat-*Parastagonospora nodorum* pathosystem. Plant Physiol 180(1):420–434

Haugrud AR, Zhang Z, Friesen TL, Faris JD (2022) Genetics of resistance to Septoria Nodorum blotch in wheat. Theor Appl Genet:1–23

Hunziker L, Tarallo M, Gough K, Guo M, Hargreaves C, Loo TS, Bradshaw RE (2021) Apoplastic effector candidates of a foliar forest pathogen trigger cell death in host and non-host plants. Sci Rep 11(1):19958

Hyde KD, Nilsson RH, Alias SA, Ariyawansa HA, Blair JE, Cai L, Zhou N (2014) One stop shop: backbones trees for important phytopathogenic genera. Fungal Divers 67(1):21–125. https://doi.org/10.1007/s13225-014-0298-1

Ibeagha AE, Hückelhoven R, Schäfer P, Singh DP, Kogel KH (2005) Model wheat genotypes as tool to uncover defense mechanisms against the hemibiotrophic fungus *Bipolaris sorokiniana*. Phytopathology 95:528–532

Izumi Y, Ohtani K, Miyamoto Y, Masunaka A, Fukumoto T, Gomi K, Akimitsu K (2012) A polyketide synthase gene, ACRTS2, is responsible for biosynthesis of host-selective ACR-toxin in the rough lemon pathotype of *Alternaria alternata*. Mol Plant-Microbe Interact 25:1419–1429. https://doi.org/10.1094/MPMI-06-12-0155-R

Jain D, Khurana J (2018) Role of pathogenesis-related (PR) proteins in plant defense mechanism. In: Singh A, Singh IK (eds) Molecular aspects of plant–pathogen interaction. Springer, Singapore, pp 265–281

Jeandet P (2015) Phytoalexins: current progress and future prospects. Molecules 20:2770–2774

Jeblick T, Leisen T, Steidele CE, Albert I, Müller J, Kaiser S, Mahler F, Sommer F, Keller S, Hückelhoven R, Hahn M, Scheuring D (2022) *Botrytis* hypersensitive response inducing protein 1 triggers non-canonical PTI to induce plant cell death. Plant Physiol 191(1):125–141. https://doi.org/10.1093/plphys/kiac476

Johal GS, Briggs SP (1992) Reductase activity encoded by the HM1 disease resistance gene in mays. Science 258:985–987

Jones DAB, Rozano L, Debler JW, Mancera RL, Moolhuijzen PM, Hane JK (2021) An automated and combinative method for the predictive ranking of candidate effector proteins of fungal plant pathogens. Sci Rep 11:19731

Jumper J, Evans R, Pritzel A, Green T, Figurnov M, Ronneberger O, Tunyasuvunakool K, Bates R, Žídek A, Potapenko A, Bridgland A, Meyer C, Kohl SAA, Ballard AJ, Cowie A, Romera-Paredes B, Nikolov S, Jain R, Adler J, Back T, Petersen S, Reiman D, Clancy E, Zielinski M, Steinegger M, Pacholska M, Berghammer T, Silver D, Vinyals O, Senior AW, Kavukcuoglu K, Kohli P, Hassabis D (2021) Applying and improving AlphaFold at CASP14.

Proteins 89(12):1711–1721. https://doi.org/10.1002/prot.26257
Juturu V, Wu JC (2018) Heterologous protein expression in *Pichia pastoris*: latest research progress and applications. Chembiochem 19(1):7–21
Kabbage M, Williams B, Dickman MB (2013) Cell death control: the interplay of apoptosis and autophagy in the pathogenicity of Sclerotinia sclerotiorum. PLoS Pathog 9(4):e1003287
Kabbage M, Kessens R, Bartholomay L, Williams B (2017) The life and death of a plant cell. Annu Rev Plant Biol 68:375–404
Kamemura K, Chihara T (2019) Multiple functions of the ER-resident VAP and its extracellular role in neural development and disease. J Biochem 165:391–400
Kanneganti TD, Huitema E, Kamoun S (2007) In planta expression of oomycete and fungal genes. Plant-Pathogen Interact Methods Mol Biol 354:35–43. https://doi.org/10.1385/1-59259-966-4:35
Kanyuka K, Rudd J (2019) Cell surface immune receptors: the guardians of the plant's extracellular spaces. Curr Opin Plant Biol 50:1–8
Kariyawasam GK, Richards JK, Wyatt NA, Running KL, Xu SS, Liu Z, Friesen TL (2022) The *Parastagonospora nodorum* necrotrophic effector SnTox5 targets the wheat gene *Snn5* and facilitates entry into the leaf mesophyll. New Phytol 233(1):409–426
Kariyawasam GK, Nelson AC, Williams SJ, Solomon PS, Faris JD, Friesen TL (2023) The necrotrophic pathogen *Parastagonospora nodorum* is a master manipulator of wheat defense. Mol Plant-Microbe Interact 36(12):764–773. https://doi.org/10.1094/MPMI-05-23-0067-IRW
Katoch S, Sharma V, Sharma D, Salwan R, Rana SK (2022) Biology and molecular interactions of *Parastagonospora nodorum* blotch of wheat. Planta 255(1):1–18
Kessler SC, Zhang X, McDonald MC, Gilchrist CL, Lin Z, Rightmyer A, Chooi YH (2020) Victorin, the host-selective cyclic peptide toxin from the oat pathogen *Cochliobolus victoriae*, is ribosomally encoded. Proc Natl Acad Sci 117(39):24243–24250
Kodama M, Rose MS, Yang G, Yun SH, Yoder OC, Turgeon BG (1999) The translocation-associated Tox1 locus of *Cochliobolus heterostrophus* is two genetic elements on two different chromosomes. Genetics 151:585–596
Kohmoto K, Otani H, Nishimura S (1982) Action sites of AM-toxins produced by the apple pathotype of *Alternaria alternata*. In: Asada Y, Bushnell WR, Ouchi S, Vance CP (eds) Plant infection: the physiological and biochemical basis. Springer, Berlin, pp 81–136
Kohmoto K, Itoh Y, Shimomura N, Kondoh Y, Otani H, Kodama M, Nishimura S, Nakatsuka S (1993) Isolation and biological activities of two host-specific toxins from the tangerine pathotype of *Alternaria alternata*. Phytopathology 83:495–502
Kombrink A (2012) Heterologous production of fungal effectors in *Pichia pastoris*. In: Bolton MD, Thomma BPHJ (eds) Plant fungal pathogens. Methods in molecular biology, vol 835. Humana Press, New York, pp 209–217
Kono Y, Kinoshita T, Takeuchi S, Daly J (1989) Structure of amino acids isolated from hydrolyzed HV-toxin M, a host-specific toxin-related compound produced by *Helminthosporium victoriae*. Agric Biol Chem 53(2):505–511
Krogh A, Larsson B, Von Heijne G, Sonnhammer EL (2001) Predicting transmembrane protein topology with a hidden Markov model: application to complete genomes. J Mol Biol 305(3):567–580
Kumar J, Schäfer P, Hückelhoven R, Kogel KH (2002) *Bipolaris sorokiniana*, a cereal pathogen of global concern: cytological and molecular approaches towards better control. Mol Plant Pathol 3:185–195
Lanoiselet V, Cother EJ, Ash GJ (2005) Risk assessment of exotic plant diseases to the Australian rice industry, with emphasis on rice blast. Farrer Centre Charles Sturt University, pp 29–30
Laurent F, Labesse G, de Wit P (2000) Molecular cloning and partial characterization of a plant VAP33 homologue with a major sperm protein domain. Biochem Biophys Res Commun 270:286–292
Lawrence DP, Park MS, Pryor BM (2012) *Nimbya* and *Embellisia* revisited, with *nov. comb* for *Alternaria celosiae* and *A. perpunctulata*. Mycol Prog 11:799–815
Lawrence DP, Gannibal PB, Peever TL, Pryor BM (2013) The sections of *Alternaria*: formalizing species-group concepts. Mycologia 105:530–546
Lawrence DP, Gannibal PB, Dugan FM, Pryor BM (2014) Characterization of *Alternaria* isolates from the infectoria species group and a new taxon from *Arrhenatherum, Pseudoalternaria arrhenatheria* sp. nov. Mycol Prog 13:257–276
Lawrence DP, Rotondo F, Gannibal PB (2016) Biodiversity and taxonomy of the pleomorphic genus *Alternaria*. Mycol Prog 15(1):1–22
Leisen T, Werner J, Pattar P, Safari N, Ymeri E, Sommer F, Hahn M (2022) Multiple knockout mutants reveal a high redundancy of phytotoxic compounds contributing to necrotrophic pathogenesis of *Botrytis cinerea*. PLoS Pathog 18(3):e1010367
Lenarčič T, Albert I, Böhm H, Hodnik V, Pirc K, Zavec AB, Podobnik M, Pahovnik D, Žagar E, Pruitt R, Nürnberger T (2017) Eudicot plant-specific sphingolipids determine host selectivity of microbial NLP cytolysins. Science 358:1431–1434
Lenz HD, Haller E, Melzer E, Kober K, Wurster K, Stahl M, Nürnberger T (2011) Autophagy differentially controls plant basal immunity to biotrophic and necrotrophic pathogens. Plant J 66(5):818–830
Levings CS, Siedow JN (1992) Molecular basis of disease susceptibility in the Texas cytoplasm of maize. Years Plant Mol Biol 10:135–147
Liakopoulou-Kyriakides M, Lagopodi AL, Thanassoulopoulos CC, Stavropoulos GS, Magafa V (1997) Isolation and synthesis of a host-selective toxin produced by *Alternaria alternata*. Phytochemistry 45:37–40. https://doi.org/10.1016/S0031-9422(96)00789-3

Liang H, Yao N, Song JT, Luo S, Lu H, Greenberg JT (2003) Ceramides modulate programmed cell death in plants. Genes Dev 17(21):2636–2641

Liang X, Rollins JA (2018) Mechanisms of broad host range necrotrophic pathogenesis in *Sclerotinia sclerotiorum*. Phytopathology 108(10):1128–1140

Lin X, Olave-Achury A, Heal R, Pais M, Witek K, Ahn HK, Jones JD (2022) A potato late blight resistance gene protects against multiple *Phytophthora* species by recognizing a broadly conserved RXLR-WY effector. Mol Plant 15(9):1457–1469

Lithourgidis AS, Roupakias DG, Damalas CA (2003) Inheritance of resistance to sclerotinia stem rot (*Sclerotinia trifoliorum*) in faba beans (*Vicia faba* L.). Field Crop Res 91:125–130

Littlejohn GR, Breen S, Smirnoff N, Grant M (2021) Chloroplast immunity illuminated. New Phytol 229(6):3088–3107

Liu Z, Faris JD, Meinhardt SW, Ali S, Rasmussen JB, Friesen TL (2004a) Genetic and physical mapping of a gene conditioning sensitivity in wheat to a partially purified host-selective toxin produced by *Stagonospora nodorum*. Phytopathology 94(10):1056–1060

Liu Z, Friesen TL, Rasmussen JB, Ali S, Meinhardt SW, Faris JD (2004b) Quantitative trait loci analysis and mapping of seedling resistance to *Stagonospora nodorum* leaf blotch in wheat. Phytopathology 94(10):1061–1067

Liu Z, Faris JD, Oliver RP, Tan KC, Solomon PS, McDonald MC, McDonald BA, Nunez A, Lu S, Rasmussen JB, Friesen TL (2009) SnTox3 acts in effector triggered susceptibility to induce disease on wheat carrying the *Snn3* gene. PLoS Pathog 5(9):e1000581

Liu Z, Zhang Z, Faris JD, Oliver RP, Syme R, McDonald MC, Friesen TL (2012) The cysteine rich necrotrophic effector SnTox1 produced by *Stagonospora nodorum* triggers susceptibility of wheat lines harboring *Snn1*. PLoS Pathog 8(1):e1002467

Liu Z, Gao Y, Kim YM, Faris JD, Shelver WL, de Wit PJGM, Xu SS, Friesen TL (2016) SnTox1, a *Parastagonospora nodorum* necrotrophic effector, is a dual-function protein that facilitates infection while protecting from wheat-produced chitinases. New Phytol 211:1052–1064

Lobstein J, Emrich CA, Jeans C, Faulkner M, Riggs P, Berkmen M (2012) SHuffle, a novel *Escherichia coli* protein expression strain capable of correctly folding disulphide bonded proteins in its cytoplasm. Microb Cell Factories 11:56

Lo Presti L, Lanver D, Schweizer G, Tanaka S, Liang L, Tollot M, Zuccaro A, Reissmann S, Kahmann R (2015) Fungal effectors and plant susceptibility. Annu Rev Plant Biol 66:513–545

Lorang J, Kidarsa T, Bradford CS, Gilbert B, Curtis M, Tzeng SC, Wolpert TJ (2012) Tricking the guard: exploiting plant defense for disease susceptibility. Science 338(6107):659–662

Lorang JM, Carkaci-Salli N, Wolpert TJ (2004) Identification and characterization of victorin sensitivity in *Arabidopsis thaliana*. Mol Plant Microbe Interact 17:577–582

Lorang JM, Sweat TA, Wolpert TJ (2007) Plant disease susceptibility conferred by a "resistance" gene. Proc Natl Acad Sci 104(37):14861–14866

Lorrain C, Gonçalves dos Santos KC, Germain H, Hecker A, Duplessis S (2019) Advances in understanding obligate biotrophy in rust fungi. New Phytol 222(3):1190–1206

Lu S, Faris JD, Sherwood R, Friesen TL, Edwards MC (2014) A dimeric PR-1-type pathogenesis-related protein interacts with ToxA and potentially mediates ToxA-induced necrosis in sensitive wheat. Mol Plant Pathol 15:650–663

Lu S, Turgeon BG, Edwards MC (2015) A ToxA-like protein from *Cochliobolus heterostrophus* induces light-dependent leaf necrosis and acts as a virulence factor with host selectivity on mays. Fungal Genet Biol 81:12–24

Luna Diez E, Pastor V, Jérôme R, Flors V, Mauch-Mani B, Ton J (2010) Callose deposition: a multifaceted plant defense response. Mol Plant-Microbe Interact 24:183–193

Lyu X, Shen C, Fu Y, Xie J, Jiang D, Li G, Cheng J (2016) A small secreted virulence-related protein is essential for the necrotrophic interactions of *Sclerotinia sclerotiorum* with its host plants. PLoS Pathog 12(2):e1005435

Macko V, Wolpert TJ, Acklin W, Jaun B, Seibl J, Meili J, Arigoni D (1985) Characterization of victorin C, the major host-selective toxin from *Cochliobolus victoriae*: structure of degradation products. Experientia 41:1366–1370

Malvestiti MC, Steentjes MBF, Beenen HG, Boeren S, van Kan JAL, Shi-Kunne X (2022) Analysis of plant cell death-inducing proteins of the necrotrophic fungal pathogens *Botrytis squamosa* and *Botrytis elliptica*. Front Plant Sci 13:993325. https://doi.org/10.3389/fpls.2022.993325

Manamgoda DS, Cai L, Bahkali AH, Chukeatirote E, Hyde KD (2011) *Cochliobolus*: an overview and current status of species. Fungal Divers 51:3–42

Mandal S, Rajarammohan S, Kaur J (2018) *Alternaria brassicae* interactions with the model Brassicaceae member *Arabidopsis thaliana* closely resembles those with Mustard (*Brassica juncea*). Physiol Mol Biol Plants 24(1):51–59

Manning VA, Hardison LK, Ciuffetti LM (2007) Ptr ToxA interacts with a chloroplast-localized protein. Mol Plant-Microbe Interact 20:168–177. https://doi.org/10.1094/MPMI-20-2-0168

Manning VA, Hamilton SM, Karplus PA, Ciuffetti LM (2008) The Arg-Gly-asp containing, solvent-exposed loop of Ptr ToxA is required for internalization. Mol Plant Microbe Interact 21:315–325

Mardanov AV, Beletsky AV, Kadnikov VV, Ignatov AN, Ravin NV (2014) The 203 kbp mitochondrial genome of the phytopathogenic fungus Sclerotinia borealis reveals multiple invasions of introns and genomic duplications. PLoS One 9(9):e107536

Markham JE, Hille J (2001) Host-selective toxins as agents of cell death in plant-fungus interactions. Mol Plant Pathol 2(4):229–239

Mayama S, Bordin APA, Morikawa T, Tanpo H, Kato H (1995) Association of avenalumin accumulation with co-segregation of victorin sensitivity and crown rust resistance in oat lines carrying the Pc-2 gene. Physiol Mol Plant Pathol 46:263–274

McDonald MC, Oliver RP, Friesen TL, Brunner PC, McDonald BA (2013) Global diversity and distribution of three necrotrophic effectors in *Phaeosphaeria nodorum* and related species. New Phytol 199(1):241–251

Mesarich CH, Ökmen B, Rovenich H, Griffiths SA, Wang C, Karimi Jashni M, de Wit PJGM (2018) Specific hypersensitive response–associated recognition of new apoplastic effectors from *Cladosporium fulvum* in wild tomato. Mol Plant-Microbe Interact 31(1):145–162

Michaelson LV, Napier JA, Molina D, Faure JD (2016) Plant sphingolipids: their importance in cellular organization and adaption. BBA Mol Cell Biol Lipids 1861:1329–1335. https://doi.org/10.1016/j.bbalip.2016.04.003

Minina EA, Bozhkov PV, Hofius D (2014) Autophagy as initiator or executioner of cell death. Trends Plant Sci 19(11):692–697

Miyashita M, Nakamori T, Murai T, Miyagawa H, Akamatsu M, Ueno T (1998) Facile syntheses of AM -toxins and analogs as cyclic depsipeptides by the solid-phase method. Biosci Biotechnol Biochem 62:1799–1801

Miyashita M, Nakamori T, Miyagawa H, Akamatsu M, Ueno T (2003) Inhibitor activity of analogs of AM-toxin, a host-specific phytotoxin from the *Alternaria alternata* apple pathotype, on photosynthetic O2 evolution in apple leaves. Biosci Biotechnol Biochem 67:635–638

Multani DS, Meeley RB, Paterson AH, Gray J, Briggs SP, Johal GS (1998) Plant pathogen microevolution: molecular basis for the origin of a fungal disease in mays. Proc Natl Acad Sci USA 95:1686–1691

Nakashima T, Ueno T, Fukami H, Taga T, Masuda H, Osaki K, Nishimura S (1985) Isolation and structure of AK-toxin I and II, host specific phytotoxic metabolites produced by *Alternaria alternata* Japanese pear pathotype. Agric Biol Chem 49(3):807–815

Nakatsuka S, Ueda K, Goto T, Yamamoto M, Nishimura S, Kohmoto H (1986) Structure of AF-toxin II, one of the host-specific toxins produced by *Alternaria alternata* strawberry pathotype. Tetrahedron Lett 27:2753–2756. https://doi.org/10.1016/S0040-4039(00)84635-3

Nejat N, Rookes J, Mantri NL, Cahill DM (2017) Plant–pathogen interactions: toward development of next-generation disease-resistant plants. Crit Rev Biotechnol 37(2):229–237

Nielsen K, Yohalem DS (2001) Origin of a polyploid *Botrytis* pathogen through interspecific hybridization between *Botrytis aclada* and *B. byssoidea*. Mycologia 93:1064–1071

Noar RD, Daub ME (2016) Transcriptome sequencing of *Mycosphaerella fijiensis* during association with *Musa acuminata* reveals candidate pathogenicity genes. BMC Genomics 17:690

Noda J, Brito N, González C (2010) The *Botrytis cinerea* xylanase Xyn11A contributes to virulence with its necrotizing activity, not with its catalytic activity. BMC Plant Biol 10(1):1–15. https://doi.org/10.1186/1471-2229-10-38

Ohtani K, Yamamoto H, Akimitsu K (2002) Sensitivity to *Alternaria alternata* toxin in citrus because of altered mitochondrial RNA processing. Proc Natl Acad Sci USA 99:2439–2444. https://doi.org/10.1073/pnas.042448499

Oh SK, Young C, Lee M, Oliva R, Bozkurt TO, Cano LM, Kamoun S (2009) In planta expression screens of *Phytophthora infestans* RXLR effectors reveal diverse phenotypes, including activation of the *Solanum bulbocastanum* disease resistance protein Rpi-blb2. Plant Cell 21(9):2928–2947

Oliver RP, Solomon PS (2010) New developments in pathogenicity and virulence of necrotrophs. Curr Opin Plant Biol 13(4):415–419

Oliver RP, Rybak K, Solomon PS, Ferguson-Hunt M (2009) Prevalence of ToxA-sensitive alleles of the wheat gene *Tsn1* in Australian and Chinese wheat cultivars. Crop Pasture Sci 60:348–352

Outram MA, Solomon PS, Williams SJ (2021a) Pro-domain processing of fungal effector proteins from plant pathogens. PLoS Pathog 17(10):e1010000

Outram MA, Sung YC, Yu D, Dagvadorj B, Rima SA, Jones DA, Williams SJ (2021b) The crystal structure of SnTox3 from the necrotrophic fungus *Parastagonospora nodorum* reveals a unique effector fold and provides insight into *Snn3* recognition and pro-domain protease processing of fungal effectors. New Phytol 231(6):2282–2296

Park P, Tsuda M, Hayashi Y, Ueno T (1977) Effect of a host-specific toxin (AM-toxin I) produced by *Alternaria mali*, an apple pathogen, on the ultrastructure of plasma membrane of cells in apple and Japanese pear leaves. Can J Bot 55:2383–2393

Park P, Nishimura S, Kohmoto K, Otani H, Tsujimoto K (1981) Two action sites of AM toxin I produced by apple pathotype of *Alternaria alternata* in host cell: an ultrastructural study. Can J Bot 59:301–310

Phan HT, Furuki E, Hunziker L, Rybak K, Tan KC (2021) GWAS analysis reveals distinct pathogenicity profiles of Australian *Parastagonospora nodorum* isolates and identification of marker-trait-associations to septoria nodorum blotch. Sci Rep 11(1):1–14

Pirc K, Clifton LA, Yilmaz N, Saltalamacchia A, Mally M, Snoj T, Žnidaršič N, Srnko M, Borišek J, Parkkila P, Anderluh G (2022) An oomycete NLP cytolysin forms transient small pores in lipid membranes. Sci Adv 8:eabj9406

Poinssot B, Vandelle E, Bentéjac M, Adrian M, Levis C, Brygoo Y, Garin J, Sicilia F, Coutos-Thévenot P, Pugin A (2003) The endopolygalacturonase 1 from

Botrytis cinerea activates grapevine defense reactions unrelated to its enzymatic activity. Mol Plant Microbe Interact 16:553–564

Ransom RF, Walton JD (1997) Histone hyperacetylation in mays in response to treatment with HC-toxin or infection by the filamentous fungus *Cochliobolus carbonum*. Plant Physiol 115:1021–1027

Reis EM (1991) Integrated disease management—the changing concepts of controlling head blight and spot blotch. In: 3rd International Conference on Wheat for the Nontraditional Warm Areas, Foz do Iguacu (Brazil), 29 Jul–3 Aug 1990. CIMMYT

Reis ER, Wunschr WA (1984) Sporulation of *Cochliobolus sativus* on residues of winter crops and its relationship to increase inoculums density in soil. Plant Dis 68:411–412

Richards JK, Kariyawasam GK, Seneviratne S, Wyatt NA, Xu SS, Liu Z, Friesen TL (2022) A triple threat: the *Parastagonospora nodorum* SnTox267 effector exploits three distinct host genetic factors to cause disease in wheat. New Phytol 233(1):427–442

Radoman B, Grünwald-Gruber C, Schmelzer B, Zavec D, Gasser B, Altmann F, Mattanovich D (2021) The degree and length of O-Glycosylation of recombinant proteins produced in *Pichia pastoris* depends on the nature of the protein and the process type. Biotechnol J 16(3):2000266

Raffaele S, Kamoun S (2012) Genome evolution in filamentous plant pathogens: why bigger can be better. Nat Rev Microbiol 10(6):417–430

Rietman H, Bijsterbosch G, Cano LM, Lee HR, Vossen JH, Jacobsen E, Visser RGF, Kamoun S, Vleeshouwers VGAA (2012) Qualitative and quantitative late blight resistance in the potato cultivar Sarpo Mira is determined by the perception of five distinct RXLR effectors. Mol Plant-Microbe Interact 25:910–919

Rocafort M, Bowen JK, Hassing B, Cox MP, McGreal B, de la Rosa S, Mesarich CH (2022) The *Venturia inaequalis* effector repertoire is dominated by expanded families with predicted structural similarity, but unrelated sequence, to avirulence proteins from other plant-pathogenic fungi. BMC Biol 20(1):1–24

Ron M, Avni A (2004) The receptor for the fungal elicitor ethylene inducing xylanase is a member of a resistance-like gene family in tomato. Plant Cell 16(6):1604–1615. https://doi.org/10.1105/tpc.022475

Ruud AK, Windju S, Belova T, Friesen TL, Lillemo M (2017) Mapping of *SnTox3–Snn3* as a major determinant of field susceptibility to Septoria nodorum leaf blotch in the SHA3/ CBRD× naxos population. Theor Appl Genet 130(7):13611374

Saharan GS, Mehta N, Meena PD (2016) Alternaria diseases of crucifers: biology, ecology and disease management. Springer Singapore, p 299. ISBN 978-981-10-0019-5

Sahu R, Sharaff M, Pradhan M, Pandey SP (2016) Elucidation of defense-related signaling responses to spot blotch infection in bread wheat (*Triticum aestivum* L.). Plant J 86(1):35–49

Saunders DGO, Win J, Cano LM, Szabo LJ, Kamoun S, Raffaele S (2012) Using hierarchical clustering of secreted protein families to classify and rank candidate effectors of rust fungi. PLoS One 7:e29847

Scheffer RP, Nelson RR, Ullstrup AJ (1967) Inheritance of toxin production and pathogenicity in *Cochliobolus carbonum* and *Cochliobolus victoriae*. Phytopathology 57(1967):1288–1291

Schouten A, van Baarlen P, van Kan JAL (2008) Phytotoxic Nep1-like proteins from the necrotrophic fungus *Botrytis cinerea* associate with membranes and the nucleus of plant cells. New Phytol 177(2):493–505

Seidl MF, van den Ackerveken G (2019) Activity and phylogenetics of the broadly occurring family of microbial Nep1-Like proteins. Annu Rev Phytopathol 57:367–386

Shi G, Friesen TL, Saini J, Xu SS, Rasmussen JB, Faris JD (2015) The wheat *Snn7* gene confers susceptibility on recognition of the *Parastagonospora nodorum* necrotrophic effector SnTox7. Plant Genome 8(2)

Shi G, Zhang Z, Friesen TL, Bansal U, Cloutier S, Wicker T, Rasmussen JB, Faris JD (2016a) Marker development, saturation mapping, and high-resolution mapping of the Septoria nodorum blotch susceptibility gene *Snn3-B1* in wheat. Mol Gen Genomics 291:107–119

Shi G, Zhang Z, Friesen TL, Raats D, Fahima T, Brueggeman RS, Lu S, Trick HN, Liu Z, Chao W, Faris JD (2016b) The hijacking of a receptor kinase-driven pathway by a wheat fungal pathogen leads to disease. Sci Adv 2(10)

Shimomura N, Otani H, Tabira H, Kodama M, Kohmoto K (1991) Two primary action sites for am-toxin produced by *Alternaria alternata* apple pathotype and their pathological significance. Jap J Phytopathol 57(2):247–255

Solomon PS, Lowe RGT, Tan KC, Waters ODC, Oliver RP (2006) *Stagonospora nodorum*: cause of *Stagonospora nodorum* blotch of wheat. Mol Plant Pathol 7:147–156

Spassieva SD, Markham JE, Hille J (2002) The plant disease resistance gene *Asc-1* prevents disruption of sphingolipid metabolism during AAL-toxin-induced programmed cell death. Plant J 32:561–572. https://doi.org/10.1046/j.1365-313X.2002.01444.x

Sperschneider J, Catanzariti AM, DeBoer K, Petre B, Gardiner DM, Singh KB, Taylor JM (2017) LOCALIZER: subcellular localization prediction of both plant and effector proteins in the plant cell. Sci Rep 7(1):1–14

Sperschneider J, Dodds PN (2021) EffectorP 3.0: prediction of apoplastic and cytoplasmic effectors in fungi and Oomycetes. Mol Plant-Microbe Interact 35:146–156

Stassen J, den Boer E, Vergeer PWJ, Andel A, Ellendorff U, Pelgrom K, Pel M, Schut J, Zonneveld O, Jeuken MJW, van den Ackerveken G (2013) Specific *in planta* recognition of two GKLR proteins of the downy mildew *Bremia lactucae* revealed in a large effector screen in lettuce. Mol Plant-Microbe Interact 26:1259–1270

Steentjes MBF, Herrera Valderrama AL, Fouillen L, Bahammou D, Leisen T, Albert I, Nürnberger T, Hahn M, Mongrand S, Scholten OE, van Kan JAL (2022) Cytotoxic activity of Nep1-like proteins on monocots. New Phytol 235:690–700

Stierle AC, Cardellina JH, Strobel GA (1988) Maculosin, a host-specific phytotoxin for spotted knapweed from *Alternaria alternata*. Proc Natl Acad Sci 85(21):8008–8011

Stukenbrock EH, McDonald BA (2007) Geographical variation and positive diversifying selection in the host-specific toxin SnToxA. Mol Plant Pathol 8:321–332

Sweat TA, Wolpert TJ (2007) Thioredoxin h 5 is required for victorin sensitivity mediated by a CC-NBS-LRR gene in *Arabidopsis*. Plant Cell 19(2):673–687

Tada Y, Spoel SH, Pajerowska-Mukhtar K, Mou Z, Song J, Wang C, Dong X (2008) Plant immunity requires conformational charges of NPR1 via S-nitrosylation and thioredoxins. Science 321(5891):952–956

Tan KC, Ferguson-Hunt M, Rybak K, Waters OD, Stanley WA, Bond CS, Stukenbrock EH, Friesen TL, Faris JD, McDonald BA, Oliver RP (2012) Quantitative variation in effector activity of ToxA isoforms from *Stagonospora nodorum* and *Pyrenophora tritici-repentis*. Mol Plant-Microbe Interact 25(4):515–522. https://doi.org/10.1094/MPMI-10-11-0273

Tarr DEK, Scott AL (2005) MSP domain proteins. Trends Parasitol 21:224–231

Teufel F, Almagro Armenteros JJ, Johansen AR, Gíslason MH, Pihl SI, Tsirigos KD, Winther O, Brunak S, von Heijne G, Nielsen H (2022) SignalP 6.0 predicts all five types of signal peptides using protein language models. Nat Biotechnol 40:1023–1025

Todd JNA, Carreón-Anguiano KG, Islas-Flores I, Canto-Canché B (2022) Fungal effectoromics: a world in constant evolution. Int J Mol Sci 23:13433

Turgeon BG, Baker SE (2007) Genetic and genomic dissection of the *Cochliobolus heterostrophus* Tox1 locus controlling biosynthesis of the polyketide virulence factor T-toxin. Adv Genet 57:219–261

Ueno T, Nakashima T, Hayashi Y, Fukami H (1975) Isolation and structure of AM-toxin III, a host specific phytotoxic metabolite produced by *Alternaria mali*. Agric Biol Chem 39:2081–2082

Umemura M (2020) Peptides derived from Kex2-processed repeat proteins are widely distributed and highly diverse in the Fungi kingdom. Fungal Biol Biotechnol 7(1):1–23

van der Hoorn RA, Laurent F, Roth R, de Wit PJGM (2000) Agroinfiltration is a versatile tool that facilitates comparative analyses of Avr 9/Cf-9-induced and Avr 4/Cf-4-induced necrosis. Mol Plant-Microbe Interact 13:439–446

Valero-Jiménez CA, Steentjes MB, Slot JC, Shi-Kunne X, Scholten OE, van Kan JAL (2020) Dynamics in secondary metabolite gene clusters in otherwise highly syntenic and stable genomes in the fungal genus *Botrytis*. Genome Biol Evol 12(12):2491–2507. https://doi.org/10.1093/gbe/evaa218

Van Baarlen P, Staats M, van Kan JAL (2004) Induction of programmed cell death in lily by the fungal pathogen *Botrytis elliptica*. Mol Plant Pathol 5(6):559–574. https://doi.org/10.1111/j.1364-3703.2004.00253.x

Van Bockhaven J, Spíchal L, Novák O, Strnad M, Asano T, Kikuchi S, De Vleesschauwer D (2015) Silicon induces resistance to the brown spot fungus *Cochliobolus miyabeanus* by preventing the pathogen from hijacking the rice ethylene pathway. New Phytol 206(2):761–773

Van der Burgh A, Joosten MHAJ (2019) Plant immunity: thinking outside and inside the box. Trends Plant Sci 24:587–601

Van der Burgh AM, Postma J, Robatzek S, Joosten MHAJ (2019) Kinase activity of SOBIR1 and BAK1 is required for immune signalling. Mol Plant Pathol 20(3):410–422

Van der Hoorn RA, Kamoun S (2008) From guard to decoy: a new model for perception of plant pathogen effectors. Plant Cell 20(8):2009–2017

van Kan JAL, van den Ackerveken GF, de Wit PJGM (1991) Cloning and characterization of cDNA of avirulence gene avr9 of the fungal pathogen *Cladosporium fulvum*, causal agent of tomato leaf mold. Mol Plant-Microbe Interact 4:52–59. https://doi.org/10.1094/mpmi-4-052

Varadi M, Anyango S, Deshpande M, Nair S, Natassia C, Yordanova G, Yuan D, Stroe O, Wood G, Laydon A, Žídek A, Green T, Tunyasuvunakool K, Petersen S, Jumper J, Clancy E, Green R, Vora A, Lutfi M, Figurnov M, Cowie A, Hobbs N, Kohli P, Kleywegt G, Birney E, Hassabis D, Velankar S (2022) AlphaFold protein structure database: massively expanding the structural coverage of protein-sequence space with high-accuracy models. Nucleic Acids Res 50(D1):D439–D444. https://doi.org/10.1093/nar/gkab1061

Veloso J, van Kan JAL (2018) Many shades of grey in *Botrytis*–host plant interactions. Trends Plant Sci 23(7):613–622

Vleeshouwers VG, Driesprong JD, Kamphuis LG, Torto-Alalibo T, Van't Slot KA, Govers F, Visser RG, Jacobsen E, Kamoun S (2006) Agroinfection-based high-throughput screening reveals specific recognition of INF elicitins in Solanum. Mol Plant Pathol 7(6):499–510. https://doi.org/10.1111/j.1364-3703.2006.00355.x

Vleeshouwers VG, Raffaele S, Vossen JH, Champouret N, Oliva R, Segretin ME, Rietman H, Cano LM, Lokossou A, Kessel G, Pel MA, Kamoun S (2011) Understanding and exploiting late blight resistance in the age of effectors. Annu Rev Phytopathol 49:507–531. https://doi.org/10.1146/annurev-phyto-072910-095326

Vleeshouwers VGAA, Oliver RP (2014) Effectors as tools in disease resistance breeding against biotrophic, hemibiotrophic, and necrotrophic plant pathogens. Mol Plant-Microbe Interact 27:196–206

Voigt C (2014) Callose-mediated resistance to pathogenic intruders in plant defense-related papillae. Front Plant Sci 5:168

Walton JD (2006) HC-toxin. Phytochemistry 67:1406–1413

Waters ODC, Lichtenzveig J, Rybak K, Friesen TL, Oliver RP (2011) Prevalence and importance of sensitivity to the *Stagonospora nodorum* necrotrophic effector SnTox3 in current western Australian wheat cultivars. Crop Pasture Sci 62:556–562

Wells K, Clark NJ (2019) Host specificity in variable environments. Trends Parasitol 35(6):452–465. https://doi.org/10.1016/j.pt.2019.04.001

Wolpert TJ, Lorang JM (2016) Victoria Blight, defense turned upside down. Physiol Mol Plant Pathol 95:8–13

Wolpert TJ, Macko V, Acklin W, Jaun B, Arigoni D (1986) Structure of minor host selective toxins from *Cochliobolus victoriae*. Experientia 42:1296–1299

Wolpert TJ, Dunkle LD, Ciuffetti LM (2002) Host-selective toxins and avirulence determinants: what's in a name? Annu Rev Phytopathol 40:251–285. https://doi.org/10.1146/annurev.phyto.40.011402.114210

Wolpert TJ, Lorang JM, Gilbert B, Chu A, Bradford II CS (2011) Analysis of a plant susceptibility response and its similarities to resistance. In: Genome-enabled integration of research in plant-pathogen systems. APS Press, St. Paul

Woudenberg JHC, Groenewald J, Binder M, Crous P (2013) *Alternaria* redefined. Stud Mycol 75(1):171–212

Woudenberg JHC, Seidl MF, Groenewald JZ, De Vries M, Stielow JB, Thomma BPHJ, Crous PW (2015) *Alternaria* section Alternaria: species, formae speciales or pathotypes? Stud Mycol 82(1):1–21

Xu SS, Friesen TL, Mujeeb-Kazi A (2004) Seedling resistance to tan spot and *Stagonospora nodorum* blotch in synthetic hexaploid wheats. Crop Sci 44:2238–2245

Yang G, Tang L, Gong Y, Xie J, Fu Y, Jiang D, Cheng J (2018b) A cerato-platanin protein SsCP1 targets plant PR1 and contributes to virulence of *Sclerotinia sclerotiorum*. New Phytol 217(2):739–755

Yang Y, Yang X, Dong Y, Qiu D (2018a) The *Botrytis cinerea* xylanase BcXyl1 modulates plant immunity. Front Microbiol 9:2535. https://doi.org/10.3389/fmicb.2018.02535

Yoder OC (1980) Toxins in pathogenesis. Annu Rev Phytopathol 18:103–129

Zhang L, Kars I, Essenstam B, Liebrand TW, Wagemakers L, Elberse J, van Kan JAL (2014) Fungal endopolygalacturonases are recognized as microbe-associated molecular patterns by the *Arabidopsis* receptor-like protein responsiveness to botrytis polygalacturonases1. Plant Physiol 164(1):352–364. https://doi.org/10.1104/pp.113.230698

Zhang Y, Qiu D, Zeng H, Guo L, Yang X (2015) BcGs1, a glycoprotein from *Botrytis cinerea*, elicits defence response and improves disease resistance in host plants. Biochem Biophys Res Commun 457(4):627–634

Zhang Z, Friesen TL, Xu SS, Shi G, Liu Z, Faris JD (2011) Two putatively homoeologous wheat genes mediate recognition of SnTox3 to confer effector-triggered susceptibility to *Stagonospora nodorum*. Plant J 65(1):27–38

Zhang Z, Running KL, Seneviratne S, Peters-Haugrud AR, Szabo-Hever A, Shi G, Faris JD (2021) A protein kinase–major sperm protein gene hijacked by a necrotrophic fungal pathogen triggers disease susceptibility in wheat. Plant J 106(3):720–732

Zhang ZC, Friesen TL, Simons KJ, Xu SS, Faris JD (2009) Identification, development and validation of markers for marker-assisted selection against the *Stagonospora nodorum* toxin sensitivity genes *Tsn1* and *Snn2* in wheat. Mol Breed 23:35–49

Zhu W, Ronen M, Gur Y, Minz-Dub A, Masrati G, Ben-Tal N, Sharon A (2017) BcXYG1, a secreted xyloglucanase from *Botrytis cinerea*, triggers both cell death and plant immune responses. Plant Physiol 175(1):438–456. https://doi.org/10.1104/pp.17.00375

Zinkernagel V, Riess F, Wendland M, Bartscherer HC (1988) Infection structures of Septoria nodorum in leaves of susceptible wheat cultivars. Zeitschrift fuer Pflanzenkrankheiten und Pflanzenschutz (Germany)

Zipfel C (2014) Plant pattern-recognition receptors. Trends Immunol 35:345–351

Zuppini A, Navazio L, Sella L, Castiglioni C, Favaron F, Mariani P (2005) An endopolygalacturonase from *Sclerotinia sclerotiorum* induces calcium-mediated signaling and programmed cell death in soybean cells. Mol Plant-Microbe Interact 18(8):849–855

Dance Between Environment and Fungal Pathogens in the Rhythm of Climate Change

7

Jaime David Acosta-España, Dolly Montaño, and Kerstin Voigt

Abstract

Climate change is an urgent global challenge with profound implications for human health. Rising temperatures and unpredictable climate patterns have contributed to the emergence of more virulent diseases, including fungal infections. These infections have devastating consequences for people and the environment. Understanding the complicated relationship between climate change and fungal diseases is critical for effective mitigation strategies.

This book chapter provides a brief overview of the impact of climate change on human health, focusing on fungal diseases. It examines how changing temperatures and precipitation patterns affect fungal pathogens' multiplication and spread, leading to severe and potentially life-threatening infections. Vulnerable populations are particularly susceptible to respiratory problems and systemic infections associated with these fungal diseases.

Urgent and sustained action is needed to address these challenges. These include reducing greenhouse gas emissions, improving resilience to extreme weather events, and increasing access to effective treatments for fungal diseases. A coordinated global effort is essential to address the far-reaching and potentially catastrophic consequences of climate change and fungal infections for human health.

In addition, this chapter highlights the importance of understanding the relationship between fungal pathogens and their environment. Variations in temperature, humidity, and land use significantly impact the spread and occurrence of fungal pathogens. Understanding this relationship is crucial for devising effective prevention and control measures.

This chapter is an updated version of "Ecological and Economical Importance of Parasitic Zoosporic True Fungi" published in Vol. 11: Agricultural Applications, 2013, ISBN 978-3-642-36820-2, https://link.springer.com/book/10.1007/978-3-642-36821-9.

J. D. Acosta-España
Jena Microbial Resource Collection, Leibniz Institute for Natural Product Research and Infection Biology – Hans Knöll Institute, Jena, Germany

Institute of Microbiology, Friedrich Schiller University Jena, Jena, Germany

Faculty of Medicine, Pontifical Catholic University of Ecuador, Quito, Ecuador

Health Sciences Faculty, Universidad Internacional SEK del Ecuador, Quito, Ecuador

D. Montaño · K. Voigt (✉)
Jena Microbial Resource Collection, Leibniz Institute for Natural Product Research and Infection Biology – Hans Knöll Institute, Jena, Germany

Institute of Microbiology, Friedrich Schiller University Jena, Jena, Germany
e-mail: Kerstin.Voigt@leibniz-hki.de

S. Grüttner et al. (eds.), *Agricultural and Industrial Applications*, The Mycota 16,
https://doi.org/10.1007/978-3-031-81904-9_7

Keywords

Climate change · Fungal pathogens · Human health · Virulent diseases · Temperature · Precipitation patterns · Respiratory problems · Systemic infections · Greenhouse gas emissions · Extreme weather events

7.1 Introduction

Climate change is one of humanity's most significant global challenges, with far-reaching and potentially catastrophic impacts on human health and well-being (United Nations 2022; World Health Organization 2022a). As temperatures continue to rise and the Earth's climate becomes increasingly unpredictable, new and more virulent diseases are emerging, many of which are caused by fungi (Nnadi and Carter 2021). These fungal diseases impact humans and cause widespread death and destruction of crops, forests, and wildlife (Fisher et al. 2020; Tibpromma et al. 2021).

The impacts of climate change on human health are complex and interrelated, with multiple pathways of exposure and potential harm (Centers for Disease Control and Prevention 2020). Rising temperatures and changing precipitation patterns can create ideal conditions for the growth and spread of fungal pathogens that can cause a range of severe and potentially life-threatening infections (Nnadi and Carter 2021). Extreme weather events such as floods and droughts can increase the likelihood of fungal diseases by allowing the growth of molds and other fungi in various environments such as homes, schools, and public spaces (Medina et al. 2014; Hernandez and Martinez 2018).

The impact of climate change on emerging and reemerging infectious diseases is becoming increasingly apparent. Fungi, particularly, are gaining attention as a potential threat due to their unique properties. Unlike viral and bacterial infections, there are no vaccines for fungi and only limited treatment options with antifungals (Casadevall 2019). They can produce infectious spores and do not necessarily require direct host-to-host contact to become infected, making them particularly difficult to control (Casadevall 2019). In addition, their ability to adapt to rising temperatures due to climate change could lead to an increase in fungal organisms that can cause disease (de Crecy et al. 2009; Casadevall 2020).

Fungal diseases can have profound and lasting effects on human health. They cause respiratory illness, skin infections, and potentially fatal systemic infections that can lead to organ failure and death (Brown et al. 2021; Gold et al. 2022). In some cases, the impairment in the immune systems of vulnerable populations, such as the elderly, young children, corticosteroid treatment, and so on, cannot effectively fight these fungal infections (Eades and Armstrong-James 2019). In addition, many fungal diseases are difficult to diagnose and treat and require extensive and often expensive medical intervention (Lass-Flörl 2017; Acosta-España and Voigt 2022a).

In addition, significant outbreaks have been observed in fungal species such as *Cryptococcus deuterogattii*, *Puccinia striiformis* (rust fungus), and *Fusarium* head blight, which may be linked to climate change (Datta et al. 2009; Goertz et al. 2010; Fernandes et al. 2016; de Vallavieille-Pope et al. 2018). These fungi have adapted to the changing temperature conditions, increasing virulence, and spreading to new regions.

Given the far-reaching and potentially catastrophic impact of climate change and fungal diseases on human health, it is imperative that we take immediate and sustained action to mitigate the effects of these threats. Achieving this goal will require a concerted global effort to reduce greenhouse gas emissions, increase resilience to extreme weather events and improve access to effective treatments for fungal diseases. Working together, we can ensure a healthy and sustainable future for all.

7.2 Overview of Fungal Pathogens in Humans

Fungal infections or mycoses can affect different parts of the body and lead to various symptoms depending on the type and severity of the infection. Fungi are found in the environment, and some of them can be seen as microbiota (*Candida* spp.) (Hernandez and Martinez 2018; Kapitan et al. 2019). While certain types of fungi are harmless, others pose a threat and can infect people, especially those with weakened immune systems, due to various risk factors (Skiada et al. 2017).

The incidence of mycoses in humans has increased in recent years due to a number of factors, including the growing number of immunocompromised individuals, the expansion of the therapeutic possibilities of immunomodulation and oncology, the use of antifungal drugs, and the increase in healthcare-associated infections (Lionakis and Hohl 2020; Mendonça et al. 2022; Gold et al. 2022).

According to the World Health Organization (WHO) (World Health Organization 2022a), the top six fungal pathogens on their priority list are:

- *Candida* species: They are responsible for several infections, including vaginal candidiasis, oral candidiasis, and invasive bloodstream infections in patients with risk factors. Risk factors include antibiotic use, immunosuppression, and the presence of medical devices such as catheters (de Santos et al. 2018).
- *Aspergillus* species: These are related to a variety of respiratory infections that can lead to lung damage. Invasive clinical diseases are also possible. Patients with aspergillosis present some risk factors, including severe asthma, cystic fibrosis, and immunosuppression (Latgé and Chamilos 2020).
- *Cryptococcus* species often cause cryptococcal meningitis, a severe and potentially life-threatening central nervous system infection. People with HIV/AIDS are particularly affected, but other risk factors and even healthy people can become infected (*C. gatti*) (Andreou et al. 2020; Chen et al. 2022).
- *Pneumocystis jirovecii*: This is responsible for *Pneumocystis* pneumonia, a severe and potentially life-threatening lung infection that often occurs in immunocompromised individuals, especially those with HIV/AIDS (Morris and Norris 2012).
- *Fusarium* species: This fungus is responsible for several infections, including keratitis (eye infections), onychomycosis (nail infections), and skin infections. Risk factors include contact lens wear, trauma, mycotoxin food poisoning, and so on (Antonissen et al. 2014; Batista et al. 2020).
- *Mucorales* (formerly *Zygomycetes*): This species of fungus is responsible for a number of infections known as mucormycosis (former zygomycosis) (Acosta-España and Voigt 2022b), a severe and potentially life-threatening infection that can affect the sinuses, brain, lungs, and so on. Risk factors include immunosuppression, uncontrolled diabetes (ketoacidosis), use of broad-spectrum antibiotics, various types of transplants, trauma, and so on (Acosta-España and Voigt 2022a).

It is essential to understand that these and other fungal pathogens have the potential to cause severe and life-threatening infections, especially in people with weakened immune systems. Early diagnosis and prompt, appropriate treatment are critical to improve outcomes and effectively control the spread of these infections (Lass-Flörl 2017).

7.3 Relationship Between Fungal Pathogens and the Environment

The relationship between fungal pathogens in humans and the environment is complex and multifaceted. Fungi are omnipresent in the environment and can be found in air, soil, and water. Some species of fungi are harmless to humans, while others can cause infections (Cavicchioli et al. 2019; Timmusk et al. 2020; Assunçã and Assunçã 2021; Zingales et al. 2022).

Fungal pathogens can be found in a variety of environmental sources. *Aspergillus* species are commonly found in soil, plants, and decaying organic material and are known to thrive in moist and warm environments (Latgé and Chamilos

2020). *Cryptococcus* species are found in soil and bird droppings and have been detected in various geographic regions (May et al. 2015).

Pneumocystis jirovecii is a fungal pathogen commonly found in the environment, especially in areas with high levels of air pollution. This fungus can survive under a wide range of environmental conditions and can be found indoors and outdoors (Walzer 2013).

Fusarium species are commonly found in soil, plant material, and water and can adapt to a wide range of environmental conditions (Abdel-Azeem et al. 2019). *Mucorales* (former *Zygomycetes*) are a fungus found in soil and decaying organic material and are known to grow in nutrient-poor environments with high moisture (Acosta-España and Voigt 2022b).

In India, researchers have discovered sporangiospores of *Mucorales* in patients suffering from mucormycosis, both in their homes and the hospital environment (Biswal et al. 2022; Ghosh et al. 2022). This finding suggests that these fungal spores may be an essential factor in the transmission and spread of the disease.

The relationship between fungal pathogens in humans and the environment is complex because fungal pathogens are found in various environmental sources. Understanding the environments in which fungal pathogens occur and the factors contributing to their growth and spread is critical for developing effective strategies to prevent and control fungal infections in humans.

7.4 Overview of Climate Change

Climate change is one of our most pressing global issues, with far-reaching impacts on the environment, human health, and the economy (Assunçã and Assunçã 2021; United Nations 2022; World Health Organization 2022b). The leading cause of climate change has been related to greenhouse gases, such as carbon dioxide, methane, and nitrous oxide, into the atmosphere. These gases trap heat and lead to rising temperatures (Cavicchioli et al. 2019; Callaghan et al. 2021; Assunçã and Assunçã 2021; World Health Organization 2022b). This leads to changing weather patterns, rising sea levels, and increased frequency of extreme weather events such as hurricanes, droughts, and floods.

As per the United Nations, the average global temperature has risen by approximately 1 °C since the preindustrial era (United Nations 2022). This may not seem like a significant increase, but the effects are already being felt worldwide: more frequent heat waves, rising sea levels, and melting polar ice caps (United Nations 2022; World Health Organization 2022b).

Regarding economic impact, the cost of climate change is expected to reach trillions of dollars in the coming decades (Tol 2009; United Nations 2022). For example, the frequency and intensity of natural disasters are expected to increase, leading to higher costs for relief and recovery efforts. Agriculture and fisheries are also likely to suffer as changing temperatures and weather patterns affect the growth of crops and fish populations (Food and Agriculture Organization of the United Nations 2018).

The impacts of climate change are not limited to the environment and the economy. Climate change also has profound implications for human health, spreading infectious diseases such as dengue fever and exacerbating existing health problems such as air pollution and heat stress (Nnadi and Carter 2021; Pacheco et al. 2021; Frías-De-León et al. 2022; World Health Organization 2022b; Zingales et al. 2022).

To combat the effects of climate change, it is essential to reduce greenhouse gas emissions and switch to cleaner energy sources, such as renewable energy. The Paris Agreement, endorsed by nearly 200 countries in 2015, aims to restrict the increase in global temperature to well below 2 °C above preindustrial levels, with additional efforts to limit the increase to 1.5 °C (Food and Agriculture Organization of the United Nations 2018).

Despite these efforts, greenhouse gas emissions continue to rise, and emissions are accelerating. In 2020, carbon dioxide emissions reached a record high, with the energy sector accounting for the largest share of emissions. This underscores the urgent need for action to reduce emissions and switch to cleaner energy sources.

Climate change is an important and urgent issue with extensive repercussions on the environment, economy, and human health. Reducing greenhouse gas emissions and transitioning to cleaner energy sources is imperative to mitigate the effects of climate change. This will require significant effort and investment, but the long-term benefits are clear and undeniable.

7.5 Impact of Climate Change on the Ecosystem and Fungi

Environmental changes significantly impact fungi and their growth, affecting human health (Cavicchioli et al. 2019; Pacheco et al. 2021; World Health Organization 2022b). Changes in temperature, humidity, and nutrient levels are just some of the environmental factors that can affect the survival, growth, and spread of fungi in the environment (Cavicchioli et al. 2019; Timmusk et al. 2020; Nnadi and Carter 2021; Frías-De-León et al. 2022; Zingales et al. 2022). As temperatures rise and become more unstable, certain species of fungi may spread more while others decline. For example, a warm and humid environment can promote the growth of fungal pathogens such as *Aspergillus*, which can cause respiratory infections in humans.

Conversely, temperature changes can also alter the behavior of fungal pathogens, making them more virulent or less virulent depending on the conditions (Cavicchioli et al. 2019; Nnadi and Carter 2021). Changes in moisture content can also have significant effects on fungal populations. Some fungal species thrive in moist environments, while others are better adapted to drier conditions. For example, the growth of *Candida*, a fungal species that can cause human infections, is favored by a moist environment (Frías-De-León et al. 2022).

On the other hand, changes in moisture levels (Ma et al. 2021) can also affect the growth and spread of *Mucorales*, which are known to cause severe and potentially life-threatening infections in humans (Acosta-España and Voigt 2022a). Loss of natural habitats and deforestation can also affect the spread of fungal pathogens (Parsons et al. 2021). Changing land use practices such as deforestation and clearing of natural habitats can promote the spread of fungal pathogens associated with soil and decaying organic matter (Fasona et al. 2018). Deforestation can lead to significant carbon dioxide emissions that contribute to climate change and facilitate the spread of fungal pathogens.

Climate change can also affect the spread of fungal pathogens, as rising temperatures and increased precipitation can create ideal conditions for the growth and spread of certain fungal species. Increases in global travel and trade have also led to the spread of fungal pathogens to new geographic regions, further exacerbating the impact of environmental changes on fungal populations and human health. Chemicals such as pesticides and fertilizers can also significantly affect fungal populations and the spread of fungal pathogens.

Pesticides and fertilizers can alter the balance of the ecosystem, leading to the decline of beneficial fungal species and the growth of harmful pathogens. Antibiotic use can also affect fungal populations, as it can promote the development of antibiotic-resistant fungal strains.

7.6 Mechanisms of Fungal Thermotolerance and Adaptation to Climate Change

Fungal thermotolerance is the ability of fungi to withstand high temperatures, which is a critical adaptation mechanism for survival under changing environmental conditions. This adaptation is primarily achieved through regulating cellular processes, including cellular respiration, protein synthesis, and membrane fluidity, which are required for fungal survival under high-temperature stress (Bakar et al. 2020).

Fungi have also evolved mechanisms to protect their DNA from damage caused by high temperatures. For example, some fungi produce heat shock proteins (HSPs) that help stabilize cell proteins and protect DNA from thermal damage. In addition, some fungi have chaperones that help fold proteins and prevent damage to cellular structures (Tereshina 2005).

Another important mechanism of fungal thermotolerance is the regulation of cellular water balance. High temperatures can cause increased water loss from cells, leading to dehydration and cell damage. Fungi have adapted to this by producing high levels of compatible solutes such as proline, trehalose, and glycerol, which help maintain cellular water balance and protect against thermal stress (Argüelles 1997; Pérez-Llano et al. 2020).

Another adaptive mechanism is the modulation of antioxidant systems that help scavenge reactive oxygen species (ROS) produced during high-temperature stress. Fungi have evolved various antioxidant enzymes such as superoxide dismutases (SODs), catalases (CATs), and peroxidases (PODs) to protect themselves from oxidative damage (Abrashev et al. 2005). In addition, some fungi also produce nonenzymatic antioxidants such as ascorbic acid and glutathione to protect themselves from ROS damage (Xiao et al. 2022).

Fungi also use adaptive phenotypic changes to cope with thermal stress. For example, some fungi produce heat-resistant spores or spores that can survive at high temperatures, which increases the chances of survival under changing environmental conditions (de Crecy et al. 2009; Xiao et al. 2022). In addition, some fungi have the ability to switch between sexual and asexual reproduction, allowing them to adapt quickly to new environmental conditions (Maresca and Carratù 1990; Xiao et al. 2022).

7.7 Thermotolerance and Its Relationship with Environmental Changes

Fungal thermotolerance is an adaptation critical for fungal survival in a changing climate, as rising temperatures can significantly impact fungal populations and ecosystems (Porras et al. 2021).

The mechanisms of fungal thermotolerance are complex and involve regulating several cellular processes, including cellular respiration, protein synthesis, and membrane fluidity, as well as producing heat shock proteins and antioxidant enzymes. These mechanisms contribute to protection against thermal stress and enable fungi to survive and grow at high temperatures (Maresca and Carratù 1990; Argüelles 1997; Porras et al. 2021; Xiao et al. 2022).

Environmental changes, such as rising temperatures and more frequent heat waves, can significantly impact fungal populations and ecosystems. These impacts can include changes in the distribution and diversity of fungal communities and the structure and function of fungus-dominated ecosystems (Větrovský et al. 2019).

In addition to temperature increases, other environmental changes, such as changes in precipitation patterns and atmospheric CO_2 levels, may also affect fungal populations (Lipson et al. 2014; Barnes et al. 2018). These effects can lead to changes in fungal growth and metabolism and in the availability of food sources and hosts for fungi (Nnadi and Carter 2021).

Despite these challenges, fungi can adapt to environmental changes through thermotolerance mechanisms, allowing them to survive in changing environments. This adaptation has essential implications for the functioning of fungus-dominated ecosystems and the provision of ecosystem services such as nutrient cycling and organic matter decomposition. Gaining insights into fungal thermotolerance and its connection to environmental changes is essential for predicting the impact of climate change on fungal populations and ecosystems.

7.8 Yeast Infections and Climate Change

Yeast infections are caused by a variety of yeast species, including *Candida albicans*, which can lead to various health problems in humans and animals (de Santos et al. 2018). Climate change is an important factor affecting the prevalence and incidence of yeast infections, as rising temperatures and changing environmental conditions can alter the growth, distribution, and virulence of yeast populations (Frías-De-León et al. 2022).

Fluctuations in temperature and humidity can increase the growth and proliferation of yeast populations, leading to an increased risk of yeast infections. For example, warmer and humid conditions can create a favorable environment for *Candida albicans'* growth, increasing the risk of vaginal yeast infections in women (Morton and Rashid 1977; de Oliveira et al. 2019).

In addition, changing precipitation patterns and the frequency of extreme weather events can lead to increased human and animal contact with yeast-contaminated water, food, and soil, increasing the risk of yeast infections (Babič et al. 2017). This is especially true in regions affected by natural disasters such as floods and hurricanes, which can lead to widespread contamination and the spread of yeast populations (Benedict and Park 2014).

Climate change can also affect the immunity of humans and animals, making them more susceptible to yeast infections (Pacheco et al. 2021). For instance, elevated temperatures and heightened stress can weaken the immune system, elevating the risk of yeast infections (Pacheco et al. 2021). In addition, changes in the distribution and abundance of other microorganisms, such as bacteria and viruses, can alter the interactions between these microorganisms and yeast, leading to differences in the prevalence and severity of yeast infections (Cavicchioli et al. 2019).

Yeast infections and climate change are closely linked, and the effects of climate change on yeast populations and the occurrence of yeast infections are complex and far-reaching. Therefore, a better understanding of the relationship between yeast infections and climate change is critical for predicting future risks and impacts and developing effective strategies to manage and prevent these infections in the face of climate change.

7.9 Mold Infections and Climate Change

Mold infections are a major public health concern. Mold exposure can lead to various health problems, including respiratory infections, allergies, and other chronic diseases. Climate change is an essential factor affecting the spread and frequency of mold infections, as rising temperatures and changing environmental conditions can alter mold populations' growth, distribution, and virulence (Hardin et al. 2003).

Changes in temperature and humidity can increase the growth and distribution of mold populations, leading to an increased risk of mold infections (Zingales et al. 2022). For example, warmer and humid conditions can create a favorable environment for the growth of common indoor mold species, such as *Aspergillus* spp. and *Fusarium* spp., which can increase the risk of respiratory infections and allergies (Timmusk et al. 2020; Assunçã and Assunçã 2021).

In addition, changing precipitation patterns and the frequency of extreme weather events can increase human exposure to mold-contaminated water, food, and soil, leading to an increased risk of mold infections (Slavin et al. 2015). This is especially true in regions affected by natural disasters such as hurricanes and floods, which result in widespread contamination of water, food, and soil, which increases the risk of mold infections (Kouadio et al. 2012). Understanding fungal thermotolerance and its correlation with environmental changes is crucial for foreseeing the consequences of climate change on fungal populations and ecosystems, particularly in regions impacted by natural disasters such as hurricanes and floods. These events can trigger extensive contamination and the dissemination of mold populations (Benedict and Park 2014).

7.10 Dimorphic Fungal Infections and Climate Change

Dimorphic fungal infections are caused by fungi that can occur in two different morphological forms: yeast (human infections) and filamentous fungi (environment) (Boyce and Andrianopoulos 2015; Hoft et al. 2022). These infections can cause significant health problems in humans and animals, especially in immunocompromised individuals (Boyce and Andrianopoulos 2015; Wilmes et al. 2022). The best-known dimorphic fungal infection is histoplasmosis, which may cause respiratory symptoms and other health problems in people who have been exposed to the fungus in contaminated soil (Boyce and Andrianopoulos 2015; Hoft et al. 2022).

Climate change may significantly impact the occurrence and spread of dimorphic fungal infections. As global temperatures rise, conditions conducive to the growth of these fungi become more favorable in many regions. This may result in an increase in the number of infections caused by these fungi and a change in their geographic distribution (van Rhijn and Bromley 2021). For example, some species of dimorphic fungi thrive in warm, humid environments. If temperatures rise in these regions, their populations may grow and spread (Menges et al. 1952; Gascón et al. 2000).

Mitigating the effects of climate change on dimorphic fungal infections will require a multifaceted approach that includes increased monitoring and surveillance, as well as efforts to decrease greenhouse gas emissions that contribute to global warming. This can consist of reducing fossil fuel use, promoting sustainable agricultural practices, and supporting the development of clean energy technologies. By taking these steps, we can protect public health and decrease the impact of climate change on dimorphic fungal infections.

References

Abdel-Azeem AM, Abdel-Azeem MA, Darwish AG, et al (2019) Fusarium: biodiversity, ecological significances, and industrial applications. In: Recent advancement in white biotechnology through fungi: volume 1: diversity and enzymes perspectives fungal biology. In: Yadav AN, Mishra S, Singh S, Gupta A (eds) Springer, Cham, pp 201–261. https://doi.org/10.1007/978-3-030-10480-1_6

Abrashev R, Dolashka P, Christova R et al (2005) Role of antioxidant enzymes in survival of conidiospores of Aspergillus niger 26 under conditions of temperature stress. J Appl Microbiol 99:902–909. https://doi.org/10.1111/J.1365-2672.2005.02669.X

Acosta-España JD, Voigt K (2022a) Mini review: risk assessment, clinical manifestation, prediction, and prognosis of mucormycosis: implications for pathogen- and human-derived biomarkers. Front Microbiol 13:1–12. https://doi.org/10.3389/FMICB.2022.895989

Acosta-España JD, Voigt K (2022b) An old confusion: Entomophthoromycosis versus mucormycosis and their main differences. Front Microbiol 13:4128. https://doi.org/10.3389/FMICB.2022.1035100/BIBTEX

Andreou M, Cogliati M, Kolonitsiou F et al (2020) Cryptococcus gattii infection in an immunocompetent host in Greece. Med Mycol Case Rep 27:1. https://doi.org/10.1016/J.MMCR.2019.12.002

Antonissen G, Martel A, Pasmans F et al (2014) The impact of Fusarium Mycotoxins on human and animal host susceptibility to infectious diseases. Toxins (Basel) 6:430. https://doi.org/10.3390/TOXINS6020430

Argüelles JC (1997) Thermotolerance and trehalose accumulation induced by heat shock in yeast cells of Candida albicans. FEMS Microbiol Lett 146:65–71. https://doi.org/10.1111/J.1574-6968.1997.TB10172.X

Assunçã R, Assunçã R (2021) Climate change influence in fungi. Eur J Pub Health 31. https://doi.org/10.1093/EURPUB/CKAB164.270

Babič MN, Gunde-Cimerman N, Vargha M et al (2017) Fungal contaminants in drinking water regulation? A tale of ecology, exposure, purification and clinical relevance. Int J Environ Res Public Health 14. https://doi.org/10.3390/IJERPH14060636

Bakar NA, Karsani SA, Alias SA (2020) Fungal survival under temperature stress: a proteomic perspective. PeerJ 8. https://doi.org/10.7717/PEERJ.10423

Barnes CJ, van der Gast CJ, McNamara NP et al (2018) Extreme rainfall affects assembly of the root-associated fungal community. New Phytol 220:1172. https://doi.org/10.1111/NPH.14990

Batista BG, de Chaves MA, Reginatto P et al (2020) Human fusariosis: an emerging infection that is difficult to treat. Rev Soc Bras Med Trop 53:1–7. https://doi.org/10.1590/0037-8682-0013-2020

Benedict K, Park BJ (2014) Invasive fungal infections after natural disasters. Emerg Infect Dis 20:349. https://doi.org/10.3201/EID2003.131230

Biswal M, Gupta P, Kanaujia R et al (2022) Evaluation of hospital environment for presence of Mucorales during COVID-19-associated mucormycosis outbreak in India - a multi-Centre study. J Hosp Infect 122:173–179. https://doi.org/10.1016/J.JHIN.2022.01.016

Boyce KJ, Andrianopoulos A (2015) Fungal dimorphism: the switch from hyphae to yeast is a specialized morphogenetic adaptation allowing colonization of a host. FEMS Microbiol Rev 39:797–811. https://doi.org/10.1093/FEMSRE/FUV035

Brown L, Leck AK, Gichangi M, et al (2021) The global incidence and diagnosis of fungal keratitis. Lancet Infect Dis https://doi.org/10.1016/S1473-3099(20)30448-5

Callaghan M, Schleussner CF, Nath S, et al (2021) Machine-learning-based evidence and attribution mapping of 100,000 climate impact studies. Nat Clim Change 11(11):966–972. https://doi.org/10.1038/s41558-021-01168-6

Casadevall A (2019) Global catastrophic threats from the fungal kingdom: fungal catastrophic threats. Curr Top Microbiol Immunol 424:21–32. https://doi.org/10.1007/82_2019_161

Casadevall A (2020) Climate change brings the specter of new infectious diseases. J Clin Invest 130:553–555. https://doi.org/10.1172/JCI135003

Cavicchioli R, Ripple WJ, Timmis KN, et al (2019) Scientists' warning to humanity: microorganisms and climate change. Nat Rev Microbiol 17(9):569–586. https://doi.org/10.1038/s41579-019-0222-5

Centers for Disease Control and Prevention (2020) Preparing for the regional health impacts of climate change in the United States

Chen Y, Shi ZW, Strickland AB, Shi M (2022) Cryptococcus neoformans infection in the central nervous system: the Battle between host and pathogen. J Fungi 8(8):1069. https://doi.org/10.3390/JOF8101069

Datta K, Bartlett KH, Baer R et al (2009) Spread of Cryptococcus gattii into Pacific northwest region of the United States. Emerg Infect Dis 15:1185. https://doi.org/10.3201/EID1508.081384

de Crecy E, Jaronski S, Lyons B et al (2009) Directed evolution of a filamentous fungus for thermotolerance. BMC Biotechnol 9:74. https://doi.org/10.1186/1472-6750-9-74/FIGURES/4

de Oliveira CC, Alves FV, de Almeida Martins PGM et al (2019) Vaginal temperature as indicative of thermoregulatory response in Nellore heifers under different microclimatic conditions. PLoS One 14:e0223190. https://doi.org/10.1371/JOURNAL.PONE.0223190

de Santos GC O, Vasconcelos CC, Lopes AJO, et al (2018) Candida infections and therapeutic strategies: mechanisms of action for traditional and alternative agents. Front Microbiol 9:1351. doi:https://doi.org/10.3389/FMICB.2018.01351/BIBTEX

de Vallavieille-Pope C, Bahri B, Leconte M et al (2018) Thermal generalist behaviour of invasive Puccinia striiformis f. sp. tritici strains under current and future climate conditions. Plant Pathol 67:1307–1320. https://doi.org/10.1111/PPA.12840

Eades CP, Armstrong-James DPH (2019) Invasive fungal infections in the immunocompromised host: mechanistic insights in an era of changing immunotherapeutics. Med Mycol 57:S307–S317. https://doi.org/10.1093/MMY/MYY136

Fasona M, Adeonipekun PA, Agboola O, et al (2018) Drivers of deforestation and land-use change in Southwest Nigeria. In: Handbook of climate change resilience, pp 1–24. https://doi.org/10.1007/978-3-319-71025-9_139-1

Fernandes KE, Dwyer C, Campbell LT, Carter DA (2016) Species in the Cryptococcus gattii complex differ in capsule and cell size following growth under capsule-inducing conditions. mSphere 1. https://doi.org/10.1128/MSPHERE.00350-16

Fisher MC, Gurr SJ, Cuomo CA, et al (2020) Threats posed by the fungal kingdom to humans, wildlife, and agriculture. MBio 11. https://doi.org/10.1128/MBIO.00449-20/ASSET/6C4AE09A-A3CA-418B-BFCD-09EB1B657093/ASSETS/GRAPHIC/MBIO.00449-20-F0001.JPEG

Food and Agriculture Organization of the United Nations (2018) Deep-ocean climate change impacts on habitat, fish and fisheries. Rome

Frías-De-León MG, García-Salazar E, Reyes-Montes MdelR et al (2022) Opportunistic yeast infections and climate change: the emergence of Candida auris. pp 161–179. https://doi.org/10.1007/978-3-030-89664-5_10

Gascón J, Torres JM, Luburich P et al (2000) Imported histoplasmosis in Spain. J Travel Med 7:89–92. https://doi.org/10.2310/7060.2000.00028

Ghosh AK, Singh R, Reddy S et al (2022) Evaluation of environmental Mucorales contamination in and around the residence of COVID-19-associated mucormycosis patients. Front Cell Infect Microbiol 12:953750. https://doi.org/10.3389/FCIMB.2022.953750/BIBTEX

Goertz A, Zuehlke S, Spiteller M et al (2010) Fusarium species and mycotoxin profiles on commercial maize hybrids in Germany. Eur J Plant Pathol 128:101–111. https://doi.org/10.1007/S10658-010-9634-9/TABLES/4

Gold JAW, Ahmad FB, Cisewski JA, et al (2022) Increased deaths from fungal infections during the COVID-19 pandemic-National Vital Statistics System, United States, January 2020–December 2021. Clin Infect Dis https://doi.org/10.1093/CID/CIAC489

Hardin BD, Kelman BJ, Saxon S (2003) Adverse human health effects associated with molds in the indoor environment. J Occup Environ Med 45:470–478. https://doi.org/10.1097/00043764-200305000-00006

Hernandez H, Martinez LR (2018) Relationship of environmental disturbances and the infectious potential of fungi. Microbiology (N Y) 164:233. https://doi.org/10.1099/MIC.0.000620

Hoft MA, Duvenage L, Hoving JC (2022) Key thermally dimorphic fungal pathogens: shaping host immunity. Open Biol 12. https://doi.org/10.1098/RSOB.210219/

Kapitan M, Niemiec MJ, Steimle A et al (2019) Fungi as part of the microbiota and interactions with intestinal bacteria. Curr Top Microbiol Immunol 422:265–301. https://doi.org/10.1007/82_2018_117

Kouadio IK, Aljunid S, Kamigaki T et al (2012) Infectious diseases following natural disasters: prevention and control measures. Expert Rev Anti-Infect Ther 10:95–104. https://doi.org/10.1586/ERI.11.155

Lass-Flörl C (2017) Current challenges in the diagnosis of fungal infections. Methods Mol Biol 1508:3–15. https://doi.org/10.1007/978-1-4939-6515-1_1

Latgé JP, Chamilos G (2020) Aspergillus fumigatus and aspergillosis in 2019. Clin Microbiol Rev 33. https://doi.org/10.1128/CMR.00140-18/ASSET/E6BB921A-A99B-4691-9FA0-4EE6A9D08B9A/ASSETS/GRAPHIC/CMR.00140-18-F0020.JPEG

Lionakis MS, Hohl TM (2020) Call to action: how to tackle emerging nosocomial fungal infections. doi:https://doi.org/10.1016/j.chom.2020.04.011

Lipson DA, Kuske CR, Gallegos-Graves LV, Oechel WC (2014) Elevated atmospheric CO2 stimulates soil fungal diversity through increased fine root production in a semiarid shrubland ecosystem. Glob Chang Biol 20:2555–2565. https://doi.org/10.1111/GCB.12609

Ma W, Zhai L, Pivovaroff A et al (2021) Assessing climate change impacts on live fuel moisture and wildfire risk using a hydrodynamic vegetation model. Biogeosciences 18:4005–4020. https://doi.org/10.5194/bg-18-4005-2021

Maresca B, Carratù L (1990) Heat shock and adaptation during temperature-activated dimorphism in the fungus Histoplasma capsulatum. Stress Proteins, pp 45–56. https://doi.org/10.1007/978-3-642-75815-7_4

May RC, Stone NRH, Wiesner DL et al (2015) Cryptococcus: from environmental saprophyte to global pathogen. Nat Rev Microbiol 14:106–117. https://doi.org/10.1038/nrmicro.2015.6

Medina A, Rodriguez A, Magan N (2014) Effect of climate change on Aspergillus flavus and aflatoxin B1 production. Front Microbiol 5:348. https://doi.org/10.3389/FMICB.2014.00348/BIBTEX

Mendonça A, Santos H, Franco-Duarte R, Sampaio P (2022) Fungal infections diagnosis—past, present and future. Res Microbiol 173:103915. https://doi.org/10.1016/J.RESMIC.2021.103915

Menges R, Furcolow M, Larsh H, Hinton A (1952) Laboratory studies on Histoplasmosis: I. The effect of humidity and temperature on the growth of Histoplasma capsulatum. J Infect Dis 90

Morris A, Norris KA (2012) Colonization by Pneumocystis jirovecii and its role in disease. Clin Microbiol Rev 25:297–317. https://doi.org/10.1128/CMR.00013-12/ASSET/C4BF1A28-A208-4FEE-BC82-E90F258A3D93/ASSETS/GRAPHIC/ZCM9990923820001.JPEG

Morton RS, Rashid S (1977) Candidal vaginitis: natural history, predisposing factors and prevention. Proc R Soc Med 70:3. https://doi.org/10.1177/00359157770700s402

Nnadi NE, Carter DA (2021) Climate change and the emergence of fungal pathogens. PLoS Pathog 17. https://doi.org/10.1371/JOURNAL.PPAT.1009503

Pacheco SE, Guidos-Fogelbach G, Annesi-Maesano I et al (2021) Climate change and global issues in allergy and immunology. J Allergy Clin Immunol 148:1366–1377. https://doi.org/10.1016/J.JACI.2021.10.011

Parsons LA, Jung J, Masuda YJ et al (2021) Tropical deforestation accelerates local warming and loss of safe outdoor working hours. One Earth 4:1730–1740. https://doi.org/10.1016/j.oneear.2021.11.016

Pérez-Llano Y, Rodríguez-Pupo EC, Druzhinina IS et al (2020) Stress reshapes the physiological response of halophile fungi to salinity. Cells 9. https://doi.org/10.3390/CELLS9030525

Porras MF, Agudelo-Cantero GA, Santiago-Martínez MG, et al (2021) Fungal infections lead to shifts in thermal tolerance and voluntary exposure to extreme temperatures in both prey and predator insects. Sci Rep 11:1–9. https://doi.org/10.1038/s41598-021-00248-z

Skiada A, Pavleas I, Drogari-Apiranthitou M (2017) Rare fungal infectious agents: a lurking enemy. F1000Res 6. https://doi.org/10.12688/F1000RESEARCH.11124.1

Slavin M, van Hal S, Sorrell TC et al (2015) Invasive infections due to filamentous fungi other than Aspergillus: epidemiology and determinants of mortality. Clin Microbiol Infect 21:490.e1–490.e10. https://doi.org/10.1016/J.CMI.2014.12.021

Tereshina VM (2005) Thermotolerance in fungi: the role of heat shock proteins and trehalose. Microbiology (N Y) 74:247–257. https://doi.org/10.1007/S11021-005-0059-Y/METRICS

Tibpromma S, Dong Y, Ranjitkar S et al (2021) Climate-fungal pathogen Modeling predicts loss of up to one-third of tea growing areas. Front Cell Infect Microbiol 11:308. https://doi.org/10.3389/FCIMB.2021.610567/BIBTEX

Timmusk S, Nevo E, Ayele F, et al (2020) Fighting Fusarium pathogens in the era of climate change: a conceptual approach. Pathogens, p 9. https://doi.org/10.3390/PATHOGENS9060419

Tol RSJ (2009) The economic effects of climate change. J Econ Perspect 23:29–51. https://doi.org/10.1257/JEP.23.2.29

United Nations (2022) Causes and Effects of Climate Change

van Rhijn N, Bromley M (2021) The consequences of our changing environment on life threatening and debilitating fungal diseases in humans. J Fungi 7:367. https://doi.org/10.3390/JOF7050367

Větrovský T, Kohout P, Kopecký M, et al (2019) A meta-analysis of global fungal distribution reveals climate-driven patterns. Nat Commun 10:1–9. https://doi.org/10.1038/s41467-019-13164-8

Walzer PD (2013) The ecology of pneumocystis: perspectives, personal recollections, and future research opportunities. J Eukaryot Microbiol 60:634. https://doi.org/10.1111/JEU.12072

Wilmes D, Mayer U, Wohlsein P et al (2022) Animal Histoplasmosis in Europe: review of the literature and molecular typing of the etiological agents. J Fungi 8:833. https://doi.org/10.3390/JOF8080833/S1

World Health Organization (2022a) WHO fungal priority pathogens list to guide research, development and public health action

World Health Organization (2022b) Climate change. Genova

Xiao W, Zhang J, Huang J, et al (2022) Response and regulatory mechanisms of heat resistance in pathogenic fungi. Appl Microbiol Biotechnol 106:17 106:5415–5431. https://doi.org/10.1007/S00253-022-12119-2

Zingales V, Taroncher M, Martino PA, et al (2022) Climate change and effects on molds and mycotoxins. Toxins 14:445. https://doi.org/10.3390/TOXINS14070445

Index

S. Grüttner et al. (eds.), *Agricultural and Industrial Applications*, The Mycota 16,
https://doi.org/10.1007/978-3-031-81904-9

GPSR Compliance

The European Union's (EU) General Product Safety Regulation (GPSR) is a set of rules that requires consumer products to be safe and our obligations to ensure this.

If you have any concerns about our products, you can contact us on ProductSafety@springernature.com

In case Publisher is established outside the EU, the EU authorized representative is:

Springer Nature Customer Service Center GmbH
Europaplatz 3
69115 Heidelberg, Germany

Batch number: 10732801

Printed by Printforce, the Netherlands